The Bio-Inspired X-Structure/Mechanism Approach for Exploring Nonlinear Benefits in Engineering

Xingjian Jing

The Bio-Inspired X-Structure/Mechanism Approach for Exploring Nonlinear Benefits in Engineering

Part III—The Nonlinear Characteristic Output Spectrum (nCOS) Function for Analysis and Design of Nonlinear Dynamics

Springer

Xingjian Jing
Department of Mechanical Engineering
City University of Hong Kong
Hong Kong, China

ISBN 978-981-96-5768-1 ISBN 978-981-96-5769-8 (eBook)
https://doi.org/10.1007/978-981-96-5769-8

This Springer imprint is published by the registered company Springer Nature Singapore Pte Ltd.
The registered company address is: 152 Beach Road, #21-01/04 Gateway East, Singapore 189721, Singapore

Preface

This book series is to present a totally new and systematic method, i.e., the X-structure/mechanism approach, for the design and analysis of beneficial nonlinearity of engineering systems, where nonlinearity is expected to take an important and critical role and thus cannot be simply ignored in structural design, dynamic analysis, and parameter selection.

The X-shaped structure or mechanism is a generic (and simple in its shape) structure or mechanism representing a class of beneficial geometric nonlinearity with versatile and designable linkage mechanism or elastic structure designs of different variants or forms (quadrilateral, diamond, polygon, K/Z/S/V/L/W-shape, or others) which all share similar geometric nonlinearity and thus similar nonlinear stiffness/damping properties, flexible in design, and easy to implement.

This book series include five parts (or volumes), i.e., Book I-Nonlinear Stiffness and Nonlinear Damping, Book II-Nonlinear Inertia and Multi-direction Vibration Isolation, Book III-The Nonlinear Characteristic Output Spectrum (nCOS) Function for Analysis and Design of Nonlinear Dynamics, Book IV-Energy Harvesting and Sensors with X-Structures, and Book V-Energy-Saving Robust Control and Other Applications with Virtual or Real X-Suspension.

Optimization of structural parameters is always a challenging topic considering complex nonlinear dynamics, strict engineering requirements and demanding performance objectives. This is Part III of this book series. It is to present a systematic frequency domain analysis, design and optimization method of critical structural or control parameters in nonlinear systems aiming at better dynamic performance and reliable structural design. The new analysis and design methods introduced in this book definitely present a totally new and powerful approach to optimization and understanding of a class of nonlinear systems from a special frequency domain point of view.

With these results, readers will be able to have a complete understanding of this series of studies and also have in-depth knowledge about nonlinear dynamic modelling, analysis, engineering design and applications, especially related to the topic of exploring and exploiting nonlinear benefits in dynamic systems.

Hong Kong, China Xingjian Jing

Acknowledgements The author would like to give his thanks to all current and previous research students and postdoc research staff in the laboratory of Nonlinear Dynamics, Vibration, and Control for their dedicated R&D efforts in the past years.

The author would like to give the greatest gratitude with heart and soul to his family for everlasting and abounding understanding and support.

Contents

Chapter 1
An Overview on the Parametric Characteristic Approach for Analysis Design and Optimization of Nonlinear Dynamics in Frequency Domain

Abstract The frequency-domain analysis and synthesis of nonlinear systems are of fundamental importance in disciplines such as signal processing, mechanical engineering, and control engineering. However, these tasks are inherently challenging due to the complex nonlinear interactions and the implicit dependencies between system outputs and model parameters. Conventional methodologies—such as harmonic balance techniques and absolute stability theory—are frequently employed, yet they often encounter limitations stemming from computational intractability and analytical nonlinearity. To overcome these constraints, a generalized and systematic frequency-domain framework, termed the nonlinear characteristic output spectrum (nCOS) function,has been developed based on Volterra series theory. This approach enables a direct parametric characterization of nonlinear system behavior in the frequency domain, offering enhanced analytical tractability and interpretability. The nCOS methodology exhibits distinct advantages over traditional techniques, providing novel insights into the manifestation of nonlinear effects and facilitating the design of nonlinear dynamic systems. Extensive research has demonstrated the applicability of the nCOS framework across a range of engineering domains, including but not limited to vibration suppression, signal analysis, fault diagnostics, and structural optimization. This chapter presents a comprehensive and up-to-date review of recent advancements in the development and application of the nCOS method.

Keywords Frequency-domain analysis · Nonlinear systems · Volterra series · Parametric characteristic analysis · Nonlinear characteristic output spectrum (nCOS)

X. Jing, *The Bio-Inspired X-Structure/Mechanism Approach for Exploring Nonlinear Benefits in Engineering*, https://doi.org/10.1007/978-981-96-5769-8_1

1.1 Introduction

1.1.1 About This Chapter

The analysis and design of nonlinear systems in the frequency domain are not only crucial to signal processing and control but also pose considerable difficulties in associated fields. Usually, harmonic balance methods or absolute stability theory etc. are employed for these purposes. However, limitations can often be encountered due to computational complexity or implicit relationship among system output response and model parameters etc. Based on the Volterra series theory, a systematic and generic frequency domain method referred to as nonlinear characteristic output spectrum (nCOS) function for analysis and design of nonlinear systems has thus been developed with a straightforward parametric characteristic approach in recent years. The method can demonstrate obvious advantages over its counterparts, and bring unique features, benefits or insights into understanding of nonlinear influence and design of nonlinear dynamics in the frequency domain. Applications of this method in vibration control, signal processing, fault detection, structural optimization and others are extensively studied in the literature as well. This chapter serves an updated review on these recent advances.

1.1.2 Research Background

The analysis and synthesis of linear systems in the frequency domain have been well established in the literature. There are many known methods and techniques developed for this purpose. However, frequency domain analysis for nonlinear systems is not straightforward due to very complicated output frequency characteristics and complex dynamic behavior such as super- or sub-harmonics, inter-modulation, chaos and bifurcation. Therefore, understanding of nonlinear dynamics in the frequency domain is far from full development.

Frequency domain analysis of nonlinear systems has been studied since 1950's. A traditional method was initiated by investigation of global stability of the stationary point with the absolute stability theory, and then frequency domain methods for the analysis of stability of stationary sets and existence of cycles and homo-clinical orbits, as well as for the estimation of dimension of attractors etc. were developed thereafter [1]. More practically, nonlinear behaviour or characteristics of a specific nonlinear part or unit in a system can be analyzed by using describing functions. It can be effectively applied for analysis of limit cycle and oscillation of nonlinear systems, and applications for controller design have also been reported [2]. However, limitations of the describing function methods are noticeable, which lead to some improved methods developed [3, 4]. Another elegant method for frequency domain nonlinear analysis is the well-known harmonic balance method. It can deal with more general problems of nonlinear systems such as sub-harmonics and jump behavior for both the time and frequency domain responses. An obvious disadvantage of the harmonic balance method is the rapidly increasing complexity as more harmonics are incorporated in the analysis. Except these methods, there are also some other results for nonlinear analysis in the frequency domain reported. For example, based on the concept of Bode diagrams, singular value decomposition, and the idea of varying eigenvalues, frequency domain methods for analysis and synthesis of uncertain systems or time-varying systems were studied in [5, 6]; and a frequency response function for convergent systems subjected to a harmonic input was also reported in [7]. A comparative discussion on these several frequency domain methods can be referred to a recent review paper in [8].

In the analysis and design of nonlinear systems, a challenging task is to determine the complicated nonlinear relationship between nonlinear output response and system characteristic parameters. Such a relationship can obviously facilitate the analysis and design of nonlinear dynamics in practice for any specific control objective.

The chapter serves as a brief review on the advances of this specific topic focusing on a Volterra series approach, and systematically introduces the nonlinear characteristic output spectrum method for nonlinear analysis and design with a unique parametric characteristic approach developed in the past more than 10 years.

1.2 The Frequency-Domain Representation of Nonlinear Systems

The input–output relationship of nonlinear systems described by nonlinear differential equations (NDEs), nonlinear autoregressive with exogenous input (NARX) models, block oriented nonlinear (BON) models including Wiener models or Hammerstein models etc., can all be cast into Volterra series up to an order N [9–11]

$$y(t) = \sum_{n=1}^{N} y_n(t), \text{ and } y_n(t) = \int_{-\infty}^{\infty} \cdots \int_{-\infty}^{\infty} h_n(\tau_1, \cdots, \tau_n) \prod_{i=1}^{n} u(t - \tau_i) d\tau_i \quad (1.1)$$

where $h_n(\tau_1, \cdots, \tau_n)$ is the nth order Volterra kernel and $u(t)$ is the input which can be any input signal with a Fourier transform $U(j\omega)$. The nth order Generalized Frequency Response Function (GFRF) is given as a multi-variate Fourier transform of Volterra kernels denoted as $H_n(j\omega_1, \ldots, j\omega_n)$ [10]. With the GFRFs, output frequency response of a nonlinear system can be obtained as [11, 12],

$$Y(j\omega) = \sum_{n=1}^{N} Y_n(j\omega) \quad \text{and}$$

$$Y_n(j\omega) = \frac{1}{(2\pi)^{n-1}} \underbrace{\int_{-\infty}^{\infty} \cdots \int_{-\infty}^{\infty}}_{n-1} \Xi d\omega_1 \cdots d\omega_{n-1} \tag{1.2}$$

where $Y_n(j\omega)$ is the nth-order output spectrum, and the Fourier transform of $y_n(t)$,

$$\begin{aligned} \Xi = {} & H_n(j\omega_1, \ldots, j\omega_{n-1}, j(\omega - \omega_1 - \cdots - \omega_{n-1})) \\ & U_n(j\omega, j\omega_1, \ldots, j\omega_{n-1}) \end{aligned} \tag{1.3}$$

$$\begin{aligned} U_n(j\omega, j\omega_1, \ldots, j\omega_{n-1}) = {} & U(j\omega_1) \cdots \\ & U(j\omega_{n-1}) U(j(\omega - \omega_1 - \cdots - \omega_{n-1})) \end{aligned} \tag{1.4}$$

The nonlinear output spectrum (NOS) above can be regarded as an equivalent frequency-domain Volterra series expansion of the original time-domain system. If the input u(t) is given and the GFRFs of the nonlinear system are known, the calculation of the output spectrum in (1.2) is straightforward via a simple integral program in computers.

1.3 The GFRFs and Nonlinear Output Spectrum (NOS)

In the analysis and design, the output spectrum in (1.2) should be computed for a given input, which further requires the computation of the GFRFs. For a parametric model given by a NDE, NARX, or BON, the harmonic probing method [9, 11] can be applied. For example, the NARX model is given by

$$\begin{aligned} y(t) = \sum_{m=1}^{M} y_m(t),\ y_m(t) = & \sum_{p=0, p+q=m}^{m} \sum_{k_1, k_{p+q}=1}^{K} c_{p,q}(k_1, \cdots, k_{p+q}) \\ & \prod_{i=1}^{p} y(t - k_i) \prod_{i=p+1}^{p+q} u(t - k_i) \end{aligned} \tag{1.5}$$

where, $y_m(t)$ is the mth-order output of the NARX model, $k_i = 1, \ldots, K$, and $\sum_{k_1, k_{p+q}=1}^{K}(\cdot) = \sum_{k_1=1}^{K}(\cdot) \cdots \sum_{k_{p+q}=1}^{K}(\cdot)$. $p + q\ (\leq M)$ is called the nonlinear degree of parameter $c_{p,q}(\cdot)$ which corresponds to the $(p + q)$th degree nonlinear terms like $\prod_{i=1}^{p} y(t - k_i) \prod_{i=p+1}^{p+q} u(t - k_i)$, *e.g.*, $y(t - 1)^p u(t - 2)^q$. $c_{0,1}(.)$ and $c_{1,0}(.)$ have degree 1 referred to as linear parameters, and all others as nonlinear parameters. The NARX model potentially include all possible linear and nonlinear combinations among input and output with different time lags. The corresponding nth-order GFRF can be determined recursively as [11, 13]

$$H_n(j\omega_1,\ldots,j\omega_n)=\frac{1}{L_n(\omega_1\cdots\omega_n)}\sum_{q=0}^{n}\sum_{p=0}^{n-q}\sum_{k_1,k_{p+q}=1}^{K}c_{p,q}(k_1,\ldots,k_{p+q})$$
$$e^{-j\sum_{i=1}^{q}(\omega_{n-q+i}k_{p+i})}H_{n-q,p}(j\omega_1,\ldots,j\omega_{n-q}) \tag{1.6a}$$

$$H_{n,p}(\cdot)=\sum_{i=1}^{n-p+1}H_i(j\omega_1,\ldots,j\omega_i)H_{n-i,p-1}(j\omega_{i+1},\ldots,j\omega_n)\exp(-j(\omega_1+\cdots+\omega_i)k_p) \tag{1.6b}$$

$$H_{n,1}(j\omega_1,\cdots,j\omega_n)=H_n(j\omega_1,\cdots,j\omega_n)\exp(-j(\omega_1+\cdots+\omega_n)k_1) \tag{1.6c}$$

$$L_n(\omega_1,\cdots,\omega_n)=1-\sum_{k_1=1}^{K}c_{1,0}(k_1)\exp(-j(\omega_1+\cdots+\omega_n)k_1) \tag{1.6d}$$

where $H_{0,0}(\cdot)=1$, $H_{n,0}(\cdot)=0$ for $n>0$, $H_{n,p}(\cdot)=0$ for $n<p$, and for $q=0$ $\exp(-j\sum_{i=1}^{q}(\omega_{n-q+i}k_{p+i}))=\begin{cases}1 & p>1\\ 0 & p\le 1\end{cases}$.

An alternative algorithm for the GFRFs of the NARX model was given in [14]. The GFRFs for the NDE model similar to the above [15], and the BON models were discussed in [16]. A different viewpoint of these GFRFs with a parametric bound approach to understanding of these GFRFs of nonlinear systems were studied in [17, 18]. For an extended SISO NARX or NDE model with a general nonlinear output function (or performance function), the GFRFs were studied in [19]. For multi-input multi-output NARX models, the GFRFs are more complicated [20].

With the GFRFs, the NOS can be determined. For example, for a multi-tone input

$$u(t)=\sum_{i=1}^{\overline{K}}|F_i|\cos(\omega_i t+\angle F_i) \tag{1.7}$$

the nonlinear output spectrum (NOS) is [11]

$$Y(j\omega)=\sum_{n=1}^{N}Y_n(j\omega),\quad \text{and}$$
$$Y_n(j\omega)=\frac{1}{2^n}\sum_{\omega_{k_1}+\cdots+\omega_{k_n}=\omega}H_n(j\omega_{k_1},\cdots,j\omega_{k_n})F(\omega_{k_1})\cdots F(\omega_{k_n}) \tag{1.8}$$

where $\overline{K}(>0)\in\mathbb{Z}$, $F_{\rm i}\in\mathbb{C}$, $F(\omega_{k_i})$ can be written explicitly as [21],

$$F(\omega_{k_i})=\left|F_{|k_i|}\right|e^{j\angle F_{|k_i|}\cdot \mathrm{sgn1}(k_i)},\quad \text{with}$$

$$\mathrm{sgn1}(a) = \begin{cases} 1 & a > 0 \\ 0 & a = 0 \\ -1 & a < 0 \end{cases} \text{ for}$$

$$a \in \mathbb{R}. \text{ and } k_i \in \left\{\pm 1, \cdots, \pm \overline{K}\right\}.$$

Nonlinear energy transfer, harmonics and intermodulation. Nonlinear output can include super-harmonics, sub-harmonics and intermodulation of input frequencies, and input energy can transfer from low frequency to high frequency or inverse. With the GFRFs and output spectrum, some fundamental nonlinear energy transfer properties were revealed in 2010 [21] and it is for the first time clearly shown that output frequencies with respect to the same input are periodically expanded from low frequency to high frequency, and can be significantly suppressed or amplified in a specific periodical frequency range with only tuning critical model parameters, demonstrating unique nonlinear harmonic cancellation properties of great potential for applications [11, 21].

The need of the parametric characteristic analysis. Although the GFRFs provide an important basis for the determination of nonlinear output frequency response, the GFRF is in fact a multivariate function defined in high dimensional frequency space, and computation of the GFRFs involves complicated recursive operations. Consequently the relationship between the model parameters and the GFRFs or NOS cannot be readily revealed from these results. In most analysis and design, the objective is to optimize output frequency response (or a nonlinear function of system states [19, 22]) with respect to model parameters. A parametric relationship can greatly facilitate the process, and provide more significant insight into the nonlinear influence.

1.4 The Parametric Characteristics Approach

To reveal the explicit relationship between model parameters and nonlinear output spectrum (NOS), the parametric characteristics approach was investigated in [12, 23, 24], and firstly clearly unveiled in 2006 [23]. With this unique approach, the GFRFs and NOS can all be written into simple polynomial functions of model parameters, corresponding to single variate or multi-variates if one or more parameters are considered.

An explicit polynomial relationship. The NOS of order N can be written as

$$Y(j\omega) = \sum_{n=1}^{N} \chi_n \cdot \phi_n(j\omega)^T \tag{1.9}$$

where χ_n is the nth-order characteristic parameter vector and each element is a monomial composed by one or several model parameters, $\phi_n(j\omega)$ is the associated complex-valued function, and both can be determined explicitly with the methods provided in [12, 23, 24]. Any model parameter corresponding to nonlinear terms in the NDE or NARX model, of interest in an analysis, will be included in χ_n

$$\chi_n = C_{n,0} \oplus \bigoplus_{\substack{k=0 \\ \sum_{i=0}^{k}(p_i+q_i)=n+k,\, 1\le p_0\le n-k,\; 2\le p_i+q_i\le n-k}}^{n-2} \left(\bigotimes_{i=0}^{k} C_{p_i,q_i} \right) \tag{1.10}$$

where $C_{p,q} = [c_{p,q}(0,\cdots,0), c_{p,q}(0,\cdots,1),\cdots,c_{p,q}(\underbrace{K,\cdots,K}_{p+q=m})]$, $\oplus$ and $\otimes$ are two operators well defined for symbolic operation on vectors [23, 24]. If a model parameter is not of interest to consider, it is set to 1 in (1.10). An element 1 in χ_n implies a pure frequency-dependent term in the polynomial (1.9). For example, if only one model parameter (e.g., c) is considered, the NOS or output frequency response function (OFRF) would be

$$Y(j\omega; c) = \overline{\phi}_0(j\omega) + c\overline{\phi}_1(j\omega) + c^2\overline{\phi}_2(j\omega) + \cdots + c^\ell\overline{\phi}_\ell(j\omega) + \cdots \tag{1.11}$$

Obviously, Eq. (1.9) is a straightforward function of system characteristic parameter c, and can significantly facilitate the analysis and design. To emphasize the parametric relationship between nonlinear output spectrum and characteristic parameters (including nonlinear parameters, excitation magnitude and frequency), Eq. (1.9) is referred to as nonlinear characteristic output spectrum (nCOS) function.

Well-defined operators. Noticeably, for two vectors C_1 and C_2 consisting of symbolic variables, $C_1 \oplus C_2$ is a vector including all the elements in C_1 and C_2 without repetition, and $C_1 \otimes C_2$ is a vector including all the elements after the Kronecker product (with/without repetition) [23, 24]. Exclusion of the repetitive terms can lead to the minimum dimension of the parameter characteristic vector χ_n. Due to this reason, at least the operator $\oplus$ should be defined to remove repetitive terms, and also it is applied to vectors and thus not simply a traditional union operator "∪" as the traditional "∪" is applied to sets/spaces and does not produce a vector or matrix.

Sufficient and necessary condition. The detailed parametric characteristics of the GFRFs and NOS were firstly discussed in [23, 24] 2006, with a sufficient and necessary condition: *The elements of the characteristic parameter vector of the nth-order GFRF χ_n include and only include all the parameters in $C_{0,n}$ and all the parameter monomials in $C_{p,q} \otimes C_{p_1,q_1} \otimes C_{p_2,q_2} \otimes \cdots \otimes C_{p_k,q_k}$ for $0 \le k \le n-2$, whose subscripts satisfy*

$$p + q + \sum_{i=1}^{k}(p_i + q_i) = n + k,\quad 2 \le p_i + q_i \le n - k,$$

$$2 \le p + q \le n - k \text{ and } 1 \le p \le n - k$$

Parametric characteristics based analysis. The parametric characteristic analysis can provide an important insight into at least the following aspects:

- System nonlinear effects on frequency response functions (including the GFRFs and output spectrum) [11:ch4–10, 19, 23, 25, 26];
- The accurate polynomial structure of the GFRFs and OFRF [11:ch7–10, 12, 23];
- Computations of the GFRFs and output spectrum [13];
- Understanding of nonlinear behaviour in the frequency domain [11:ch4–6, 21, 26];
- Design of output response by employing nonlinearities [11:ch9–10, 12, 26–28].

1.5 The Nonlinear Characteristic Output Spectrum (nCOS)

As already discussed, the nCOS function is an explicit expression for the nonlinear relationship between NOS and system characteristic parameters (model parameters, frequency, and excitation magnitude) [12]. This relationship can be accurately obtained with less simulation trials and computation cost compared with a pure simulation based study or traditional theoretical computation, and can provide a significant and straightforward method for analysis and design of nonlinear systems with the developed algorithms [11, 12]. Moreover, linear and nonlinear components in the NOS of a given nonlinear system by NDE, NARX or NBO models can be accurately obtained, with some simulation or experiment data [12, 16, 29]. These output spectrum components can then be used for system analysis and design for different purposes such as signal processing, vibration control and fault detection, system identification or nonlinear damping design [12, 16, 22, 29–31].

For example, considering a 2DOF nonlinear vehicle suspension system, the estimated nCOS function is obtained as below [12], in terms of a damping coefficient b_1,

$$
\begin{aligned}
\hat{Y}(j\omega) = \rho \hat{Y}_1(j\omega) + \rho^2 \hat{Y}_2(j\omega) + \rho^3 \underbrace{\left[\hat{Y}_3(j\omega)_{|b_1=0} + b_1 \cdot \hat{\phi}_3(b_1; j\omega)\right]}_{\hat{Y}_3(j\omega, b_1)} \\
+ \rho^4 \underbrace{\left[\hat{Y}_4(j\omega)_{|b_1=0} + b_1 \cdot \hat{\phi}_4(b_1; j\omega)\right]}_{\hat{Y}_4(j\omega, b_1)} \\
+ \rho^5 \underbrace{\left[\hat{Y}_5(j\omega)_{|b_1=0} + b_1 \cdot \hat{\phi}_5(b_1; j\omega) + b_1^2 \cdot \hat{\phi}_5(b_1^2; j\omega)\right]}_{\hat{Y}_5(j\omega, b_1)} + \cdots
\end{aligned}
$$

It is an explicit function of frequency ω, model parameter b_1 and input magnitude ρ. It was thus revealed that a proper nonlinear damping together with a small linear damping can quickly suppress transient vibration and simultaneously keep a small linear damping performance at steady state.

Some qualitative results can directly be obtained with the nCOS function. For instance [11]:ch7, [25], considering (1.11), the following is obvious:

(a) $\frac{\partial|Y(j\omega;c)|}{\partial c} < 0$ for some $c \Rightarrow \exists$some $n > 0$, $\sum_{i=0}^{n} sign(c^{n-1})\overline{\phi}_i \cdot \overline{\phi}_{n-i}^{*} < 0$

(b) $p_1 = \mathrm{Re}(\overline{\phi}_0(j\omega) \cdot \overline{\phi}_1(-j\omega)) < 0 \Rightarrow$ there exists $\varepsilon > 0$ such that $\frac{\partial|Y(j\omega;c)|}{\partial c} < 0$ for $|c| < \varepsilon$. Here $sign(x) = \begin{cases} 1 & x \geq 0 \\ -1 & \text{else} \end{cases}$, $\mathrm{Re}(\cdot)$ is to get the real part of (.). If a nonlinear parameter c satisfies $p_1 = \mathrm{Re}(\overline{\phi}_0(j\omega) \cdot \overline{\phi}_1(-j\omega)) < 0$, then it can be utilized for the vibration suppression objective [26].

Importantly, given a desired output magnitude Y^*, an optimal c^* can be found such that

$$\min_{c}(|Y(j\omega; c)| - Y^*) \tag{1.12}$$

Some systematic methods have been developed for this purpose and applied [12, 22, 25].

1.6 Parametric Convergence Bounds

With the parametric characteristic idea, bound characteristics of the GFRFs and NOS can all be evaluated conveniently [17, 18], which result in parametric insight into the bound of nonlinear frequency responses in terms of the linear component of the system and its model parameters. Based on this technical line, parametric convergence bounds of Volterra series expansion of nonlinear systems were investigated systematically in [32, 33]. The results present for the first time a closed and parametric form of the bound of output response of a nonlinear system in the frequency domain. Importantly, these results for the first time clearly show at what parametric/frequency range and/or for what input amplitude a given NDE or NARX system has a convergent Volterra series expansion by solving a simple parametric equation with less conservatism. Moreover, the results can also indicate the appearance of complicated non-Volterra nonlinear dynamics such as chaos or bifurcation with a specific and explicit parameter range. These results solved the long existing fundamental issue related to the application of Volterra series modelling and its associated methods.

1.7 Some Applications

Based on the parametric characteristics, the potential influence that a specific nonlinear term can bring to system output response can be conveniently investigated [21, 26, 28]. The determination of the GFRFs and output spectrum can also be directly obtained analytically from the parametric characteristics [13]. Importantly, a series of algorithms were developed to numerically determine the nCOS accurately from a given NDE or NARX model or from experimental data [11, 12, 29]. Therefore, nonlinear output frequency response (NOFR) can easily be designed with respect to any model parameter (corresponding to linear or nonlinear components) [12, 19, 22, 25, 34].

Control of nonlinear systems. Some recent studies with this approach reveal uniquely that a nonlinear vibration system can be well controlled with only a simple linear or nonlinear feedback, and the resulting control law is much easier to implement or can reduce energy cost [22, 34]. Any nonlinear function or performance function to be optimized in terms of system output or states can be regarded as a new output and thus be optimized with such a nCOS approach [19, 22].

Understanding and utilization of nonlinear effects. It is for the first time revealed that output frequency response would take an alternating series with respect to some model parameters for convergence or vibration suppression in 2008 [26], and some nonlinear damping can achieve very beneficial effect in vibration control—high damping for strong vibration but low damping for others [26, 28]. This further leads to a series of recent studies on bio-inspired vibration isolation structures [35, 36] and utilization of nonlinear dynamics in active vibration control for better performance with less energy cost simultaneously [37, 38]. A general technique was developed for easily considering nonlinear damping effect [28], followed by many others later. Neural networks or exponential function can also be conducted.

Fault detection with nCOS. Estimation of the nCOS function or the *n*th-order nCOS function with the traditional least square method may lead to significant estimation error. This would be a serious issue if the estimated output spectrum is used for fault detection. To solve this problem, a series of novel and easy-to-operate algorithms were developed in [12, 29] and this leads to unique nonlinear features for fault detection—SOOS and nonlinear output regulation methods [30, 31, 39] with only experimental data.

1.8 Conclusions

A systematic parametric characteristic approach to the analysis and design of nonlinear systems in the frequency domain has been developed in the past years, referred to as the nCOS function method. The nCOS is an explicit and analytic function of system characteristic parameters in a polynomial form, and can be easily obtained with theoretical computation or experimental data, which thus provides an alternative and powerful tool for characterization, analysis and design of nonlinear systems and nonlinear features. The method can be well applied in control, vibration, fault detection, system identification, signal processing and structural optimization etc.

Obvious benefits of this parametric characteristic approach based on the GFRFs and nCOS function lie in that it can provide a convenient and straightforward tool for analysis and design of output frequency response in terms of characteristic parameters (model parameters and input amplitude etc.), present a special insight into nonlinear influence and nonlinear output frequency characteristics in the frequency domain based on which utilization of nonlinear benefits can thus be done readily with a clear technical guidance, and also produce new understanding of nonlinear dynamics from the viewpoint of Volterra series.

References

1. Leonov G.A., Ponomarenko D.V. Smirnova V.B., Frequency-domain methods for nonlinear analysis, theory and applications. World Scientific Publishing, Singapore, 1996
2. Taylor J. H. and Strobel K. L., Nonlinear Compensator Synthesis via Sinusoidal-Input Describing Functions, Proc. American Control Conference, Boston MA, 1242–1247, 1985
3. Nuij P.W.J.M., Bosgra O.H., Steinbuch M. Higher-order sinusoidal input describing functions for the analysis of non-linear systems with harmonic responses. Mechanical Systems and Signal Processing, 20, pp1883–1904, 2006
4. Engelberg, S. Limitations of the Describing Function for Limit Cycle Prediction, IEEE trans. Automatic Control, 47(11), pp 1887–1890, 2002
5. Orlowski P., Frequency domain analysis of uncertain time-varying discrete-time systems, Circuits Systems Signal Processing, Vol. 26, No. 3, 2007, PP. 293–310, 2007
6. Shah M. A., Franchek M. A. Frequency-based controller design for a class of nonlinear systems, International Journal of Robust and Nonlinear Control, 9(12), 825–840, 1999
7. Pavlov A., Van De Wouw N., and Nijmeijer H., Frequency Response Functions for Nonlinear Convergent Systems, IEEE Trans. Automatic Control, Vol 52, No 6, 1159–1165, 2007
8. David Rijlaarsdam, Pieter Nuij, Johan Schoukens, Maarten Steinbuch, A comparative overview of frequency domain methods for nonlinear systems, Mechatronics 42 (2017) 11–24
9. Rugh W.J., Nonlinear System Theory: the Volterra/Wiener Approach, Baltimore, Maryland, U.S.A.: Johns Hopkins University Press, 1981
10. George D.A., Continuous nonlinear systems, Technical Report 355, MIT Research Laboratory of Electronics, Cambridge, Mass. Jul. 24, 1959
11. Jing XJ and Lang ZQ, Frequency domain analysis and design of nonlinear systems based on Volterra series expansion: A parametric characteristic approach, Springer Int. Publishing, Switzerland, 2015
12. Jing X.J., Nonlinear characteristic output spectrum for nonlinear analysis and design, IEEE/ASME Transaction on Mechatronics, 19(1), 171–183, 2014
13. Jing X. J., Lang Z.Q., On the Generalized Frequency Response Functions of Volterra Systems, Journal of Dynamic Systems, Measurement and Control, 131(6), 2009
14. Peyton Jones JC, Billings SA, Recursive algorithm for computing the frequency response of a class of nonlinear difference equation models. Int J Control 50(5):1925–1940 (1989)
15. Billings SA, Peyton-Jones JC, Mapping nonlinear integro-differential equation into the frequency domain. Int J Control 54:863–879 (1990)
16. Jing X. J., Frequency Domain Analysis and Identification of Block-Oriented Nonlinear Systems, Journal of Sound and Vibration, 330 (22), 5427–5442, 2011

17. Jing X. J., Lang Z.Q., and Billings S.A., New Bound Characteristics of NARX Model in the Frequency Domain. International Journal of Control, Vol 80, No 1, 140–149, 2007
18. Jing X. J., et al., Magnitude bounds of generalized frequency response functions for nonlinear Volterra systems described by NARX model. Automatica, 44, 838–845, 2008
19. Jing X. J., Lang Z.Q. and Billings S.A., Frequency domain analysis for nonlinear Volterra systems with a general nonlinear output function. Inter. J. of Control, 81:2, 235–251, 2008
20. Swain A. K., Billings S. A., Generalized frequency response function matrix for MIMO nonlinear systems, International Journal of Control, 74(8), Pages 829–844, 2001
21. Jing X.J., Lang Z.Q. and Billings S.A., Output Frequency Properties of Nonlinear Systems, International Journal of Non-Linear Mechanics, 45(7), 2010, p 681–690
22. Xiao Z.L., Jing X.J., An SIMO Nonlinear System Approach to Analysis and Design of Vehicle Suspensions, IEEE/ASME Trans on Mechatronics, 20 (6), 2015
23. Jing X. J., Lang Z. Q., Billings S. A. and Tomlinson G. R., The parametric characteristic of frequency response functions for nonlinear systems. Int. J. of Control, 79(12), 2006
24. Jing X. J., Lang Z.Q. et.al., Parametric Characteristic Analysis for Generalized Frequency Response Functions of Nonlinear Systems, Circ., Syst. and Sign. Proc., 28: 699–733, 2009
25. Jing X. J., Lang Z.Q. and Billings S.A., Output Frequency Response Function based Analysis for Nonlinear Volterra Systems. Mech. Syst. and Sign. Process., 22, 102–120, 2008
26. Jing X. J., Lang Z.Q. and Billings S.A., Nonlinear influence in the frequency domain: Alternating series, Systems and Control Letters, 60 (5), 295–309, 2011
27. Jing X.J., Lang Z.Q. et.al., Frequency Domain Analysis for Suppression of Output Vibration from Periodic Disturbance using Nonlinearities. J. of sound and vibration 314, 536–557, 2008
28. Jing X. J. and Lang Z.Q., Frequency Domain Analysis of a Dimensionless Cubic Nonlinear Damping System Subject to Harmonic Input, Nonlinear Dynamics, 58: 469–485, 2009
29. Jing X. J., Truncation order and its effect in a class of nonlinear systems,Automatica,48(11), 2012
30. Jing X.J., Li QK, A Nonlinear Decomposition and Regulation Method for Nonlinearity Characterization, Nonlinear Dynamics, 83(3): 1355–1377, 2016
31. Q Li, Jing XJ, A second-order output spectrum approach for fault detection of bolt loosening in a satellite-like structure with a sensor chain, Nonlinear Dynamics, 89(1): 587–606, 2017
32. Jing XJ, Xiao ZL, On Convergence of Volterra Series Expansion of a Class of Nonlinear Systems, Asian Journal of Control, 19 (3), 1089–1102, 2017
33. Xiao Z.L., Jing X.J., Cheng L., Estimation of Parametric Convergence Bounds for Volterra Series Expansion of Nonlinear Systems, Mech. Syst. and Sign. Process., 45(1), 28–48, 2014
34. Xiao Z.L., Jing X.J., Frequency-domain analysis and design of linear feedback of nonlinear systems and applications in vehicle suspensions, IEEE/ASME Trans on Mechatr., 21(1), 2016
35. Bian J, Jing XJ, Superior Nonlinear Passive Damping Characteristics of the Bio-inspired Limb-Like or X-shaped Structure, accepted by Mech. Syst. and Sign. Proc., 2018
36. Wu Z, Jing XJ*, Bian J, Li F, and Allen R, Vibration isolation by exploring bio-inspired structural nonlinearity, Bioinspiration and Biomimetics, 10 (5), 056015–056015, 2015
37. Huihui Pan, Xingjian Jing, Weichao Sun, and Huijun Gao, A Bio-inspired Dynamics-Based Adaptive Tracking Control for Nonlinear Suspension Systems, IEEE Transactions on Control Systems Technology, 26(3), 903–914, 2018
38. Jingying Li , Xingjian Jing, Zhengchao Li and Xianlin Huang, Fuzzy Adaptive Control for Nonlinear Suspension Systems Based on A Bio-inspired Reference Model with Deliberately Designed Nonlinear Damping. IEEE Trans on Industrial Electronics, 66(11): 8713–8723, 2019. https://doi.org/10.1109/TIE.2018.2884219
39. Quankun Li, Xingjian Jing, Frequency Domain Nonlinear Vibration Features for Fault Diagnosis in Complex Structures, Springer, 2025

Chapter 2
General Frequency Response Functions and Applications in System Identification

Abstract The analysis and design of nonlinear systems remain a persistent challenge in the literature, while also being critically important across a wide range of engineering applications—including dynamic modeling, response prediction, performance enhancement, and fault diagnosis. Over the years, numerous methodologies have been proposed to address these challenges. Among them, the harmonic balance method stands out as a widely adopted and extensively studied technique for nonlinear dynamic analysis, with diverse applications. A concise overview of such conventional methods is provided in the preceding chapter. In the following chapters, we focus on a frequency-domain approach grounded in Volterra series theory—namely, the nonlinear Characteristic Output Spectrum (nCOS) function method. The nCOS framework is formulated based on the concept of Generalized Frequency Response Functions (GFRFs), which serve as a natural extension of the classical transfer function paradigm used in linear system theory.This formulation enables a systematic and parametric representation of nonlinear system behavior in the frequency domain, offering enhanced analytical capabilities for both theoretical investigation and practical design. This chapter is to present how to calculate the GFRFs of a nonlinear system and thus serves to provide fundamentals of the nCOS framework.

Keyword Frequency domain analysis · Nonlinear systems · General frequency response functions · Block-oriented models

2.1 Introduction

2.1.1 About This Chapter

Analysis and design of nonlinear systems have been a challenging topic in the literature and are also greatly demanded in most engineering systems for dynamics modelling, response analysis, performance improvement, fault diagnosis and so on. There are quite a few different methods which have been proposed in the past years. A well-known one for nonlinear dynamics analysis is the harmonic balance method,

X. Jing, *The Bio-Inspired X-Structure/Mechanism Approach for Exploring Nonlinear Benefits in Engineering*, https://doi.org/10.1007/978-981-96-5769-8_2

extensively studied and used for different purposes. A brief introduction to these methods can be referred to the last chapter. Here, we introduce the Volterra series based frequency domain method. As discussed in the following chapters, the new method is referred to as the nCOS function approach. The nCOS function is established on the basis of the concept of General Frequency Response Functions (GFRFs), which can be understood as an extension of the transfer function concept of linear systems.

In this chapter, the GFRF is introduced and the method about how to derive a GFRF is given and illustrated with detailed examples. For a class of block-oriented nonlinear models, the GFRFs are given based on these methods. For other more complicated nonlinear systems described by nonlinear differential equations or nonlinear autoregressive with exogenous input models (NARX), please also refer to my previous book [1] and papers (e.g., [2]).

2.1.2 Research Background

Nonlinear analysis in the frequency domain has been an interesting topic in the literature. This is likely due to the fact that frequency domain methods often offer a description of system dynamics that is both coordinate-free and equivalent, making it more comprehensible for engineers. A variety of methods have been devised for this very reason. These include methods based on absolute stability theory [3], extensions from Bode diagrams, singular value decomposition, and the idea of fluctuating eigenvalues [4–6], as well as describing functions [7–9] and harmonic balance methods [10], among others.

Frequency domain analysis of nonlinear systems can also be conducted using the concept of the generalized frequency response function (GFRF), which is a generalization of the transfer function for linear systems to nonlinear systems. The GFRF was first introduced in [11] based on the Volterra series expansion of nonlinear systems, which provides a powerful technique for the analysis and design of nonlinear systems in the frequency domain. Thereafter, by using the probing method [12], the GFRFs for a nonlinear system described by nonlinear differential equations (NDE) or nonlinear auto-regressive model with exogenous input (NARX) can directly be obtained from its model parameters [13–18]. The system output spectrum was studied in [19]. The bound characteristics of the frequency response functions were investigated in [20–23]. The bound results were greatly generalized in [22–24] where the bound expressions are described into an elegant and concise form which is a polynomial of the first order GFRF with model parameters as coefficients. The parametric characteristics of the GFRFs and nonlinear output spectrum of NDE or NARX systems were studied in [25] and some novel output frequency characteristics were revealed in [26] recently. These results demonstrate a great progress in the frequency domain analysis of nonlinear systems and provide a significant basis for the analysis and design of nonlinear systems in the frequency domain. The advantages of the GFRF based methods could lie in that it is mathematically elegant for a considerably large class of nonlinear systems (e.g., the GFRFs can be obtained for any NDE or NARX system directly), not restrict to any specific input signals such as harmonic or triangle or step inputs, and provides similar techniques to those for linear systems etc. Applications of these results can be found in many areas such as sound and vibration control, fault detection and biological data interpretation etc. [27–29].

Block-oriented nonlinear systems such as Hammerstein and Wiener models are composed by a cascade combination of a linear dynamic model and a static (memoryless) nonlinear function. Theoretically, any nonlinear systems which have a Volterra or Wiener functional expansion can be represented by a finite sum of Wiener models with a sufficient degree of accuracy [30, 31]. It was shown that the Wiener model is a reasonable model for many chemical and biological processes [32–34]. The MR damping systems can also be well approximated by a Hammerstein model [35]. Applications of these block-oriented models can be found in many areas such as mechanical systems [35], control systems [36], communication systems [37], chemical processes [33], and biological systems [34]. Although great progress has been made in the frequency domain analysis of nonlinear systems and the block-oriented nonlinear models have also been extensively studied in the literature, few results are reported about the analytical analysis of the frequency response of block-oriented nonlinear systems. Note that several frequency domain identification methods for block-oriented nonlinear systems are discussed in [35, 38–41]. However, no analytical nonlinear frequency response functions and output spectrum are derived and the identification can only be conducted under predefined harmonic excitations and requires sufficient structural information or invertibility of the nonlinear part. In [42], nonlinear transfer functions are discussed for Wiener–Hammerstein models. However, nonlinear output spectrum is still not addressed and the static nonlinear function in the model is only restricted to a univariate polynomial function.

This research develops analytical frequency response functions, including Generalized Frequency Response Functions (GFRFs) and the nonlinear output spectrum of block-oriented nonlinear systems. This is achieved by utilizing the probing method and leveraging the structural information of system models. The resulting GFRFs and nonlinear output spectrum clearly illustrate the relationship between frequency response functions and model parameters, as well as the dependence of frequency response functions on the model's linear part. The model's nonlinear part can be a more generalized multivariate polynomial function. These findings form a crucial foundation for the analysis and design of block-oriented nonlinear systems.

The study also explores the frequency domain identification of block-oriented models. An efficient algorithm is presented for estimating different orders of nonlinear output spectra. This algorithm can accurately generate the different order output spectrum caused by different order nonlinearity in the system, without needing any structural information of the system. This makes it applicable to any nonlinear systems. The estimated output spectrum can be used for the nonlinear analysis and design of the system under consideration, and can also indicate the system's nonlinear severity for non-destructive evaluation purposes.

A general procedure for frequency domain identification of a block-oriented model is subsequently discussed, and both parametric and nonparametric models can be generated. Compared to existing frequency domain identification methods, this new procedure does not necessarily require extensive model structure information, does not need the invertibility of the nonlinear part, and is not restricted to harmonic input signals. Simulation examples are provided to demonstrate these new findings.

2.2 The Generalized Frequency Response Functions and Output Spectrum

2.2.1 *Basis of Volterra Series Related Theory*

The input–output relationship for a wide class of nonlinear systems can be approximated by a Volterra series of order N as [12, 31]

$$y(t) = \sum_{n=1}^{N} y_n(t) \tag{2.1a}$$

$$y_n(t) = \int_{-\infty}^{\infty} \cdots \int_{-\infty}^{\infty} h_n(\tau_1, \ldots, \tau_n) \prod_{i=1}^{n} u(t - \tau_i) \mathrm{d}\tau_i \tag{2.1b}$$

where $h_n(\tau_1, \ldots, \tau_n)$ is a real valued function of $\tau_1, \ldots, \tau_n$ known as the nth order Volterra kernel. The nth order generalized frequency response function (GFRF) is defined as

$$H_n(j\omega_1, \ldots, j\omega_n) = \int_{-\infty}^{\infty} \cdots \int_{-\infty}^{\infty} h_n(\tau_1, \ldots, \tau_n) \exp(-j(\omega_1\tau_1 + \cdots + \omega_n\tau_n)) \mathrm{d}\tau_1 \cdots \mathrm{d}\tau_n \tag{2.2}$$

which is the multidimensional Fourier transform of $h_n(\tau_1, \cdots, \tau_n)$. With the inverse Fourier transform of the nth order GFRF, (2.1b) can be written as

$$y_n(t) = \frac{1}{(2\pi)^n} \int_{-\infty}^{\infty} \cdots \int_{-\infty}^{\infty} H_n(j\omega_1, \ldots, j\omega_n) \prod_{i=1}^{n} U(j\omega_i) e^{j(\omega_1 + \cdots + \omega_n)t} \mathrm{d}\omega_1 \cdots \mathrm{d}\omega_n$$

denoting $\omega_n = \omega - \omega_1 - \cdots - \omega_{n-1}$, which can be further written as

$$y_n(t) = \frac{1}{2\pi} \int_{-\infty}^{\infty} \left[\frac{1}{(2\pi)^{n-1}} \underbrace{\int_{-\infty}^{\infty} \cdots \int_{-\infty}^{\infty}}_{n-1} H_n(j\omega_1, \cdots, j\omega_{n-1}, \right.$$

$$j(\omega - \omega_1 - \cdots - \omega_{n-1}))$$
$$U_n(j\omega_1, \cdots j\omega_{n-1})\mathrm{d}\omega_1 \cdots \mathrm{d}\omega_{n-1}\big]e^{j\omega t}\mathrm{d}\omega$$

where $U_n(j\omega_1, \ldots, j\omega_{n-1}) = U(j\omega_1)\cdots U(j\omega_{n-1})U(j(\omega - \omega_1 - \cdots - \omega_{n-1}))$. Therefore the Fourier transform of $y_n(t)$ is obtained as

$$Y_n(j\omega) = \frac{1}{(2\pi)^{n-1}} \underbrace{\int_{-\infty}^{\infty} \cdots \int_{-\infty}^{\infty}}_{n-1} H_n(j\omega_1, \ldots, j\omega_{n-1}, j(\omega - \omega_1 - \cdots - \omega_{n-1}))$$
$$U_n(j\omega_1, \ldots, j\omega_{n-1})\mathrm{d}\omega_1 \cdots \mathrm{d}\omega_{n-1} \tag{2.3}$$

which is referred to as the nth-order output spectrum. The output spectrum of the nonlinear system in (2.1) can then be computed by

$$Y(j\omega) = \sum_{n=1}^{N} Y_n(j\omega) \tag{2.4}$$

Obviously, the output spectrum involves the computation of the GFRFs. For a practical nonlinear system, the GFRFs can be derived by using the probing method [12]. Consider the excitation of system (2.1) with an input consisting of n complex exponentials defined as

$$u(t) = \sum_{r=1}^{n} e^{j\omega_r t} \tag{2.5}$$

The output y(t) is given as

$$y(t) = \sum_{n=1}^{N} \sum_{r_1, r_n=1}^{n} H_n(j\omega_{r_1}, \cdots, j\omega_{r_n}) e^{j(\omega_{r_1} + \cdots + \omega_{r_n})t} \tag{2.6}$$

Then for a nonlinear system, replacing the input and output by (2.5–2.6), the nth-order GFRF can be obtained by extracting the coefficients of the term $e^{j(\omega_{r_1} + \cdots + \omega_{r_n})t}$. With this method, the following result can be obtained.

Lemma 1. *Consider a static polynomial function*

$$y(t) = f(u(t)) = c_1 u(t) + c_2 u(t)^2$$
$$+ c_3 u(t)^3 + \cdots + c_n u(t)^n + \cdots \tag{2.7}$$

The nth-order GFRF between the input $u(t)$ and output $y(t)$ is given by

$$H_n(j\omega_1, \cdots, j\omega_n) = c_n \tag{2.8}$$

Proof. See Appendix 1. □

The example above, although simple, is useful for the following derivation of the GFRFs of block-oriented nonlinear models. Note that many complex nonlinear functions allow a polynomial approximation to any degree of accuracy with sufficiently high order [43]. For this reason, polynomial functions are widely adopted as the nonlinear part of block-oriented nonlinear models.

2.2.2 The Harmonic Probing Method

To demonstrate the probing method for deriving the GFRFs of a nonlinear system, examples are given below for a better demonstration of this method.

Example 01. For a nonlinear system with single input $u(t)$ and single output $x(t)$ shown below,

$$\ddot{x} + c_0\dot{x} + c_1x^2\dot{x} + k_0x + k_1x^2 = u + \dot{u}^2 \tag{2.9}$$

its general frequency response functions can be calculated with the harmonic probing method as follows.

Firstly, considering a periodic excitation with a single frequency,

$$u(t) = e^{j\omega t} \tag{2.10}$$

The resulting output response $x(t)$ can be expressed by

$$x(t) = H_1(\omega)e^{j\omega t} + \cdots \tag{2.11}$$

Substituting (2.10) and (2.11) into (2.9), and extracting coefficients of $e^{j\omega t}$ gives,

$$-\omega^2H_1(\omega) + c_0j\omega H_1(\omega) + k_0H_1(\omega) = 1$$

Thus, $H_1(\omega)$ can be calculated as

$$H_1(\omega) = \frac{1}{\left(-\omega^2 + c_0j\omega + k_0\right)}$$

Secondly, considering a periodic excitation with two frequencies,

$$u(t) = e^{j\omega_1 t} + e^{j\omega_2 t} \tag{2.12}$$

The resulting output response $x(t)$ can be expressed by

$$\begin{aligned} x(t) &= C_1^1 H_1(\omega_1) e^{j\omega_1 t} + C_1^1 H_1(\omega_2) e^{j\omega_2 t} \\ &\quad + C_2^1 H_2(\omega_1, \omega_2) e^{j(\omega_1+\omega_2)t} + \cdots \end{aligned} \tag{2.13}$$

where C_n^i means combination operation.

Substituting (2.12) and (2.13) into (2.9), and extracting coefficients of $e^{j(\omega_1+\omega_2)t}$ gives,

$$\begin{aligned} &-(\omega_1+\omega_2)^2 2H_2(\omega_1,\omega_2) + c_0 j(\omega_1+\omega_2) 2H_2(\omega_1,\omega_2) \\ &\qquad + k_0 2H_2(\omega_1,\omega_2) + k_1 2H_1(\omega_1)H_1(\omega_2) = -2\omega_1\omega_2 \end{aligned}$$

Thus, $H_2(\omega_1, \omega_2)$ can be calculated, as

$$H_2(\omega_1,\omega_2) = \frac{-\omega_1\omega_2 - k_1 H_1(\omega_1)H_1(\omega_2)}{\left(-(\omega_1+\omega_2)^2 + c_0 j(\omega_1+\omega_2) + k_0\right)}$$

Thirdly, considering a periodic excitation with triple frequencies,

$$u(t) = \sum_{i=1}^{3} e^{j\omega_i t} \tag{2.14}$$

The resulting output response $x(t)$ can be expressed by

$$\begin{aligned} x(t) &= C_1^1 H_1(\omega_1) e^{j\omega_1 t} + C_1^1 H_1(\omega_2) e^{j\omega_2 t} \\ &\quad + C_1^1 H_1(\omega_3) e^{j\omega_3 t} + C_2^1 H_2(\omega_1,\omega_2) e^{j(\omega_1+\omega_2)t} \\ &\quad + C_2^1 H_2(\omega_2,\omega_3) e^{j(\omega_2+\omega_3)t} + C_2^1 H_2(\omega_1,\omega_3) e^{j(\omega_1+\omega_3)t} \\ &\quad + C_3^1 C_2^1 H_3(\omega_1,\omega_2,\omega_3) e^{j(\omega_1+\omega_2+\omega_3)t} + \cdots \end{aligned} \tag{2.15}$$

Substituting (2.14) and (2.15) into (2.9), and extracting coefficients of $e^{j(\omega_1+\omega_2+\omega_3)t}$ gives,

$$\begin{aligned} &-(\omega_1+\omega_2+\omega_3)^2 6H_3(\omega_1,\omega_2,\omega_3) \\ &+ c_0 j(\omega_1+\omega_2+\omega_3) 6H_3(\omega_1,\omega_2,\omega_3) \\ &+ c_1 \begin{pmatrix} 3H_1(\omega_1)H_1(\omega_2) j\omega_3 H_1(\omega_3) \\ +3H_1(\omega_1) j\omega_2 H_1(\omega_2) H_1(\omega_3) \\ +j\omega_1 3H_1(\omega_1)H_1(\omega_2)H_1(\omega_3) \end{pmatrix} \\ &+ k_0 6H_3(\omega_1,\omega_2,\omega_3) + k_1 \begin{pmatrix} 2H_1(\omega_1) 2H_2(\omega_2,\omega_3) \\ +2H_1(\omega_2) 2H_2(\omega_1,\omega_3) \\ +2H_1(\omega_3) 2H_2(\omega_1,\omega_2) \end{pmatrix} = 0 \end{aligned}$$

Thus, $H_3(\omega_1, \omega_2, \omega_3)$ can be calculated as

$$H_3(\omega_1,\omega_2,\omega_3) = \frac{c_1\begin{pmatrix}3H_1(\omega_1)H_1(\omega_2)j\omega_3H_1(\omega_3)\\+3H_1(\omega_1)j\omega_2H_1(\omega_2)H_1(\omega_3)\\+j\omega_1 3H_1(\omega_1)H_1(\omega_2)H_1(\omega_3)\end{pmatrix} + k_1\begin{pmatrix}2H_1(\omega_1)2H_2(\omega_2,\omega_3)\\+2H_1(\omega_2)2H_2(\omega_1,\omega_3)\\+2H_1(\omega_3)2H_2(\omega_1,\omega_2)\end{pmatrix}}{-\begin{pmatrix}-6(\omega_1+\omega_2+\omega_3)^2\\+c_0 j6(\omega_1+\omega_2+\omega_3)+6k_0\end{pmatrix}}$$

Similarly, considering a periodic excitation with multiple frequencies,

$$u(t) = \sum_{i=1}^{n} e^{j\omega_i t} \tag{2.16}$$

The resulting output response $x(t)$ can be expressed by

$$\begin{aligned} x(t) = &\sum_{a=1}^{n} C_1^1 H_1(\omega_a)e^{j\omega_a t} + \sum_{a=1,b=1,a\neq b}^{n} C_2^1 H_2(\omega_a,\omega_b)e^{j(\omega_a+\omega_b)t} \\ &+ \sum_{a=1,b=1,c=1,a\neq b\neq c}^{n} C_3^1 C_2^1 H_3(\omega_a,\omega_b,\omega_c)e^{j(\omega_a+\omega_b+\omega_c)t} \\ &+ \sum_{a=1,b=1,c=1,d=1,a\neq b\neq c\neq d}^{n} C_4^1 C_3^1 C_2^1 H_4(\omega_a,\omega_b,\omega_c,\omega_d)e^{j(\omega_a+\omega_b+\omega_c+\omega_d)t} \\ &\ldots \\ &+ C_n^1 C_{n-1}^1 \ldots C_2^1 H_n(\omega_1,\omega_2,\ldots,\omega_n)e^{j(\omega_1+\omega_2+\ldots+\omega_n)t} \end{aligned} \tag{2.17}$$

Substituting (2.16) and (2.17) into (2.9), and extracting coefficients of $e^{j(\omega_1+\omega_2+\ldots+\omega_n)t}$ gives the n-th order $H_n(\omega_1,\omega_2,\ldots,\omega_n)$.

Example 02. Consider a SIMO system as follows:

$$\begin{cases} \sum_{l=0}^{L} c_{1,0}^{10}(l;1)\frac{d^l x_1}{dt^l} + \sum_{l=0}^{L} c_{1,0}^{01}(l;1)\frac{d^l x_2}{dt^l} = \sum_{l=0}^{L} c_{0,1}^{00}(l)\frac{d^l u}{dt^l} \\ \quad + \sum_{m=p+q=2}^{N}\sum_{p=0}^{m}\sum_{\substack{s_1=0\\ s_1+s_2=p}}^{p}\sum_{l_1,\cdots,l_{p+q}=0}^{L} c_{p,q}^{s_1 s_2}(l_1,\cdots,l_{p+q};1) \\ \quad \times \prod_{i=1}^{s_1}\frac{d^{l_i}x_1}{dt^{l_i}}\prod_{i=s_1+1}^{s_1+s_2}\frac{d^{l_i}x_2}{dt^{l_i}}\prod_{i=p+1}^{p+q}\frac{d^{l_i}u}{dt^{l_i}} \\ \dot{x}_2 = \sum_{m=p+q=2}^{N}\sum_{p=0}^{m}\sum_{\substack{s_1=0\\ s_1+s_2=p}}^{p}\sum_{l_1,\cdots,l_{p+q}=0}^{L} c_{p,q}^{s_1 s_2}(l_1,\cdots,l_{p+q};2) \\ \quad \times \prod_{i=1}^{s_1}\frac{d^{l_i}x_1}{dt^{l_i}}\prod_{i=s_1+1}^{s_1+s_2}\frac{d^{l_i}x_2}{dt^{l_i}}\prod_{i=p+1}^{p+q}\frac{d^{l_i}u}{dt^{l_i}} \end{cases} \tag{2.18}$$

where x_i is the ith output for $i = 1, 2$ with $x_1 = x$; N is the maximum nonlinearity order in terms of output $\boldsymbol{x}$ and input $\boldsymbol{u}$; $c_{p,q}^{s_1 s_2}(l_1, \ldots, l_{p+q}; j)$ is the model parameter of the corresponding term in the jth equation, which has a nonlinear degree $p + q$ (p order in terms of the outputs x_1, x_2, and q order in terms of the input u), referred to as nonlinear parameter if $\mathrm{p} + \mathrm{q} > 1$ otherwise as linear parameter if $\mathrm{p} + \mathrm{q} = 1$; l_i is the differential order with the maximum order L; $s_1, s_2 \in \mathbb{N}$, which are the nonlinearity degrees of x_1, x_2 in the nonlinear term (s_1 corresponds to the output x_1, and s_2 corresponds to the output x_2), e.g.,

$$\prod_{i=1}^{s_1}\frac{d^{l_i}x_1}{dt^{l_i}}\prod_{i=s_1+1}^{s_1+s_2}\frac{d^{l_i}x_2}{dt^{l_i}}\prod_{i=p+1}^{p+q}\frac{d^{l_i}u}{dt^{l_i}},$$

respectively. When $s < s_0$, $\prod_{i=s_0}^{s}(*) = 1$.

It is supposed that the equilibrium of the nonlinear system above is zero and thus DC terms are not considered. To obtain the GFRFs, the principle is based on the fact that a growing exponential input signal $u = e^{j\omega_1 t}$ is simply scaled by a homogeneous system

$$y(t) = \int_{-\infty}^{\infty} h_n(\tau_1, \ldots, \tau_n)u(t-\tau_1)\cdots u(t-\tau_n)d\tau_1\cdots d\tau_n$$

to produce the output signal $y = H_n(\omega_1, \ldots, \omega_1)e^{jn\omega_1 t}$.

For the first-order GFRF, assuming that the input is given as $u = e^{j\omega_1 t}$, the output of system (2.18) can be expressed as

$$\begin{aligned} x_i &= H_0^i + \frac{1}{2\pi}\int_{-\infty}^{+\infty} H_1^i(j\Omega_1)\cdot U(j\Omega_1)e^{j\Omega_1 t}d\Omega_1 \\ &= H_0^i + \frac{1}{2\pi}\int_{-\infty}^{+\infty} H_1^i(j\Omega_1)\cdot 2\pi\delta(\Omega_1-\omega_1)e^{j\Omega_1 t}d\Omega_1 \\ &= H_0^i + H_1^i(j\omega_1)e^{j\omega_1 t},\ i=1,2. \end{aligned} \tag{2.19}$$

Substituting (2.19) into (2.18) and equating the coefficients of $e^{j\omega_1 t}$ to be zero, the results are as follows.

$$\begin{aligned} &\sum_{l=0}^{L} c_{1,0}^{10}(l;1)(j\omega_1)^l\cdot H_1^1(j\omega_1) + \sum_{l=0}^{L} c_{1,0}^{01}(l;1)(j\omega_1)^l\cdot H_1^2(j\omega_1) = \sum_{l=0}^{L} c_{0,1}^{00}(l;1)(j\omega_1)^l \\ &\quad + \sum_{\substack{p=1,\\ s_1+s_2=p}}^{N-1}\sum_{l_{p+1}=0}^{L} c_{p,1}^{s_1s_2}(0,\ldots,0,l_{p+1};1)\left(H_0^1\right)^{s_1}\left(H_0^2\right)^{s_2}(j\omega_1)^{l_{p+1}} \\ &\quad + \sum_{\substack{p=2,\\ t_2=p-1}}^{N}\sum_{l_1=0}^{L} c_{p,0}^{1t_2}(l_1,0,\ldots,0;1)(j\omega_1)^{l_1}(H_0^2)^{t_2}\cdot H_1^1(j\omega_1) \\ &\quad + \sum_{\substack{p=2,\\ t_1=p-1}}^{N}\sum_{l_2=0}^{L} c_{p,0}^{t_11}(0,\ldots,0,\ldots,l_2;1)(j\omega_1)^{l_2}(H_0^1)^{t_1}\cdot H_1^2(j\omega_1) \\ &\quad + \sum_{p=2}^{N}\sum_{s_1=2}^{p}\sum_{i=1}^{s_1}\sum_{l_i=0}^{L} c_{p,0}^{s_1s_2}(0,l_i,0;1)\left(H_0^1\right)^{s_1-1}\left(H_0^2\right)^{s_2}(j\omega_1)^{l_i}\cdot H_1^1(j\omega_1) \\ &\quad + \sum_{p=2}^{N}\sum_{s_2=2}^{p}\sum_{i=1}^{s_2}\sum_{l_{p-i+1}=0}^{L} c_{p,0}^{s_1s_2}(0,l_{p-i+1},0;1)\left(H_0^1\right)^{s_1}\left(H_0^2\right)^{s_2-1}(j\omega_1)^{l_{p-i+1}} \\ &\quad \cdot H_1^2(j\omega_1), \end{aligned}$$

$$\begin{aligned} (j\omega_1)\cdot H_1^2(j\omega_1) &= \sum_{\substack{p=1,\\ s_1+s_2=p}}^{N-1}\sum_{l_{p+1}=0}^{L} c_{p,1}^{s_1s_2}(0,\ldots,0,l_{p+1};2)\left(H_0^1\right)^{s_1}\left(H_0^2\right)^{s_2}(j\omega_1)^{l_{p+1}} \\ &\quad + \sum_{\substack{p=2,\\ t_2=p-1}}^{N}\sum_{l_1=0}^{L} c_{p,0}^{1t_2}(l_1,0,\ldots,0;2)(j\omega_1)^{l_1}(H_0^2)^{t_2}\cdot H_1^1(j\omega_1) \\ &\quad + \sum_{\substack{p=2,\\ t_1=p-1}}^{N}\sum_{l_2=0}^{L} c_{p,0}^{t_11}(0,\ldots,0,\ldots,l_2;2)(j\omega_1)^{l_2}(H_0^1)^{t_1}\cdot H_1^2(j\omega_1) \\ &\quad + \sum_{p=2}^{N}\sum_{s_1=2}^{p}\sum_{i=1}^{s_1}\sum_{l_i=0}^{L} c_{p,0}^{s_1s_2}(0,l_i,0;2)\left(H_0^1\right)^{s_1-1}\left(H_0^2\right)^{s_2}(j\omega_1)^{l_i}\cdot H_1^1(j\omega_1) \end{aligned}$$

$$+\sum_{p=2}^{N}\sum_{s_2=2}^{p}\sum_{i=1}^{s_2}\sum_{l_{p-i+1}=0}^{L} c_{p,0}^{s_1 s_2}(0, l_{p-i+1}, 0; 2)\left(H_0^1\right)^{s_1}\left(H_0^2\right)^{s_2-1}(j\omega_1)^{l_{p-i+1}}$$

$$\cdot H_1^2(j\omega_1), \tag{2.20}$$

Writing the equation above into the matrix form,

$$\begin{bmatrix} l_{1,1} & l_{1,2} \\ l_{2,1} & l_{2,2} \end{bmatrix}\begin{bmatrix} H_n^1(j\omega_1, \cdots, j\omega_n) \\ H_n^2(j\omega_1, \cdots, j\omega_n) \end{bmatrix} = \begin{bmatrix} A_{1,n} \\ A_{2,n} \end{bmatrix}$$

and the related terms are

$$l_{1,1} = \sum_{l=0}^{L} c_{1,0}^{10}(l; 1)(j\omega_1)^l -$$

$$\sum_{\substack{p=2,\\ t_2=p-1}}^{N}\sum_{l_1=0}^{L} c_{p,0}^{1t_2}(l_1, 0, \cdots, 0; 1)(j\omega_1)^{l_1}(H_0^2)^{t_2},$$

$$l_{1,2} = \sum_{l=0}^{L} c_{1,0}^{01}(l; 1)(j\omega_1)^l -$$

$$\sum_{\substack{p=2,\\ t_1=p-1}}^{N}\sum_{l_2=0}^{L} c_{p,0}^{t_1 1}(0, \cdots, 0, \cdots, l_2; 1)(j\omega_1)^{l_2}(H_0^1)^{t_1},$$

$$l_{2,1} = -\sum_{\substack{p=2,\\ t_2=p-1}}^{N}\sum_{l_1=0}^{L} c_{p,0}^{1t_2}(l_1, 0, \cdots, 0; 2)(j\omega_1)^{l_1}(H_0^2)^{t_2},$$

$$l_{2,2} = (j\omega_1) -$$

$$\sum_{\substack{p=2,\\ t_1=p-1}}^{N}\sum_{l_2=0}^{L} c_{p,0}^{t_1 1}(0, \cdots, 0, \cdots, l_2; 2)(j\omega_1)^{l_2}(H_0^1)^{t_1},$$

$$A_{1,1} = \sum_{l=0}^{L} c_{0,1}^{00}(l; 1)(j\omega_1)^l$$

$$+\sum_{\substack{p=1,\\ s_1+s_2=p}}^{N-1}\sum_{l_{p+1}=0}^{L} c_{p,1}^{s_1 s_2}(0, \ldots, 0, l_{p+1}; 1)\left(H_0^1\right)^{s_1}\left(H_0^2\right)^{s_2}(j\omega_1)^{l_{p+1}}$$

$$+\sum_{p=2}^{N}\sum_{s_1=2}^{p}\sum_{i=1}^{s_1}\sum_{l_i=0}^{L} c_{p,0}^{s_1 s_2}(0, l_i, 0; 1)\left(H_0^1\right)^{s_1-1}\left(H_0^2\right)^{s_2}(j\omega_1)^{l_i} \cdot H_1^1(j\omega_1)$$

$$+\sum_{p=2}^{N}\sum_{s_2=2}^{p}\sum_{i=1}^{s_2}\sum_{l_{p-i+1}=0}^{L} c_{p,0}^{s_1s_2}(0,l_{p-i+1},0;1)\left(H_0^1\right)^{s_1}\left(H_0^2\right)^{s_2-1}(j\omega_1)^{l_{p-i+1}}\cdot H_1^2(j\omega_1),$$

$$A_{2,1}=\sum_{\substack{p=1,\\ s_1+s_2=p}}^{N-1}\sum_{l_{p+1}=0}^{L} c_{p,1}^{s_1s_2}(0,\ldots,0,l_{p+1};2)\left(H_0^1\right)^{s_1}\left(H_0^2\right)^{s_2}(j\omega_1)^{l_{p+1}}$$

$$+\sum_{p=2}^{N}\sum_{s_1=2}^{p}\sum_{i=1}^{s_1}\sum_{l_i=0}^{L} c_{p,0}^{s_1s_2}(0,l_i,0;2)\left(H_0^1\right)^{s_1-1}\left(H_0^2\right)^{s_2}(j\omega_1)^{l_i}\cdot H_1^1(j\omega_1)$$

$$+\sum_{p=2}^{N}\sum_{s_2=2}^{p}\sum_{i=1}^{s_2}\sum_{l_{p-i+1}=0}^{L} c_{p,0}^{s_1s_2}(0,l_{p-i+1},0;2)\left(H_0^1\right)^{s_1}\left(H_0^2\right)^{s_2-1}(j\omega_1)^{l_{p-i+1}}\cdot H_1^2(j\omega_1),$$

For the nth-order GFRF, following previous procedures and giving the input as $u=e^{j\omega_1 t}+\cdots+e^{j\omega_n t}$, the output can then be given as

$$x_i=H_0^i+\sum_{\substack{\text{all combinations}\\ \text{of } \omega_{i1} \text{ in } \boldsymbol{\omega}}} H_1^i(j\omega_{i1})e^{j\omega_{i1}t}+\cdots+$$
$$+\sum_{\substack{\text{all permutations}\\ \text{of } (\omega_{i1},\ldots,\omega_{in})}}\ \sum_{\substack{\text{all combinations}\\ \text{of } (\omega_{i1},\ldots,\omega_{in}) \text{ in } \boldsymbol{\omega}}} H_n^i(j\omega_{i1},\ldots,j\omega_{in})e^{j\omega_{i1}t+\cdots+j\omega_{in}t} \tag{2.21}$$

where $\boldsymbol{\omega}=(\omega_1,\ldots,\omega_n)$. Substituting (2.21) into (2.18) and equating the coefficients of $e^{j\omega_1 t+\cdots+j\omega_n t}$ to be zero, a recursive calculation can then be obtained as follows,

The nth-order GFRF can be calculated as

$$H_n=\begin{bmatrix} H_n^1(j\omega_1,\cdots,j\omega_n)\\ H_n^2(j\omega_1,\cdots,j\omega_n)\end{bmatrix}=\begin{bmatrix} l_{1,1} & l_{1,2}\\ l_{2,1} & l_{2,2}\end{bmatrix}^{-1}\begin{bmatrix} A_{1,n}\\ A_{2,n}\end{bmatrix}=L_n^{-1}\begin{bmatrix} A_{1,n}\\ A_{2,n}\end{bmatrix} \tag{2.22}$$

where

$$l_{1,1}=\sum_{l=0}^{L} c_{1,0}^{10}(l;1)(j\omega_1+\cdots+j\omega_n)^l$$
$$-\sum_{\substack{p=2,\\ t_2=p-1}}^{N}\sum_{l_1=0}^{L} c_{p,0}^{1t_2}(l_1,0,\ldots,0;1)(j\omega_1+\cdots+j\omega_n)^{l_1}(H_0^2)^{t_2},$$

$$l_{1,2}=\sum_{l=0}^{L} c_{1,0}^{01}(l;1)(j\omega_1+\cdots+j\omega_n)^l$$
$$-\sum_{\substack{p=2,\\ t_1=p-1}}^{N}\sum_{l_2=0}^{L} c_{p,0}^{t_11}(0,\ldots,0,\ldots,l_2;1)(j\omega_1+\cdots+j\omega_n)^{l_2}(H_0^1)^{t_1},$$

$$l_{2,1} = -\sum_{\substack{p=2,\\ t_2=p-1}}^{N}\sum_{l_1=0}^{L} c_{p,0}^{1t_2}(l_1, 0, \cdots, 0; 2)(j\omega_1 + \cdots + j\omega_n)^{l_1}(H_0^2)^{t_2},$$

$$l_{2,2} = (j\omega_1 + \cdots + j\omega_n)$$
$$-\sum_{\substack{p=2,\\ t_1=p-1}}^{N}\sum_{l_2=0}^{L} c_{p,0}^{t_1 1}(0, \ldots, 0, \ldots, l_2; 2)(j\omega_1 + \cdots + j\omega_n)^{l_2}(H_0^1)^{t_1}, \quad (2.23)$$

and

$$A_{i,n} = \sum_{l_1,\cdots,l_n=0}^{L} c_{0,n}^{00}(l_1, \ldots, l_n; i)\prod_{i=1}^{n}(j\omega_i)^{l_i}$$
$$+\sum_{m=p+q=2}^{N}\sum_{p=1}^{m}\sum_{\substack{t=t_1+t_2=0\\ t_1\le s_1, t_2\le s_2}}^{p-1}\sum_{l_1,\cdots,l_{p+q}=0}^{L} c_{p,q}^{s_1 s_2}(l_1, \ldots, l_{p+q}; i)$$
$$\times (H_0^1)^{t_1}(H_0^2)^{t_2}\prod_{i=1}^{q}(j\omega_{n-q+i})^{l_{n-q+i}} H_{n-q,p-t}^{t_1\prime t_2\prime}(j\omega_1, \ldots, j\omega_{n-q}), \quad (2.24)$$

$$H_{n-q,p-t}^{t_1\prime t_2\prime}(j\omega_1, \ldots, j\omega_{n-q}) = \sum_{i=1}^{t_1\prime}(j\omega_1 + \cdots + j\omega_i)^{l_{t_1+1}}$$
$$H_i^1(j\omega_1, \ldots, j\omega_i)H_{n-q-i,p-t-1}^{(t_1\prime-i)t_2\prime}(j\omega_{i+1}, \ldots, j\omega_{n-q})$$
$$+\sum_{i=1}^{t_2\prime}(j\omega_1 + \cdots + j\omega_i)^{l_{s_1+t_2+1}}$$
$$H_i^2(j\omega_1, \ldots, j\omega_i)H_{n-q-i,p-t-1}^{t_1\prime(t_2\prime-i)}(j\omega_{i+1}, \ldots, j\omega_{n-q}), \quad (2.25)$$

where $\sum_{i=0}^{0}(\cdot) = 0$, $H_{0,0}(.) = 1$, $H_{n,0}(.) = 0$ for $n > 0$, $H_{n,p}(.) = 0$ for $n < p$, and $s_1 + s_2 = p$, $t_1 + t_2 = t$, $t_1\prime = s_1 - t_1$, $t_2\prime = s_2 - t_2$. The recursion in (2.25) ends when $p - t = 1$ (as such, $t_1\prime = 0$ or 1 and $t_2\prime = 1$ or 0) holds. In such case, note also that and $s_1 + s_2 = p$, $t_1 + t_2 = t$, $t_1\prime = s_1 - t_1$, $t_2\prime = s_2 - t_2$, and thus $t_1\prime = 0$ or 1 and $t_2\prime = 1$ or 0 hold. Further with (2.25) and noting that $\sum_{i=0}^{0}(\cdot) = 0$, $H_{0,0}(.) = 1$, $H_{n,0}(.) = 0$ for $n > 0$, $H_{n,p}(.) = 0$ for $n < p$, (2.26) can be obtained.

That is,

$$H_{\overline{n},1}^{10}(j\omega_1, \ldots, j\omega_{\overline{n}}) = (j\omega_1 + \cdots + j\omega_{\overline{n}})^{l_{s_1}} H_{\overline{n}}^1(j\omega_1, \ldots, j\omega_{\overline{n}}),$$
$$H_{\overline{n},1}^{01}(j\omega_1, \ldots, j\omega_{\overline{n}}) = (j\omega_1 + \cdots + j\omega_{\overline{n}})^{l_{s_1+s_2}} H_{\overline{n}}^2(j\omega_1, \ldots, j\omega_{\overline{n}}), \quad (2.26)$$

where $\overline{n} = 1,2,\ldots$,or n. For the first-order GFRF,

$$
\begin{aligned}
A_{1,1} &= \sum_{l=0}^{L} c_{0,1}^{00}(l;\,1)(j\omega_1)^l \\
&+ \sum_{\substack{p=1,\\ s_1+s_2=p}}^{N-1} \sum_{l_{p+1}=0}^{L} c_{p,1}^{s_1 s_2}(0,\ldots,0,l_{p+1};\,1)\left(H_0^1\right)^{s_1}\left(H_0^2\right)^{s_2}(j\omega_1)^{l_{p+1}} \\
&+ \sum_{p=2}^{N}\sum_{s_1=2}^{p}\sum_{i=1}^{s_1}\sum_{l_i=0}^{L} c_{p,0}^{s_1 s_2}(0,l_i,0;\,1)\left(H_0^1\right)^{s_1-1}\left(H_0^2\right)^{s_2}(j\omega_1)^{l_i}\cdot H_1^1(j\omega_1) \\
&+ \sum_{p=2}^{N}\sum_{s_2=2}^{p}\sum_{i=1}^{s_2}\sum_{l_{p-i+1}=0}^{L} c_{p,0}^{s_1 s_2}(0,l_{p-i+1},0;\,1)\left(H_0^1\right)^{s_1}\left(H_0^2\right)^{s_2-1}(j\omega_1)^{l_{p-i+1}}\cdot H_1^2(j\omega_1), \\
A_{2,1} &= \sum_{\substack{p=1,\\ s_1+s_2=p}}^{N-1} \sum_{l_{p+1}=0}^{L} c_{p,1}^{s_1 s_2}(0,\ldots,0,l_{p+1};\,2)\left(H_0^1\right)^{s_1}\left(H_0^2\right)^{s_2}(j\omega_1)^{l_{p+1}} \\
&+ \sum_{p=2}^{N}\sum_{s_1=2}^{p}\sum_{i=1}^{s_1}\sum_{l_i=0}^{L} c_{p,0}^{s_1 s_2}(0,l_i,0;\,2)\left(H_0^1\right)^{s_1-1}\left(H_0^2\right)^{s_2}(j\omega_1)^{l_i}\cdot H_1^1(j\omega_1) \\
&+ \sum_{p=2}^{N}\sum_{s_2=2}^{p}\sum_{i=1}^{s_2}\sum_{l_{p-i+1}=0}^{L} c_{p,0}^{s_1 s_2}(0,l_{p-i+1},0;\,2)\left(H_0^1\right)^{s_1}\left(H_0^2\right)^{s_2-1}(j\omega_1)^{l_{p-i+1}}\cdot H_1^2(j\omega_1),
\end{aligned}
\tag{2.27}
$$

To show the above calculation, a simple example is given here. The first and second order GFRFs of a SIMO nonlinear system is calculated accordingly.

$$
\begin{aligned}
&\ddot{x}_1 + 0.2\dot{x}_1 + 1.1x_1 + 1.4x_2 = u + 0.1x_1^2 + 0.2x_1x_2 + 0.1x_2^2 \\
&\dot{x}_2 = 0.1x_1^2 + 0.2x_1x_2 + 0.1x_2^2
\end{aligned}
$$

The coefficients in the equation above are

$$
\begin{aligned}
&c_{1,0}^{10}(2;\,1) = 1,\ c_{1,0}^{10}(1;\,1) = 0.2,\ c_{1,0}^{10}(0;\,1) = 1.1,\ c_{1,0}^{01}(0;\,1) = 1.4, \\
&c_{0,1}^{00}(0;\,1) = 1, \\
&c_{2,0}^{20}(0,0;\,1) = 0.1,\ c_{2,0}^{11}(0,0;\,1) = 0.2,\ c_{2,0}^{02}(0,0;\,1) = 0.1, \\
&c_{2,0}^{20}(0,0;\,2) = 0.1,\ c_{2,0}^{11}(0,0;\,2) = 0.2,\ c_{2,0}^{02}(0,0;\,2) = 0.1,
\end{aligned}
$$

(a) **The Calculation of the First Order GFRF**

$$H_1 = \begin{bmatrix} H_1^1(j\omega_1) \\ H_2^2(j\omega_1) \end{bmatrix} = \begin{bmatrix} l_{1,1} & l_{1,2} \\ l_{2,1} & l_{2,2} \end{bmatrix}^{-1} \begin{bmatrix} A_{1,1} \\ A_{2,1} \end{bmatrix} = L_1^{-1} \begin{bmatrix} A_{1,1} \\ A_{2,1} \end{bmatrix}$$

where

$$\begin{aligned} l_{1,1} &= (jw)^2 + 0.2(jw) + 1.1 - 0.2H_0^2, l_{1,2} = 1.4 - 0.2H_0^1, \\ l_{2,1} &= -0.2H_0^2, l_{2,2} = (jw) - 0.2H_0^1, \end{aligned}$$

$$\begin{aligned} A_{1,1} &= 1 + 0.2H_0^1 H_1^1(jw) + 0.2H_0^2 H_1^2(jw), \\ A_{1,2} &= 0.2H_0^1 H_1^1(jw) + 0.2H_0^2 H_1^2(jw), \end{aligned}$$

Then the first order GFRF can be calculated by following the matrix operation.

(b) **The Calculation of the Second Order GFRF**

$$H_2 = \begin{bmatrix} H_2^1(j\omega_1, j\omega_2) \\ H_2^2(j\omega_1, j\omega_2) \end{bmatrix} = \begin{bmatrix} l_{1,1} & l_{1,2} \\ l_{2,1} & l_{2,2} \end{bmatrix}^{-1} \begin{bmatrix} A_{1,2} \\ A_{2,2} \end{bmatrix} = L_2^{-1} \begin{bmatrix} A_{1,2} \\ A_{2,2} \end{bmatrix}$$

where

$$\begin{aligned} l_{1,1} &= (jw_1 + jw_2)^2 + 0.2(jw_1 + jw_2) + 1.1 - 0.2H_0^2, \\ l_{1,2} &= 1.4 - 0.2H_0^1, \\ l_{2,1} &= -0.2H_0^2, \\ l_{2,2} &= (jw_1 + jw_2) - 0.2H_0^1, \end{aligned}$$

$$\begin{aligned} A_{1,1} = 0.2H_0^1 H_2^1&(jw_1 + jw_2) + 0.1H_1^1(jw_1)H_1^1(jw_2) \\ &+ 0.2H_1^1(jw_1)H_1^2(jw_2) \\ &+ 0.2H_0^2 H_2^2(jw_1 + jw_2) + 0.1H_1^2(jw_1)H_1^2(jw_2), \\ A_{1,2} = 0.2H_0^1 H_2^1&(jw_1 + jw_2) + 0.1H_1^1(jw_1)H_1^1(jw_2) \\ &+ 0.2H_1^1(jw_1)H_1^2(jw_2) \\ &+ 0.2H_0^2 H_2^2(jw_1 + jw_2) + 0.1H_1^2(jw_1)H_1^2(jw_2), \end{aligned}$$

Then the second order GFRF can be calculated by following the matrix operation.

2.2.3 *The GFRFs and Output Spectrum for the Block-Oriented Nonlinear Models*

Block-oriented nonlinear models, such as the Wiener model, Hammerstein model, and Wiener–Hammerstein model, are commonly utilized in practical applications.

In this section, we first establish nonlinear frequency response functions for these models. This is done under the assumption that the nonlinear component in all these models can be approximated by a polynomial, as indicated in Eq. (2.7). Following this, we extend the nonlinear function to accommodate a broader scenario.

A. **Frequency Response Functions of Wiener Systems**

Consider the Wiener model which can be described by

$$u(t) = g \circ r(t) \quad \text{and} \quad y(t) = f(u(t)) \tag{2.28ab}$$

where "∘" represents the convolution operator, $g(t)$ is the impulse response of the linear part, and $f(u(t))$ is the nonlinear part of the system, which is a static nonlinear function. The linear part is defined as a stable single input single output system, which can be described by parametric FIR/IIR models or nonparametric models, and the nonlinear part is described by the polynomial in (2.7).

Note that the GFRFs for (2.28b) are obtained in (2.8), and Eq. (2.28a) can be written in the frequency domain as

$$U(\mathrm{j}\omega) = G(\mathrm{j}\omega)R(\mathrm{j}\omega)$$

where $U(\mathrm{j}\omega)$,$G(\mathrm{j}\omega)$ and $R(\mathrm{j}\omega)$ are the corresponding Fourier transforms of $u(t)$, $g(t)$ and $r(t)$ respectively. Using (2.3–2.4), the output spectrum of (2.28ab) can be obtained with the nth-order output spectrum as

$$\begin{aligned}
Y_n(j\omega) &= \frac{1}{(2\pi)^{n-1}}\underbrace{\int_{-\infty}^{\infty}\cdots\int_{-\infty}^{\infty}}_{n-1} c_n U_n(j\omega_1,\cdots,j\omega_{n-1})\mathrm{d}\omega_1\cdots\mathrm{d}\omega_{n-1} \\
&= \frac{1}{(2\pi)^{n-1}}\underbrace{\int_{-\infty}^{\infty}\cdots\int_{-\infty}^{\infty}}_{n-1} c_n\prod_{i=1}^{n}(G(j\omega_i)R(j\omega_i))\mathrm{d}\omega_1\cdots\mathrm{d}\omega_{n-1} \\
&= \frac{1}{(2\pi)^{n-1}}\underbrace{\int_{-\infty}^{\infty}\cdots\int_{-\infty}^{\infty}}_{n-1}\left(c_n\prod_{i=1}^{n}G(j\omega_i)\right)\prod_{i=1}^{n}R(j\omega_i)\mathrm{d}\omega_1\cdots\mathrm{d}\omega_{n-1} \\
&= \frac{c_n}{(2\pi)^{n-1}}\underbrace{\int_{-\infty}^{\infty}\cdots\int_{-\infty}^{\infty}}_{n-1}\left(\prod_{i=1}^{n}G(j\omega_i)\right)\prod_{i=1}^{n}R(j\omega_i)\mathrm{d}\omega_1\cdots\mathrm{d}\omega_{n-1}
\end{aligned} \tag{2.29}$$

where $\omega_n = \omega - \omega_1 - \cdots - \omega_{n-1}$. Comparing (2.29) with (2.3) gives the nth-order GFRF of the Wiener model in (2.28ab) as

$$H_n(j\omega_1,\ldots,j\omega_n) = c_n\prod_{i=1}^{n}G(j\omega_i) \tag{2.30}$$

With the nonlinear frequency response functions given in (2.29) and (2.30), the output spectrum of the Wiener system in (2.28ab) can therefore be computed under any input signal $R(j\omega)$. That is, the nth-order output spectrum can be obtained from (2.3) to (2.30) as,

$$Y_n(j\omega) = \frac{c_n}{(2\pi)^{n-1}} \underbrace{\int_{-\infty}^{\infty} \cdots \int_{-\infty}^{\infty}}_{n-1} \left(\prod_{i=1}^{n-1} G(j\omega_i) R(j\omega_i) \right) G(j(\omega - \omega_1 - \cdots - \omega_{n-1})) R(j(\omega - \omega_1 - \cdots - \omega_{n-1})) \mathrm{d}\omega_1 \cdots \mathrm{d}\omega_{n-1}$$

Let $\Pi(j\omega_i) = G(j\omega_i)R(j\omega_i)$, then $Y_1(j\omega_i) = c_1\Pi(j\omega_i)$, which represents the linear behavior of the Wiener system. The equation above can be further written as

$$Y_n(j\omega) = \frac{c_n}{(2\pi)^{n-1}} \underbrace{\int_{-\infty}^{\infty} \cdots \int_{-\infty}^{\infty}}_{n-1} \left(\prod_{i=1}^{n-1} \Pi(j\omega_i) \right) \Pi(j(\omega - \omega_1 - \cdots - \omega_{n-1})) \mathrm{d}\omega_1 \cdots \mathrm{d}\omega_{n-1} \tag{2.31}$$

Hence, it can be observed that the nth-order nonlinear output spectrum of the Wiener model (2.28ab) is entirely reliant on the frequency response of the system's linear part. If the frequency response of the system's linear part (which could be nonparametric) is known, the higher order output spectrum can be instantly calculated. Conversely, if the higher order output spectrum is known, the linear part of the system can be subsequently estimated. These aspects will be further elaborated in the subsequent sections. The total output spectrum is a combination of all these nth order output spectra.

The nonlinear frequency response functions derived above can offer valuable insights into the analytical analysis and design of Wiener systems in the frequency domain. It's worth noting that the magnitude bound of the system output spectrum often provides a useful understanding of the system dynamics at various frequencies and a direct understanding of the relationship between frequency response functions and system parameters [22–24]. With the GFRFs developed above, the bound characteristics of the output spectrum of the Wiener system (2.28ab) can be readily examined.

It's also recognized that output frequencies of nonlinear systems are always more complex than those of linear systems, which include sub- or super- harmonics and inter-modulations [26]. The GFRFs and output spectrum developed above explicitly illuminate the analysis of output frequency characteristics of Wiener systems.

B. **The GFRFs of Wiener–Hammerstein Systems**

Similar results can be established for the other block-oriented nonlinear models. The Wiener–Hammerstein model can be described by

$$u(t) = g \circ r(t); \quad x(t) = f(u(t)) \text{ and}$$
$$y(t) = p \circ x(t) \tag{2.32abc}$$

where p is the impulse response of the linear part that follows the nonlinear element $f(u(t))$, and g denotes the preceding linear part. Consider the subsystem from $r(t)$ to $x(t)$, which is a Wiener model in (2.28ab). According to (2.29), the nth-order output spectrum of this subsystem is

$$X_n(j\omega) = \frac{1}{(2\pi)^{n-1}} \underbrace{\int_{-\infty}^{\infty} \cdots \int_{-\infty}^{\infty}}_{n-1} \left(c_n \prod_{i=1}^{n} G(j\omega_i) \right)$$
$$\prod_{i=1}^{n} R(j\omega_i) \mathrm{d}\omega_1 \cdots \mathrm{d}\omega_{n-1}$$

where $\omega_n = \omega - \omega_1 - \cdots - \omega_{n-1}$. Then the nth-order output spectrum of (2.32abc) is

$$Y_n(j\omega) = X_n(j\omega)P(j\omega)$$
$$= \frac{c_n P(j\omega)}{(2\pi)^{n-1}} \underbrace{\int_{-\infty}^{\infty} \cdots \int_{-\infty}^{\infty}}_{n-1} \left(\prod_{i=1}^{n} G(j\omega_i) \right)$$
$$\prod_{i=1}^{n} R(j\omega_i) \mathrm{d}\omega_1 \cdots \mathrm{d}\omega_{n-1} \tag{2.33}$$

where $P(j\omega)$ is the Fourier transform of $p(t)$. Therefore, the nth-order GFRF for (2.32abc) is

$$H_n(j\omega_1, \ldots, j\omega_n) = c_n P(j\omega) \prod_{i=1}^{n} G(j\omega_i) \tag{2.34a}$$

Noting that $\omega = \omega_1 + \cdots + \omega_n$ in (2.34a), the equation above can be written as

$$H_n(j\omega_1, \ldots, j\omega_n) = c_n P(j\omega_1 + \cdots + j\omega_n) \prod_{i=1}^{n} G(j\omega_i) \tag{2.34b}$$

C. **The GFRFs of Hammerstein Systems**

The Hammerstein model can be described by

$$x(t) = f(r(t)) \text{ and } y(t) = p \circ x(t) \tag{2.35ab}$$

The nth-order GFRF for the nonlinear element from the input $r(t)$ to the output $x(t)$ is given in (9), i.e.,

$$H_n(j\omega_1, \cdots, j\omega_n) = c_n$$

Then the nth-order output spectrum of the nonlinear part is

$$X_n(j\omega) = \frac{1}{(2\pi)^{n-1}} \underbrace{\int_{-\infty}^{\infty} \cdots \int_{-\infty}^{\infty}}_{n-1} c_n \prod_{i=1}^{n} R(j\omega_i) \mathrm{d}\omega_1 \cdots \mathrm{d}\omega_{n-1}$$

Therefore, the nth-order output spectrum of (2.35ab) is given by

$$\begin{aligned} Y_n(j\omega) &= X_n(j\omega)P(j\omega) \\ &= \frac{c_n P(j\omega)}{(2\pi)^{n-1}} \underbrace{\int_{-\infty}^{\infty} \cdots \int_{-\infty}^{\infty}}_{n-1} \prod_{i=1}^{n} R(j\omega_i) \mathrm{d}\omega_1 \cdots \mathrm{d}\omega_{n-1} \end{aligned} \tag{2.36}$$

Hence, the nth-order GFRF of (2.35ab) is achieved as

$$H_n(j\omega_1, \ldots, j\omega_n) = c_n P(j\omega_1 + \cdots + j\omega_n) \tag{2.37}$$

D. **Output Spectrum Subject to Different Input Signals**

With the GFRFs obtained for the block-oriented models, the output spectrum of these nonlinear systems can readily be evaluated by using (2.3) and (2.4). Note that although the input signal $R(j\omega)$ mentioned above can be any appropriate signals, in practice a multi-tone signal, i.e.,

$$\eta(t) = \sum_{i=1}^{K} |F_i| \cos(\omega_i t + \angle F_i) \tag{2.38}$$

is usually adopted for system analysis, where F_i is a complex number, $\angle F_i$ is the argument and $|F_i|$ is the modulus. In this case, with the GFRFs obtained, the output spectrum under the input signal (2.38) can be computed by following the method in [19] as

$$Y(j\omega) = \sum_{n=1}^{N} \frac{1}{2^{n-1}} \sum_{\omega_{k_1} + \cdots + \omega_{k_n} = \omega} H_n(j\omega_{k_1}, \ldots, j\omega_{k_n}) F(\omega_{k_1}) \cdots F(\omega_{k_n}) \tag{2.39}$$

where $F(\omega_{k_i}) = |F_{|k_i|}|e^{j\angle F_{|k_i|}\cdot \mathrm{sig}(k_i)}$ for $k_i \in \{\pm 1, \cdots, \pm \overline{K}\}$, $\mathrm{sgn}(a) = \begin{cases} 1 & a \geq 0 \\ -1 & a < 0 \end{cases}$,

and $\omega_{k_i} \in \{\pm\omega_1, \cdots, \pm\omega_{\overline{K}}\}$.

E. **Summary**

In summary, the Generalized Frequency Response Functions (GFRFs) and output spectrum of the block-oriented nonlinear systems discussed above are derived using model structure information and are formulated into analytical functions of model parameters. These frequency response functions are not limited to a specific input (such as sinusoidal signals) but can accommodate any form of input signals, i.e., R(jω). A lot of existing frequency-domain results for nonlinear analysis necessitate a specific sinusoidal input signal [7–10, 35, 38–41]. In the GFRFs, the relationships between the output spectrum, the GFRFs, the system's nonlinear parameters, and the system's linear dynamics are clearly illustrated. With these frequency response functions, the bound characteristics of the output spectrum and output frequency characteristics can be examined following the methods in [22–26]. Consequently, these findings can significantly aid the analysis and design of nonlinear systems in the frequency domain and extend the frequency domain theory from linear systems to nonlinear scenarios.

It's worth mentioning that while the Generalized Frequency Response Functions (GFRFs) for Nonlinear Dynamic Error (NDE) systems can be calculated using the recursive algorithm provided in [14], it may be challenging or complex to apply this recursive algorithm to the block-oriented nonlinear models presented in (2.28ab), (2.32abc), and (2.35ab). A clear drawback is that to utilize the recursive algorithm in [14], all the block-oriented models must be expanded into a polynomial form. This process disrupts the straightforward model structure information, which in turn complicates the relationship between the frequency response functions and the system's linear and nonlinear elements.

2.2.4 Extensions to a More General Polynomial Case

Note that the nonlinear part of the block-oriented models discussed above is only a univariate polynomial function. To represent a more general class of nonlinear systems, this static nonlinear function can be extended to a more general case as follows. Consider the following Wiener model

$$u(t) = g \circ r(t) \quad \text{and} \quad y(t) = f(u(t), r(t)) \tag{2.40ab}$$

where the nonlinear function is defined as a more general multivariate polynomial function as

$$f(u(t), r(t)) = \sum_{m=1}^{M} \sum_{p=0}^{m} \sum_{k_1, k_m=0}^{K} c_{p,m-p}(k_1, \cdots, k_m) \prod_{i=1}^{p} \frac{\mathrm{d}^{k_i} u(t)}{\mathrm{d}t^{k_i}} \prod_{i=p+1}^{m} \frac{\mathrm{d}^{k_i} r(t)}{\mathrm{d}t^{k_i}} \tag{2.40c}$$

where M is the maximum nonlinear degree of the polynomial, K is the maximum order of the derivative and $c_{p,m-p}(k_1, \cdots, k_m)$ is the coefficient of the term $\prod_{i=1}^{p} \frac{\mathrm{d}^{k_i} u(t)}{\mathrm{d}t^{k_i}} \prod_{i=p+1}^{m} \frac{\mathrm{d}^{k_i} r(t)}{\mathrm{d}t^{k_i}}$ in the polynomial.

It can be seen that the univariate polynomial function (2.28b) is only a special case of the general form (2.40c). That is, if letting $c_{p,0}(\underbrace{0, \ldots, 0}_{p}) = c_p$ for p = 1, 2, ... and the other coefficients in (2.40c) are zero, then (2.40c) will become (2.28b). Obviously, (2.40abc) can represent a wider class of nonlinear systems. For example, if the polynomial function $f(.)$ is of sufficiently high degree and includes all the possible linear and nonlinear combinations of input $u(t)$ of sufficiently high orders, it will be an equivalent nonlinear infinite impulse response model of the Volterra class of nonlinear systems [44].

To derive the GFRFs for Wiener system (2.40ab), the model can be regarded as a nonlinear differential equation model with two outputs $u(t)$ and $y(t)$, and one input $r(t)$. The probing method mentioned in Sect. 2.2.1 can be employed again to this objective. Note that the frequency response function from the input $r(t)$ to the intermediate output $u(t)$ is the Fourier transform of the impulse response function g(t), i.e., $G(\mathrm{j}\omega)$, which is a linear dynamics; while the frequency responses from the input $r(t)$ to the output $y(t)$ involve nonlinear dynamics. The latter are the GFRFs to be derived. The terms in the polynomial function $f(u(t),r(t))$ can be categorized into three groups, i.e., pure input terms $c_{0,m}(k_1, \cdots, k_m) \prod_{i=1}^{m} \frac{\mathrm{d}^{k_i} r(t)}{\mathrm{d}t^{k_i}}$, pure output terms $c_{p,0}(k_1, \cdots, k_p) \prod_{i=1}^{p} \frac{\mathrm{d}^{k_i} u(t)}{\mathrm{d}t^{k_i}}$, and input–output cross terms $c_{p,m-p}(k_1, \cdots, k_m) \prod_{i=1}^{p} \frac{\mathrm{d}^{k_i} u(t)}{\mathrm{d}t^{k_i}} \prod_{i=p+1}^{m} \frac{\mathrm{d}^{k_i} r(t)}{\mathrm{d}t^{k_i}}$ $(0 < p < m)$. Therefore, by applying the probing method, each group of terms corresponds to a specific form of contributions to the nth-order GFRF of Wiener system (2.40ac), which can be written as

$$H_n(j\omega_1, \ldots, j\omega_n) = H_{n_r}(j\omega_1, \ldots, j\omega_n) + H_{n_{ru}}(j\omega_1, \ldots, j\omega_n) + H_{n_u}(j\omega_1, \ldots, j\omega_n) \tag{2.41a}$$

where $H_{n_r}(j\omega_1, \ldots, j\omega_n)$ represents the contribution from the pure input terms and similar notations are used for the other terms. This is a special case of the system studied in [45]. Therefore following the method there, it can be obtained that

$$H_{n_r}(j\omega_1,\ldots,j\omega_n)=\sum_{k_1,k_n=0}^{K}c_{0,n}(k_1,\ldots,k_n)(j\omega_1)^{k_1}\cdots(j\omega_n)^{k_n} \tag{2.41b}$$

$$H_{n_{ru}}(j\omega_1,\ldots,j\omega_n)=\sum_{q=1}^{n-1}\sum_{p=1}^{n-q}\sum_{k_1,k_{p+q}=0}^{K}c_{p,q}(k_1,\ldots,k_{p+q})(j\omega_{n-q+1})^{k_{p+1}} \cdots(j\omega_n)^{k_{p+q}}H_{n-q,p}(j\omega_1,\ldots,j\omega_{n-q}) \tag{2.41c}$$

$$H_{n_u}(j\omega_1,\ldots,j\omega_n)=\sum_{p=1}^{n}\sum_{k_1,k_p=0}^{K}c_{p,0}(k_1,\ldots,k_p)H_{n,p}(j\omega_1,\ldots,j\omega_n) \tag{2.41d}$$

$$H_{n,p}(\cdot)=G(j\omega_1)H_{n-1,p-1}(j\omega_2,\ldots,j\omega_n)(j\omega_1)^{k_p} \tag{2.41e}$$

$$H_{n,1}(j\omega_1,\ldots,j\omega_n)=\begin{cases}G(j\omega_1)(j\omega_1)^{k_1} & \text{n}=1\\ 0 & \text{else}\end{cases} \tag{2.41f}$$

Since the polynomial function (2.28b) is included in (2.40c), it can be verified that the nth-order GFRF in (2.31) for the Wiener model in (2.28ab) can be derived from (2.41a–f). As mentioned above, letting $c_{p,0}(\underbrace{0,\ldots,0}_{p})=c_p$ and the others zero in (2.40b) will result in (2.28b). That is, only the parameter $c_{n,0}(\underbrace{0,\ldots,0}_{n})=c_n$ is not zero and the others are zero in (2.41b–f). Therefore,

$$H_n(j\omega_1,\ldots,j\omega_n)=H_{n_r}(j\omega_1,\ldots,j\omega_n) +H_{n_{ru}}(j\omega_1,\ldots,j\omega_n)+H_{n_u}(j\omega_1,\ldots,j\omega_n) \tag{2.42a}$$

$$H_{n_r}(j\omega_1,\ldots,j\omega_n)=0 \tag{2.42b}$$

$$H_{n_{ru}}(j\omega_1,\ldots,j\omega_n)=0 \tag{2.42c}$$

$$H_{n_u}(j\omega_1,\ldots,j\omega_n)=c_{n,0}(0,\ldots,0)H_{n,n}(j\omega_1,\ldots,j\omega_n) \tag{2.42d}$$

$$\begin{aligned}H_{n,n}(\cdot)&=G(j\omega_1)H_{n-1,n-1}(j\omega_2,\ldots,j\omega_n)(j\omega_1)^0\\&=G(j\omega_1)G(j\omega_2)\cdots G(j\omega_{n-1})H_{1,1}(j\omega_n)\end{aligned} \tag{2.42e}$$

$$H_{1,1}(j\omega_n)=G(j\omega_n) \tag{2.42f}$$

The nth-order GFRF for (2.28ab) can now be obtained from (2.42a–f) which is exactly Eq. (2.31).

Similarly, the Hammerstein model and Wiener–Hammerstein model may also be extended with the general polynomial function defined in (2.40b), and the corresponding GFRFs can be derived respectively. Note that only input nonlinearity is involved in the Hammerstein model. The extended polynomial function for the Hammerstein model can be written as

$$x(t) = f(r(t)) = \sum_{m=1}^{M} \sum_{k_1,k_m=0}^{K} c_{0,m}(k_1, \ldots, k_m) \prod_{i=1}^{m} \frac{d^{k_i} r(t)}{dt^{k_i}} \tag{2.43}$$

Following the same line as before, the nth-order GFRF for the extended Hammerstein model (2.43) and (2.35b) can be obtained as

$$\begin{aligned} H_n(j\omega_1, \ldots, j\omega_n) = {} & P(j\omega_1 + \cdots + j\omega_n) \\ & \sum_{k_1,k_n=0}^{K} c_{0,n}(k_1, \ldots, k_n)(j\omega_1)^{k_1} \cdots (j\omega_n)^{k_n} \end{aligned} \tag{2.44}$$

The nth-order GFRF for the extended Wiener–Hammerstein model (2.32ac) and (2.40b) can be obtained as

$$\begin{aligned} H_n(j\omega_1, \ldots, j\omega_n) = {} & P(j\omega_1 + \cdots + j\omega_n) \\ & \big(H_{n_r}(j\omega_1, \ldots, j\omega_n) + H_{n_{ru}}(j\omega_1, \ldots, j\omega_n) \\ & + H_{n_u}(j\omega_1, \ldots, j\omega_n)\big) \end{aligned} \tag{2.45}$$

where $H_{n_r}(j\omega_1, \ldots, j\omega_n)$, $H_{n_{ru}}(j\omega_1, \ldots, j\omega_n)$ and $H_{n_u}(j\omega_1, \ldots, j\omega_n)$ are given by (2.41b–f).

From the GFRFs of different block-oriented models, it can be seen that the nth-order GFRF for the Wiener–Hammerstein could be more powerful to represent the dynamics of a system than the other two models, and the Hammerstein model is the simplest one. This is basically consistent with practical experience. It shall be noted that the nonlinear frequency response functions above are all developed for continuous time system models. However, it would be easy to extend them to the corresponding discrete time system models. This is not discussed in details in this study.

2.3 Nonparametric Identification of Different Orders of Output Spectrum

The frequency response functions developed in the previous section demonstrate that the system output frequency response of a nonlinear system can invariably be divided into different orders of output spectrum, as per Eq. (2.4). This holds true regardless of the specifics of the Generalized Frequency Response Functions (GFRFs), or the nonlinear or linear components in a given system. Consequently, an alternative approach for the identification and analysis of a nonlinear system would be to initially estimate the different orders of output spectrum, rather than directly identifying the nonlinear model.

From the parametric characteristic analysis of the GFRFs and output spectrum of nonlinear systems, it is established that the nth-order output spectrum can consistently be expressed as a polynomial function of model parameters [25]. This is in the "linear-in-parameters" form. Therefore, once the different orders of output spectrum are determined, numerous existing system identification methods (for instance, Least Square) can be utilized to estimate the model parameters involved in each order output spectrum. This ultimately leads to the acquisition of the nonlinear system model.

It's important to note that the identification of block-oriented nonlinear systems in the frequency domain has been explored in existing literature. Current methods typically assume a known structure and invertibility of nonlinearity, necessitate the linear part to be a rational transfer function, and involve numerous repeated experiments with varying magnitudes and frequencies, restrictively using sinusoidal input signals [35, 38–42].

In this section, to estimate the different orders of output spectrum, there is no need for prior knowledge of the model structure and orders. The linear part of the system is merely expected to be stable, and the input signal could be any signal with a Fourier transformation R(jω), which can be easily obtained by performing a fast Fourier transform (FFT) on the real input signals.

Considering the general nonlinear output spectrum in (2.3–2.4) with an input signal whose Fourier transform is $\rho R(j\omega)$, where ρ is a constant which is used to represent different magnitude of the input, the output spectrum can then be rewritten as

$$\begin{aligned} Y(j\omega) &= \sum_{n=1}^{\infty} \rho^n Y_n(j\omega) \\ &= \rho Y_1(j\omega) + \rho^2 Y_2(j\omega) + \rho^3 Y_3(j\omega) + \cdots \end{aligned} \tag{2.46}$$

where $Y_n(j\omega)$ is the nth-order output spectrum given in (2.3) with $U(j\omega_i) = R(j\omega_i)$ (i = 1, 2, …, n). The task is to determine $Y_n(j\omega)$ given the input $R(j\omega)$ and the measured output $Y(j\omega)$. To this aim, define a residue function as (for n = 1, 2, 3, …)

$$\begin{aligned} \varepsilon_{[n,\rho]}(j\omega) &= Y(j\omega) - \sum_{i=1}^{n} \rho^i Y_i(j\omega) \\ &= \rho^{n+1} Y_{n+1}(j\omega) + \rho^{n+2} Y_{n+2}(j\omega) \\ &\quad + \rho^{n+3} Y_{n+3}(j\omega) + \cdots \end{aligned} \tag{2.47}$$

That is, Eq. (2.46) can be written as

$$Y(j\omega) = \rho Y_1(j\omega) + \rho^2 Y_2(j\omega) + \cdots + \rho^N Y_N(j\omega) + \varepsilon_{[N,\rho]}(j\omega) \tag{2.48}$$

where N is the maximum truncated order such that $\varepsilon_{[n,\rho]}(j\omega)$ is a sufficiently small value, i.e.,

$$N = \arg \max_{\substack{1<n<\infty \\ 0<\rho<\overline{\rho}}} \left| \varepsilon_{[n,\rho]}(j\omega) \right| \leq \varepsilon \tag{2.49}$$

where ε is a small positive threshold value that is determined heuristically and $\overline{\rho}$ corresponds to the maximum input magnitude. To estimate $Y_n(j\omega)$ for $n = 1, 2, 3, \ldots$ N, the system can be excited by the same input $R(j\omega)$ of different magnitudes 1, ρ_1, $\rho_2, \ldots, \rho_{N-1}$, and there will be a series of output that can be obtained accordingly, which are denoted by $Y(j\omega)$, $Y(j\omega)_{\rho 1}$, $Y(j\omega)_{\rho 2}, \ldots, Y(j\omega)_{\rho N-1}$. It can be obtained that

$$\begin{bmatrix} 1 & 1 & \cdots & 1 \\ \rho_1 & \rho_1^2 & \cdots & \rho_1^N \\ \vdots & \vdots & \ddots & \vdots \\ \rho_{N-1} & \rho_{N-1}^2 & \cdots & \rho_{N-1}^N \end{bmatrix} \begin{bmatrix} \hat{Y}_1(j\omega) \\ \hat{Y}_2(j\omega) \\ \vdots \\ \hat{Y}_N(j\omega) \end{bmatrix} = \begin{bmatrix} Y(j\omega)_{\rho_1} \\ Y(j\omega)_{\rho_2} \\ \vdots \\ Y(j\omega)_{\rho_{N-1}} \end{bmatrix}$$

From the equation above, different order output spectrum can be estimated as

$$\begin{bmatrix} \hat{Y}_1(j\omega) \\ \hat{Y}_2(j\omega) \\ \vdots \\ \hat{Y}_N(j\omega) \end{bmatrix} = \begin{bmatrix} 1 & 1 & \cdots & 1 \\ \rho_1 & \rho_1^2 & \cdots & \rho_1^N \\ \vdots & \vdots & \ddots & \vdots \\ \rho_{N-1} & \rho_{N-1}^2 & \cdots & \rho_{N-1}^N \end{bmatrix}^{-1} \begin{bmatrix} Y(j\omega)_1 \\ Y(j\omega)_{\rho_1} \\ \vdots \\ Y(j\omega)_{\rho_{N-1}} \end{bmatrix} \tag{2.50}$$

Note that $\rho_1, \rho_2, \ldots, \rho_{N-1}$ should be properly designed such that the square matrix in (2.50) is nonsingular. The procedure for the identification of different order output spectrum can be summarized as follows (referred to as Output Spectrum Estimation (OSE) algorithm):

Step 1: Given an initial truncation order N, a residue threshold ε, input signal $R(j\omega)$, and determine the input magnitude range.

Step 2: Find $N - 1$ values for $\rho_1, \rho_2, \ldots, \rho_{N-1}$ considering the input magnitude range such that the square matrix in (2.50) is nonsingular. This can be obtained by searching in an appropriate range stochastically. Use input signals $R(j\omega)$, $\rho_1 R(j\omega)$, $\rho_2 R(j\omega), \ldots, \rho_{N-1} R(j\omega)$ to activate the system N times and obtain N series of output $Y(j\omega)_1$, $Y(j\omega)_{\rho 1}$, $Y(j\omega)_{\rho 2}, \ldots, Y(j\omega)_{\rho N-1}$ after Fourier transform.

Step 3: Use (2.50) to find the estimates of different order output spectrum and use (2.49) to check the effectiveness of the truncation order N. If the desired threshold residue can be achieved, then the procedure terminates otherwise increase N and repeat step 2 and step 3.

It is important to highlight that the estimation of different order output spectrum mentioned above requires only some input–output data and is not dependent on any specific model structures of the linear and nonlinear elements of the system being studied. Furthermore, with the estimated different order output spectra of a nonlinear system characterized by a NDE or NARX model, system identification can be performed more conveniently by adhering to a parametric identification procedure in [46] and a parametric characteristic analysis in [25]. The method in [46] can be significantly enhanced with reduced computational cost. This will be explored in a subsequent study. In what follows, the focus is solely on the identification of block-oriented models, and a general procedure is proposed for this purpose.

2.4 Frequency Domain Identification of Block-Oriented Models

With the estimates of different order output spectrum, effective methods could be developed to identify a detailed parametric model for a practical block-oriented nonlinear system. As an example, consider the identification of Wiener systems in this section. Similar results can be extended to the other block-oriented models.

For a Wiener system, it can be known from (2.3) that

$$Y_1(j\omega) = \kappa G(j\omega)R(j\omega) \tag{2.51}$$

where κ is a constant value. Without specialty it can be supposed that $\kappa = 1$. Therefore, given $R(j\omega)$ and $\hat{Y}_1(j\omega)$, the complex valued function $G(j\omega)$ can be estimated if $Y_1(j\omega) \neq 0$, denoted by $\hat{G}(j\omega)$. Then several methods can be adopted to fit a linear transfer function for $\hat{G}(j\omega)$ [47, 48]. In this study, let

$$\hat{G}(j\omega) = \frac{B(j\omega)}{A(j\omega)} = \frac{b_0(j\omega)^m + b_1(j\omega)^{m-1} + \cdots + b_m}{a_0(j\omega)^n + a_1(j\omega)^{n-1} + \cdots + a_{n-1}(j\omega) + 1} \tag{2.52}$$

which can be further written as

$$\begin{aligned} &-\left(a_0(j\omega)^n + a_1(j\omega)^{n-1} + \cdots + a_{n-1}(j\omega)\right)\hat{G}(j\omega) \\ &+ b_0(j\omega)^m + b_1(j\omega)^{m-1} + \cdots + b_m = \hat{G}(j\omega) \end{aligned} \tag{2.53}$$

This is a linear-in-parameter regressive form and the standard least squares can be applied to estimate the parameters $\{a_0, a_1, \ldots, a_{n-1}, b_0, \ldots b_m\}$. In case that there is noise in $\hat{G}(j\omega)$, this estimation will be biased. However, the instrumental variable method [48] could be applied to achieve an unbiased estimation. This is not discussed in detail. With the estimated $\hat{G}(j\omega)$, it would be easy to estimate the nonlinear part of the Wiener system both parametrically and non-parametrically. It shall be noted that the nonparametric determination of the nonlinear part in this identification process involves no knowledge on the structural information, nonlinear order and invertibility of the nonlinear part. However, those assumptions are commonly used in the literature [35, 38–41].

To estimate a parametric Wiener model in (2.28ab), noting that $Y_1(j\omega) = c_1 G(j\omega_i)R(j\omega_i)$ and again supposing without specialty that $c_1 = 1$, it can be obtained from (2.31b) that

$$Y_n(j\omega) = \frac{c_n}{(2\pi)^{n-1}} \underbrace{\int_{-\infty}^{\infty} \cdots \int_{-\infty}^{\infty}}_{n-1} \left(\prod_{i=1}^{n-1} Y_1(j\omega_i) \right)$$
$$Y_1(j(\omega - \omega_1 - \cdots - \omega_{n-1}))\mathrm{d}\omega_1 \cdots \mathrm{d}\omega_{n-1} \tag{2.54}$$

It can be seen that, given the estimated different order output spectrum $\hat{Y}_1(j\omega)$, $\hat{Y}_2(j\omega)$, ..., $\hat{Y}_N(j\omega)$ (using the OSE algorithm in Sect. 2.2.3), it is straightforward to find estimates of the nonlinear parameters c_n for $n = 2, 3, \ldots$ from (2.54).

In the identification algorithm above, it is supposed that $Y_1(j\omega) \neq 0$, which implies that $c_1 \neq 0$ for the Wiener model (2.28ab). In case that $Y_1(j\omega) = 0$, it would not be straightforward to estimate the linear part of the Wiener system under study. Considering again the Wiener model (2.28ab), suppose m is the smallest nonlinear degree such that $\hat{Y}_m(j\omega) \neq 0$ and additionally suppose $c_m = 1$ without specialty. Applying the inverse Fourier transform to both sides of (2.54) yields

$$\hat{y}_n(t) = f^{-1}\left[\hat{Y}_n(j\omega)\right] = c_n \hat{y}_1(t)^n = c_n \hat{u}(t)^n \tag{2.55}$$

Therefore, it can be obtained that

$$\hat{y}_m(t) = \hat{u}(t)^m \tag{2.56}$$

There would be two cases to be considered.

Case I: m is the smallest odd integer satisfying $\hat{Y}_m(j\omega) \neq 0$. Then $u(t)$ can be unique determined from (2.56) as $\hat{u}(t) = \sqrt[m]{\hat{y}_m(t)}$. Therefore, given the estimate for $Y_m(j\omega)$, an estimate for the intermediate state $u(t)$ can be obtained from (2.55) and (2.56), and then the Wiener model (2.28ab) can be identified immediately.

Case II: there are no terms of odd degree in (2.28b), and thus m is an even integer which can be written as $m = 2d$, d is a positive integer. It gives from (2.56) that $\hat{u}(t)^2 = \sqrt[d]{\hat{y}_m(t)}$. Since (2.28b) includes only the terms of even degree, (2.55) gives

$$c_n = \hat{y}_n(t)/\hat{u}(t)^n = \hat{y}_n(t)/\hat{u}(t)^{2d} \tag{2.57}$$

Therefore, given the estimates $\hat{Y}_n(j\omega)$ for $n = 2d$ (d are some positive integer), the estimates for $u(t)^2$ and $y_n(t)$ with $n = 2d$ (d are some positive integer) can be obtained from (2.55) to (2.56), and then the estimates for c_n's can be obtained from (2.57) immediately. With the estimate of the nonlinear part, the linear part could be identified consequently in this case.

Similar results can be obtained for Wiener model (2.40abc) using the parametric characteristic analysis results in [25]. To this aim, the linear transfer function can be obtained firstly using the methods above, then the nth-order output spectrum ($n > 1$) can be expressed into a polynomial function of nonlinear parameters in (2.40c), which is also in a linear-in-parameters form such that the nonlinear model parameters involved can be identified eventually. This is not discussed in detail in this study.

As a summary, the following procedure can be used to identify a Wiener model (referred to as Wiener Model Estimation (WiME) procedure).

Step 1: Using the OSE algorithm in Sect. 2.2.3 to obtain the estimates of different order output spectrum $\hat{Y}_1(j\omega)$, $\hat{Y}_2(j\omega)$, …, $\hat{Y}_N(j\omega)$.

Step 2: If $Y_1(j\omega) \neq 0$, then the linear part can be estimated through the regressive Eq. (2.53), and the nonlinear part can readily be estimated both parametrically and non-parametrically after the linear part is known. This could result in an estimated Wiener model without knowing the structure and invertibility of the nonlinear part.
Otherwise, go to Step 3.

Step 3: Consider the identification of the Wiener model (2.28ab).
If $Y_m(j\omega) \neq 0$, and m is the smallest odd integer,
Using (2.55) to find the mth-order nonlinear output $\hat{y}_{\mathrm{m}}(\mathrm{t})$, the intermediate state $u(t)$ can therefore be estimated through (2.56), i.e., $\hat{u}(\mathrm{t}) = \sqrt[m]{\hat{y}_m(t)}$. With this estimated intermediate state and the known input–output signals $r(t)$ and $y(t)$, the linear part and the nonlinear part of the Wiener model (2.28ab) can both be estimated.
If there are no terms of odd degree in (2.28b),
It can be obtained from (2.56) that $\hat{u}(t)^2 = \sqrt[d]{\hat{y}_m(t)}$. Using (2.57) to estimate the nonlinear coefficients c_{n} for $n = 2d$ and d are positive integers. With the known nonlinear part, the linear part could be identified consequently.

The identification algorithm for Wiener systems, as discussed above, takes into account several scenarios based on the specific nonlinear orders present in the system. These can be determined using the OSE algorithm outlined in Sect. 2.2.3. If the first order output spectrum is present or if there is an odd order nonlinearity in the nonlinear part, the WiME procedure can accurately estimate the linear part of the Wiener system without any assumptions about the structure and invertibility of the nonlinear part. The nonlinear part can then be estimated subsequently. If not, the WiME procedure can initially identify the nonlinear part as a polynomial and then determine the linear part. When compared to other methods such as those in [40, 41], the WiME procedure is simpler to comprehend and implement, with fewer assumptions about the system.

It's worth mentioning that the WiME procedure can be readily adapted for the identification of other nonlinear models, such as Hammerstein models, Wiener–Hammerstein models, and NARX models, among others. As per the WiME procedure, the OSE algorithm is initially used to generate estimates of the different order output spectrum of a nonlinear system. This step doesn't necessitate any structural information about the system's nonlinear dynamics. The subsequent step in system identification involves estimating the corresponding model parameters relative to a specific model structure. For instance, when identifying a Hammerstein model with a polynomial nonlinear component, the coefficients of the nonlinear polynomial function can be easily estimated through Eq. (2.36), using the estimated linear part transfer function P(jω).

2.5 Simulations

To validate and illustrate the practicality of the aforementioned results, two examples are presented in this section. The first example explores a Wiener system. The findings from this study can be directly applied, leading to the demonstration of some intriguing results. The second example involves a system that isn't precisely a Wiener system, but rather a nonlinear mass-spring-damper system. It is demonstrated that the results from this study can still be utilized to uncover the nonlinear dynamics of the system.

1. **Analysis and Identification of a Wiener System**

Consider a Wiener system described by (2.28ab) with

$$\mathrm{G}(s) = \frac{26}{s^2 + 2s + 26} \quad \text{and} \quad f(u) = u + 2u^2 - u^4 \tag{2.58ab}$$

(a) **Frequency Domain Analysis**

Using the results in Sect. 2.2.2, the *n*th-order GFRF of the Wiener system above can be readily obtained from (2.31) as

$$H_1(j\omega_1) = c_1 G(j\omega_1) = G(j\omega_1) \tag{2.59a}$$

$$H_2(j\omega_1, j\omega_2) = c_2 \prod_{i=1}^{2} G(j\omega_i) = 2G(j\omega_1)G(j\omega_2) \tag{2.59b}$$

$$\begin{aligned} H_4(j\omega_1, \ldots, j\omega_4) &= c_4 \prod_{i=1}^{4} G(j\omega_i) \\ &= -G(j\omega_1)G(j\omega_2)G(j\omega_3)G(j\omega_4) \end{aligned} \tag{2.59c}$$

where $G(j\omega_i) = \frac{26}{26-\omega_i^2+j2\omega_i}$, and $H_n(j\omega_1, \ldots, j\omega_n) = 0$ for all the other orders. From (2.30), the nth-order output spectrum subject to an input $R(j\omega)$ can therefore be computed accordingly as

$$Y_1(j\omega) = c_1 G(j\omega)R(j\omega) = G(j\omega)R(j\omega) \tag{2.60a}$$

$$Y_2(j\omega) = \frac{c_2}{2\pi}\int_{-\infty}^{\infty} G(j\omega_1)R(j\omega_1)G(j(\omega-\omega_1))R(j(\omega-\omega_1))\mathrm{d}\omega_1 \tag{2.60b}$$

$$Y_4(j\omega) = \frac{c_4}{(2\pi)^3}\underbrace{\int_{-\infty}^{\infty}\cdots\int_{-\infty}^{\infty}}_{3} \left(\prod_{i=1}^{3} G(j\omega_i)R(j\omega_i)\right)G(j(\omega-\omega_1-\cdots-\omega_3)) R(j(\omega-\omega_1-\cdots-\omega_3))\mathrm{d}\omega_1\cdots\mathrm{d}\omega_3 \tag{2.60c}$$

where $Y_n(j\omega) = 0$ for all the other orders. Thus the nonlinear output spectrum of the Wiener system is given by

$$Y(\mathrm{j}\omega) = Y_1(\mathrm{j}\omega) + Y_2(\mathrm{j}\omega) + Y_4(\mathrm{j}\omega) \tag{2.61}$$

Using (2.59–2.61), the nonlinear output spectrum of the Wiener system (2.58ab) can be evaluated readily. This could provide an explicit insight into the nonlinear dynamics of the system and demonstrate an alternative and effective approach to nonlinear analysis in the frequency domain. Importantly, it can be seen from (2.59–2.61) that the nonlinear output spectrum and GFRFs reveal clearly the dependence of these nonlinear frequency response functions on model parameters (including the nonlinear coefficients c_1, c_2 and c_4 and the linear impulse response function G(s)). This potentially provides an important and powerful basis for the analysis, design and control of the Wiener system under study. For example, the influence of the nonlinear parameters c_1, c_2 and c_4 on the system output spectrum under a sinusoidal input $r(t) = A\cos(2\pi \times \Omega \times t)$ can be easily studied through (2.61). In this case, the output spectrum can be computed by directly following (2.39) as

$$\begin{aligned} Y(\mathrm{j}\omega) &= Y_1(\mathrm{j}\omega) + Y_2(\mathrm{j}\omega) + Y_4(\mathrm{j}\omega) \\ &= c_1 G(j\omega_{k_1})A(\omega_{k_1}) \\ &\quad + \frac{c_2}{2}\sum_{\omega_{k_1}+\omega_{k_2}=\omega} G(j\omega_{k_1})G(j\omega_{k_2})A(\omega_{k_1})A(\omega_{k_2}) \\ &\quad + \frac{c_4}{8}\sum_{\omega_{k_1}+\cdots+\omega_{k_4}=\omega} G(j\omega_{k_1})\cdots G(j\omega_{k_4})A(\omega_{k_1})\cdots A(\omega_{k_4}) \\ &= c_1 G(j\omega) + \frac{c_2}{2}\sum_{\omega_{k_1}+\omega_{k_2}=\omega} G(j\omega_{k_1})G(j\omega_{k_2}) \\ &\quad + \frac{c_4}{8}\sum_{\omega_{k_1}+\cdots+\omega_{k_4}=\omega} G(j\omega_{k_1})\cdots G(j\omega_{k_4}) \end{aligned} \tag{2.62}$$

where $\omega_{k_i} \in \{+\Omega, -\Omega\}$. With (2.62), the output spectrum at any frequency subject to a sinusoidal input can be evaluated readily, and the effect of different model parameters on system output spectrum can also be investigated. For example, the harmonic response at 2Ω can be obtained as

$$Y(j2\Omega) = c_1 G(j2\Omega) + \frac{c_2}{2} G(j\Omega)G(j\Omega) + \frac{c_4}{8} \cdot 4 \cdot G(j\Omega)G(j\Omega)G(j\Omega)G(-j\Omega) \tag{2.63}$$

If setting $\Omega = 0.8115$ Hz which is the natural frequency of the linear part, the magnitude of the harmonic response is shown in Fig. 2.1. It can be seen clearly how the harmonic response changes with different parameters c_1, c_2 and c_4 ($c_1 = 2$ in Fig. 2.1a and c1 = 20 in Fig. 2.1b).

A fascinating observation is that the harmonic response can be reduced to nearly zero (0.0212 in Fig. 2.1a and 0.2231 in Fig. 2.1b) when the nonlinear parameters c2 and c4 are set to suitable values. It's also worth noting that the harmonic response is heavily influenced by the linear dynamics (i.e., G(s)) in the system, indicating a strong dependency of the Wiener system's nonlinear dynamics on its linear part. This could be particularly relevant for sound and vibration control, as the system's harmonic response can be designed or controlled by intentionally introducing suitable linear or nonlinear dynamics. Clearly, the nonlinear output spectrum and GFRFs developed in this study offer a promising potential for techniques aimed at this objective. This will be the focus of a future study.

(b) **System Identification in the Frequency Domain**

In case that the system model is not known, the WiME procedure can be employed to identify the model easily. To demonstrate this, the WiME procedure could be conducted as follows:

The first Step is to apply the OSE algorithm to estimate different order output spectrum. To this aim, the initial truncation order N is firstly guessed to be 5 since it is not known; the input signal $r(t)$ is purposefully chosen as

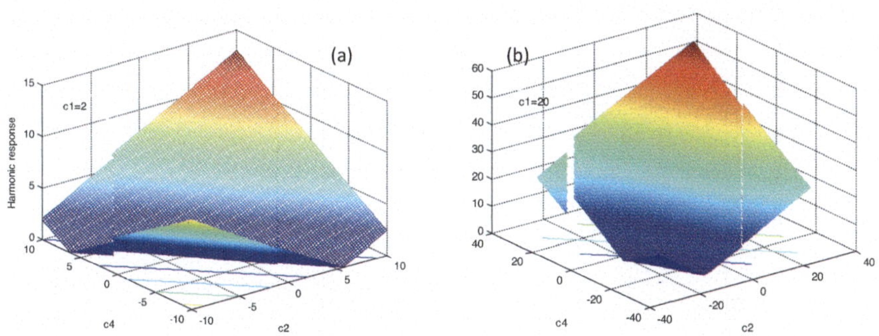

Fig. 2.1 Harmonic response with respect to different model parameters, showing clearly the harmonic response can be suppressed greatly with an appropriate relationship between c_2 and c_4 (the x and y coordinates represent the values of c_2 and c_4 and the z coordinate represents the magnitude of the output frequency response)

$$r(t) = \rho \frac{3}{2\pi} \frac{\sin(2\pi \times 200 \times t) - \sin(2\pi \times 0.001 \times t)}{t} \tag{2.64}$$

for $-25 < t < 25$ with a sampling rate 0.001 Hz, whose Fourier transform can be obtained via FFT shown in Fig. 2.2 for $\rho = 1$; the $N - 1 = 4$ values for ρ can be generated by searching stochastically in [0, 10], which are obtained as 1.9625, 3.1748, 3.1643, 2.1756 and can guarantee that the square matrix in (2.50) is nonsingular. With this algorithm setting, the OSE algorithm can yield the estimates of the nth-order output spectrum $Y_n(j\omega)$ for $n = 1,2,\ldots,5$, which are shown in Fig. 2.3.

It can be seen from Fig. 2.3 that the 3rd and 5th order output spectrum are obviously trivial and insignificant compared with the other orders. This implies that both $Y_3(j\omega)$ and $Y_5(j\omega)$ could be 0, which is consistent with the real system. Note that the initial nonlinear order is guessed to $N = 5$ in this estimation. Actually, different initial nonlinear order $N \geq 4$ can be tried to check the consistency of the OSE algorithm. When the guessing initial nonlinear order is set to $N = 4$, the estimation results are illustrated in Fig. 2.4 compared with those under $N = 5$. It can be seen that the estimates under these two cases match very well. That is, both of them produce the same estimate on the nonlinear order and exactly the same estimates on output spectra.

To identify the Wiener model with the estimated different order output spectrum, the second step of the WiME procedure can be conducted since $Y_1(j\omega) \neq 0$. With Eq. (2.53), it would be straightforward to estimate a linear model for the linear part using a least square algorithm. What should be emphasized here is that the estimation of the Wiener model with (2.53) and (2.54) is completely dependent on the estimate of the first order output spectrum. Therefore, the estimation accuracy of the first

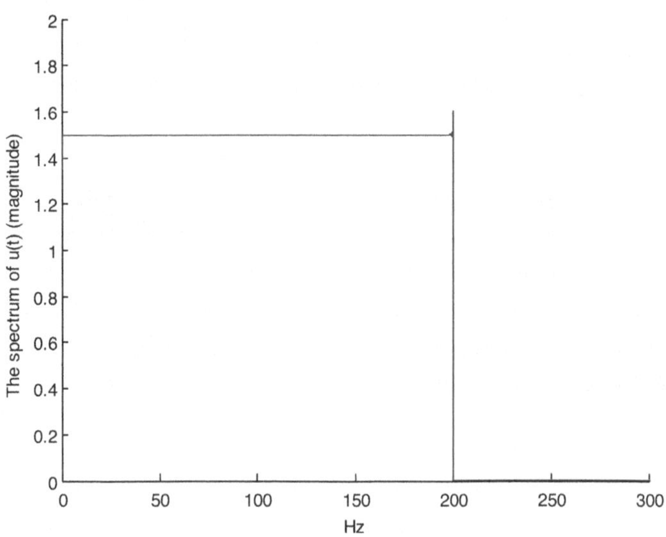

Fig. 2.2 FFT of the input signal (the unit of x-axis is Hz, and the y-axis represents the magnitude of the input signal)

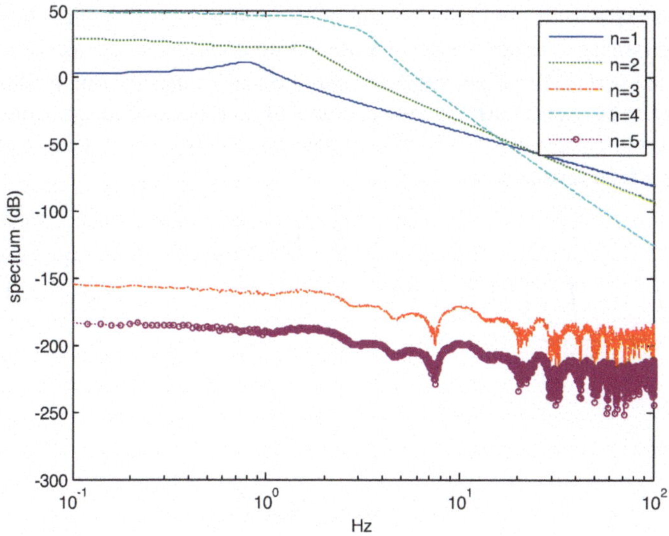

Fig. 2.3 The estimates of the nth-order output spectrum (the unit of x-axis is Hz, and the y-axis is the magnitude in dB)

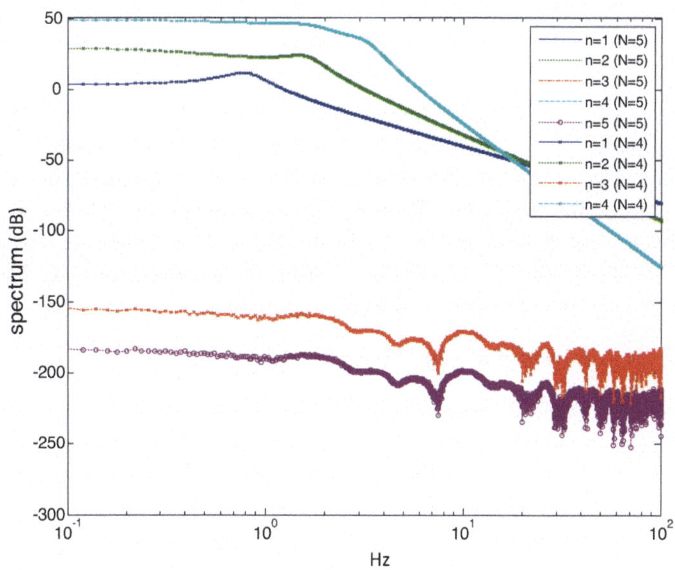

Fig. 2.4 Comparisons between two different estimation of the output spectrum with the first one using the guessing nonlinear order $N = 5$ and the second one $N = 4$ (the unit of x-axis is Hz, and the y-axis is the magnitude in dB)

order output spectrum predominates the estimation accuracy of the whole model. To demonstrate the accuracy of the first order output spectrum, comparison between the estimated first order output spectrum and the real theoretical one is illustrated in Fig. 2.5. It can be seen clearly that the estimated first order output spectrum matches the true theoretical one very well. To estimate the nonlinear part, Eq. (2.54) can be employed, which results in

$$\hat{c}_n = \frac{(2\pi)^{n-1}\hat{Y}_n(j\omega)}{\underbrace{\int_{-\infty}^{\infty}\cdots\int_{-\infty}^{\infty}}_{n-1}\left(\prod_{i=1}^{n-1}\hat{Y}_1(j\omega_i)\right)\hat{Y}_1(j(\omega-\omega_1-\cdots-\omega_{n-1}))\mathrm{d}\omega_1\cdots\mathrm{d}\omega_{n-1}} \tag{2.65}$$

Therefore,

$$|\hat{c}_n| = \frac{(2\pi)^{n-1}\left|\hat{Y}_n(j\omega)\right|}{\left|\underbrace{\int_{-\infty}^{\infty}\cdots\int_{-\infty}^{\infty}}_{n-1}\left(\prod_{i=1}^{n-1}\hat{Y}_1(j\omega_i)\right)\hat{Y}_1(j(\omega-\omega_1-\cdots-\omega_{n-1}))\mathrm{d}\omega_1\cdots\mathrm{d}\omega_{n-1}\right|} \tag{2.66}$$

Note that the estimate of the parameter $\mathrm{c_n}$ given in (2.65) or (2.66) is obtained with the estimates of nth-order output spectrum $Y_\mathrm{n}(\mathrm{j}\omega)$ at the frequency ω, which can be denoted by $\hat{c}_n(\omega)$. The estimates of the parameter c_n at different frequencies should be around the true value. Therefore, a better estimate could be obtained by taking an average of those estimates. To illustrate this, the estimates of the parameter c_2 and c_4 are shown in Fig. 2.6, where the averaged estimate for c_2 is 2.0035 (the true value of c_2 is 2) and for c_4 is -1.0065 (the true value is -1).

2. **Nonlinear Analysis of a Non-Wiener System**

Consider a mass-spring-damper system described by

$$240\ddot{x} + 196\dot{x} + 16000x + kx^2 = F(t) \tag{2.67a}$$

$$y = 196\dot{x} + 16000x + kx^2 \tag{2.67b}$$

There is a nonlinear term in the system with $k = 5\cdot10^6$, which could be considered as a fault in the spring characteristic.

To understand the nonlinear dynamics in the system and to evaluate the nonlinear effect on system dynamic behavior, the OSE algorithm can be applied firstly to achieve the estimates of different order output spectrum of the system. Following the similar procedure in Example 1, the first 5 orders of output spectra are estimated under the actuation signal (2.64) with $\rho = 1, 1.7065, 1.4082, 0.0357, 1.0114$, which

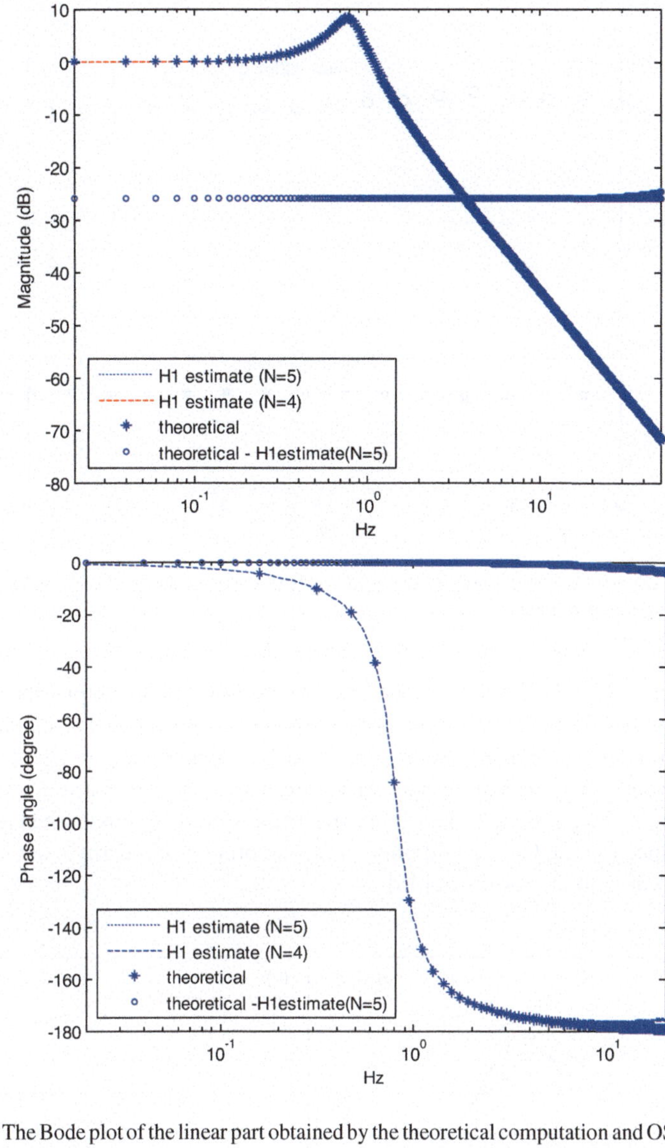

Fig. 2.5 The Bode plot of the linear part obtained by the theoretical computation and OSE algorithm (under two guessing initial nonlinear orders $N = 5$ and 4), and the errors between the theoretical computation and the OSE results (the unit of x-axis is Hz, and the y-axis is the magnitude in dB and phase angle in degree, respectively)

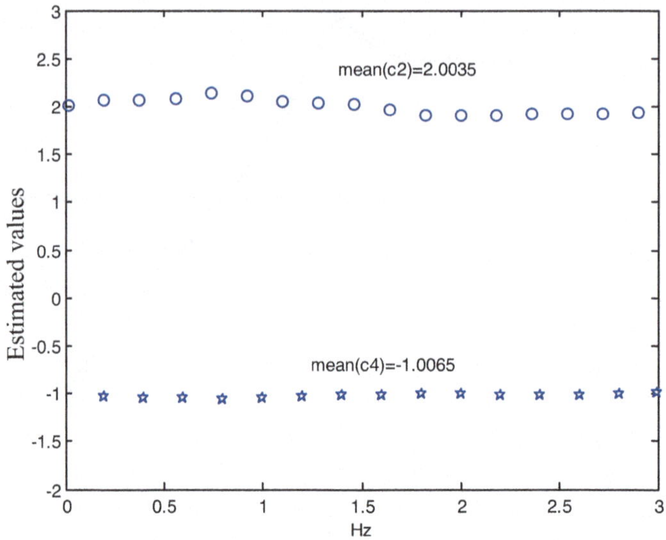

Fig. 2.6 The estimation of the parameters c_2 and c_4 (the x-axis is the frequency in Hz, and the y-axis is the estimated values)

are shown in Fig. 2.7a. It can be seen that the overall system output spectrum is basically dominated by the first three orders of output spectra (the 4th and 5th order output spectra can be regarded to be minor since their magnitudes are about −50 dB below). The high order (>1) output spectra demonstrate clearly the harmonic response in the system and the second order output spectrum is more dominant then the other high order spectra. The latter could imply a second order nonlinearity existing in the system according to the results in [26].

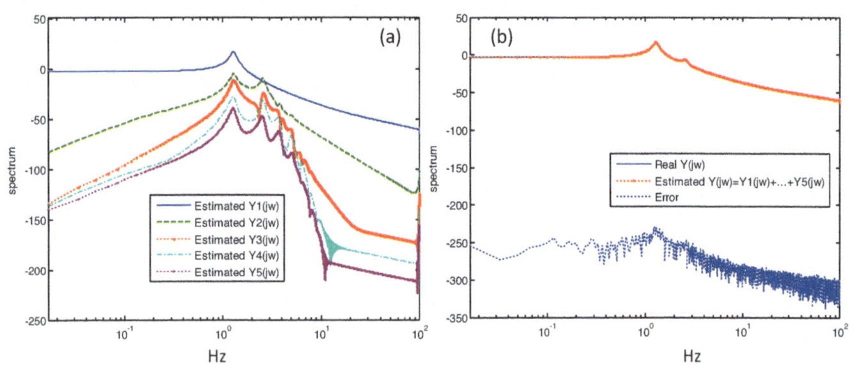

Fig. 2.7 **a** The estimated different order output spectrum; **b** The estimated and real system total output spectrum and the error (the unit is dB in y-axis and Hz in x-axis)

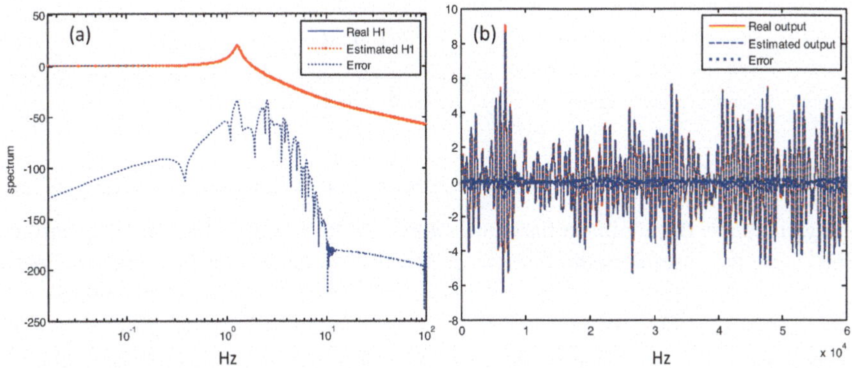

Fig. 2.8 **a** The estimated and real linear part transfer functions; **b** The estimated [by a Wiener model (2.28ab)] and real system outputs (the unit is dB in y-axis and Hz in x-axis)

The total system output spectrum can be approximated by adding up these five orders of output spectra (primarily the first three orders). This is compared with the actual system output spectrum and is depicted in Fig. 2.7b. A secondary resonance peak is clearly visible at a frequency that is twice the linear resonance frequency, due to the significant contribution from the second order nonlinear dynamics. The error line is essentially below −250 dB, signifying the precision of the estimation. To further illustrate the accuracy of the estimation, the first order Generalized Frequency Response Function (GFRF), i.e., the transfer function of the system's linear part, can be estimated by dividing the first order spectrum by the input spectrum. This is displayed in Fig. 2.8a, alongside a comparison to the actual transfer function of the linear part. The error is essentially below −50 dB, indicating a strong correlation between the estimated and actual transfer functions.

Although the system does not have a Wiener model structure, the input output dynamics can be approximated by a Wiener model (2.28ab) with the nonlinear part as a polynomial. The polynomial coefficients can be computed by (2.65), which are $c_1 = 1$, $c_2 = 0.009$, $c_3 = 0.0001$.The linear part can also be estimated from (2.53), which is straightforward (this is omitted here). The predicted output with the estimated Wiener model and the real output is shown in Fig. 2.8b (subject to a white Gaussian input with zero mean and STD 10). Obviously, the predicted output of the estimated Wiener model can match the real output of the system very well. It should be noted that with the estimated different order output spectrum, a more accurate NDE or NARX model could be fitted using a parametric characteristic analysis method in [25]. Also, note that the identification of localized fault nonlinearity in structural dynamics is conducted by using a Wiener series approach with parametric kernels in [49]. Using a Wiener model or Wiener–Hammerstein model could be an alternative and more efficient approach to the same objective. These topics will be focused in a future study.

With the estimated different order output spectrum $\hat{Y}_i(j\omega)$(i = 1, 2, …), the nonlinear severity of the system can be easily evaluated by

$$\delta = \frac{\int_{-\infty}^{\infty}\left|\sum_{i=2}^{N}\hat{Y}_i(j\omega)\right|^2 \mathrm{d}\omega}{\int_{-\infty}^{\infty}|Y(j\omega)|^2 \mathrm{d}\omega} \tag{2.68}$$

Note that $Y(j\omega) = \sum_{i=1}^{N}\hat{Y}_i(j\omega)$. The severity above is defined as a ratio of the energy taken by the nonlinear harmonics and the total energy in the output spectrum. This could be a useful method in non-destructive evaluation for fault severity. With (2.68), the nonlinear severity for system (2.67ab) with $k = 5{\cdot}10^6$ can be computed as $\delta = 0.58\%$. For comparison, when $k = 10^6$ in system (2.67ab), the different orders of output spectra are also estimated, which is shown in Fig. 2.9a and the nonlinear severity is computed with (2.68) as $\delta = 0.033\%$. It can be seen from Figs. 2.7a and 2.9ab that the high order (n > 1) output spectra when $k = 1{\cdot}10^6$ are obviously weaker than those when $k = 5{\cdot}10^6$. This is completely consistent with the computation results by the severity function. The nonlinear severities computed above indicate that the nonlinear part takes only a minor part in system output behaviors although $k = 5{\cdot}10^6$ in the first case and $k = 1{\cdot}10^6$ in the latter case. Therefore, under small excitation signals, the system can basically be regarded as a linear system. Moreover, the estimated first order output spectra under two cases match very well. This further demonstrates the consistency of the OSE algorithm in the estimation of system output spectrum.

It is worth mentioning that in the identification of a nonlinear system, whether using block-oriented models or others, a crucial task could be to precisely estimate the linear component of the system, regardless of the models used. Typically, the system nonlinearity could vary significantly. However, a precise estimation of the linear part will greatly assist in accurately estimating the involved nonlinearity. From the examples provided above, it's evident that the identification of the first order output spectrum is remarkably consistent and accurate when using the proposed

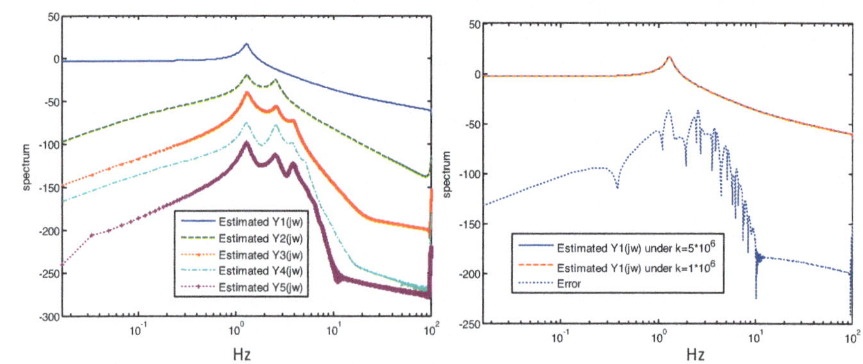

Fig. 2.9 **a** The estimated different order output spectrum; **b** The estimated first order output spectrum (the unit is dB in y-axis and Hz in x-axis)

OSE algorithm. This underscores the utility and potential of the new results for frequency domain analysis and design of nonlinear systems.

2.6 Conclusions

This study develops and discusses analytical frequency response functions, including the generalized frequency response functions (GFRFs) and nonlinear output spectrum, with examples demonstrating the derivation of GFRFs for a general nonlinear system, and gives more discussions about block-oriented nonlinear systems. These block-oriented nonlinear models are increasingly utilized across various fields, these new findings offer a systematic approach and a crucial foundation for the analysis, design, and identification of these nonlinear systems in the frequency domain.

The potential significance of these new results could be that these frequency response functions provide analytical expressions for the relationship between frequency response functions and model parameters, and for the dependence of frequency response functions on the linear part of the model. The effectiveness of the resulting estimation algorithms could be attributed to their ability to operate without requiring extensive model structure information, without needing the invertibility of the nonlinear part (which could be a more general multivariate polynomial function), and without being restricted to harmonic input signals, among other things.

Future research will concentrate on applying these new frequency domain methods to sound and vibration control systems, non-destructive evaluation, and biological systems, among others. The idea of output spectrum estimation will also be further explored for identification problems of more general Nonlinear Dynamic Error (NDE) systems and Nonlinear AutoRegressive with eXogenous inputs (NARX) systems.

Appendix 1: Proof of Lemma 1

The GFRFs for (2.7) can be obtained by directly applying the probing method. Note that (2.6) can be expanded as

$$\begin{aligned} y(t) &= n!H_n(j\omega_1, \ldots, j\omega_n)e^{j(\omega_1+\cdots+\omega_n)t} \\ &\quad + (n-1)!H_{n-1}(j\omega_1, \ldots, j\omega_{n-1})e^{j(\omega_1+\cdots+\omega_{n-1})t} \\ &\quad + \cdots + H_1(j\omega_1)e^{j\omega_1 t} + (\text{the other terms}) \end{aligned} \tag{2.69}$$

Using (2.5) and (2.69) with n = 3 in (2.7), it can be obtained that

$$y(t) = 3!H_3(j\omega_1, \ldots, j\omega_3)e^{j(\omega_1+\cdots+\omega_3)t}$$

$$
\begin{aligned}
&+ 2H_2(j\omega_1, j\omega_2)e^{j(\omega_1+\omega_2)t} \\
&+ H_1(j\omega_1)e^{j\omega_1 t} + \text{(the other terms)} \\
=& c_1\left(e^{j\omega_1 t} + e^{j\omega_2 t} + e^{j\omega_3 t}\right) + c_2\left(e^{j\omega_1 t} + e^{j\omega_2 t} + e^{j\omega_3 t}\right)^2 \\
&+ c_3\left(e^{j\omega_1 t} + e^{j\omega_2 t} + e^{j\omega_3 t}\right)^3 + \cdots \\
&+ c_{\mathrm{n}}\left(e^{j\omega_1 t} + e^{j\omega_2 t} + e^{j\omega_3 t}\right)^{\mathrm{n}} + \cdots
\end{aligned}
$$

Extracting the coefficients of the term $e^{j\omega_1 t}$ from the equation above, it can be obtained that

$$H_1(j\omega_1) = c_1$$

Extracting the coefficients of $e^{j(\omega_1+\omega_2)t}$, it can be obtained that

$$H_2(j\omega_1, j\omega_2) = c_2$$

Similarly, from the coefficients of $e^{j(\omega_1+\cdots+\omega_3)t}$, it can be obtained that

$$H_3(j\omega_1, \cdots, j\omega_3) = c_3$$

Following the same method, it can be obtained the nth-order GFRF as

$$H_n(j\omega_1, \cdots, j\omega_n) = c_n$$

This completes the proof.

References

1. Jing XJ and Lang ZQ, Frequency domain analysis and design of nonlinear systems based on Volterra series expansion: A parametric characteristic approach, Springer Int. Publishing, Switzerland, 2015
2. Jing XJ, Frequency domain analysis and identification of block-oriented nonlinear systems Journal of Sound and Vibration 330 (22), 5427–5442, 2011
3. Leonov G.A.,Ponomarenko D.V.&Smirnova V.B., Frequency-domain methods for nonlinear analysis, theory and applications. World Scientific Publishing Co. Pte. Ltd., Singapore, 1996
4. Orlowski P., Frequency domain analysis of uncertain time-varying discrete-time systems, Circuits Systems Signal Processing, Vol. 26, No. 3, 2007, PP. 293–310, 2007
5. Glass J. W. and Franchek M. A., Frequency Based Nonlinear Controller Design of Regulating Systems Subjected to Time Domain Constraints, Proceedings of the American Control Conference San Diego, California, pp 2082–2086, 1999
6. Shah M. A., Franchek M. A. Frequency-based controller design for a class of nonlinear systems, International Journal of Robust and Nonlinear Control, 9(12), 825–840, 1999
7. Alleyne A. and Hedrick J.K., Nonlinear adaptive control of active suspensions, IEEE Transactions on Control Systems Technology, Vol 3, No 1, pp 94–101, 1995

8. Gelb A. and Velde W. v.d., Multiple input describing functions and nonlinear system design. McGraw Hill, NJ, 1968
9. Nuij, P.W.J.M., Bosgra O.H. and Steinbuch M., Higher order sinusoidal input describing functions for the analysis of nonlinear systems with harmonic responses. Mech Syst Signal Process, 20(8): 1883–1904, 2006
10. Schmidt G., Tondl A., Non-linear Vibrations, Cambridge University Press, Cambridge, 1986.
11. George D.A., Continuous nonlinear systems, Technical Report 355, MIT Research Laboratory of Electronics, Cambridge, Mass. Jul. 24, 1959
12. Rugh W.J., Nonlinear System Theory: the Volterra/Wiener Approach, Baltimore, Maryland, U.S.A.: Johns Hopkins University Press, 1981
13. Peyton Jones J.C. and Billings S.A. Recursive algorithm for computing the frequency response of a class of nonlinear difference equation models. International Journal of Control, Vol. 50, No. 5, 1925–1940. 1989
14. Billings S.A. and Peyton-Jones J.C., Mapping nonlinear integro-differential equation into the frequency domain, International Journal of Control, Vol 54, 863–879, 1990
15. Swain A.K. and Billings S.A.. Generalized frequency response function matrix for MIMO nonlinear systems. International Journal of Control. Vol. 74. No. 8, 829–844, 2001
16. Jing X.J., Lang Z.Q., Billings S.A., Tomlinson G.R., The parametric characteristic of frequency response functions for nonlinear systems. International Journal of Control, 79 (12) (2006), pp. 1552–1564
17. Zhang H., Billings S.A, and Zhu Q.M., "Frequency response functions for nonlinear rational models". International Journal of Control, 61, 1073–1097, 1995
18. Jing X. J., Lang Z.Q., On the Generalized Frequency Response Functions of Volterra Systems, Transactions of the ASME, Journal of Dynamic Systems, Measurement and Control, 131(6), 2009
19. Lang Z.Q., and Billings S. A. Output frequency characteristics of nonlinear systems. International Journal of Control, Vol. 64, 1049–1067, 1996
20. Zhang H. and Billings S.A., Gain bounds of higher order nonlinear transfer functions, International Journal of Control, Vol 64, No 4, 767–773, 1996
21. Billings S.A. and Lang Z.Q., A bound of the magnitude characteristics of nonlinear output frequency response functions, International Journal of Control, Part 1 Vol 65, No. 2, 309–328 and Part 2, Vol 65, No. 3, 365–384, 1996
22. Jing X.J., Lang Z.Q., and Billings S. A., New Bound Characteristics of NARX Model in the Frequency Domain, International Journal of Control, 80(1), pp140–149, 2007
23. Jing X. J., Lang Z.Q. and Billings S.A., Magnitude bounds of generalized frequency response functions for nonlinear Volterra systems described by NARX model. Automatica, 44, 838–845, 2008
24. Jing X. J., Lang Z.Q. and Billings S.A., Frequency-Dependent Magnitude Bounds of the Generalized Frequency Response Functions for NARX model, European Journal of Control, 15(1), 2009
25. Jing X. J., Lang Z.Q. and Billings S.A., Parametric Characteristic Analysis for Generalized Frequency Response Functions of Nonlinear Systems, Circuits, Systems and Signal Processing, 28: 699–733, 2009
26. Jing X.J., Lang Z.Q., Billings S.A., Output Frequency Properties of Nonlinear Systems, International Journal of Non-Linear Mechanics, 45(7), September 2010, pp 681–690
27. X.J. Jing, Z.Q. Lang and S.A. Billings, Frequency Domain Analysis for Suppression of Output Vibration from Periodic Disturbance using Nonlinearities. Journal of sound and vibration 314, 536–557, 2008
28. Jing X. J. and Lang Z.Q., Frequency Domain Analysis of a Dimensionless Cubic Nonlinear Damping System Subject to Harmonic Input, Nonlinear Dynamics, 58: 469–485, 2009
29. Jing X.J. et al, Coding characteristics of spiking local interneurons during imposed limb movements in the locust, Journal of Neurophysiology, 103:603–615, 2010
30. Korenberg M.J., Statistical identification of parallel cascades of linear and nonlinear systems. IFAC Symp. Ident. Sys. Param Est, 1: 580–585, 1982

31. Boyd S. and Chua L.O., Fading memory and the problem of approximating nonlinear operators with Volterra series. IEEE Trans. On Circuits and Systems, CAS-32(11), 1150–1161, 1985
32. Zhu, Y. Distillation column identification for control using Wiener model. American control conference, 1999
33. Kalafatis A. et al, A new approach to the identification of pH processes based on the Wiener model. Chemical Engineering Science, 50(23), 3693–3701, 1995
34. Hunter I.W. and Korenberg M.J., The identification of nonlinear biological systems: Wiener and Hammerstein cascade models. Biological Cybernetics, 55, 135–144, 1986
35. Huang A. Tanskanen J.M.A., Hartimo I.O., Design of optimum power estimator based on Wiener model applied to mobile transmitter power control. Proceedings of the IEEE Inter. Symposium on Circuits and Systems, 5:249–252, 1998
36. Bloemen H.H.J., et al, Wiener model identification and predictive control for dual composition control of a distillation column. Journal of Process Control, 11: 601–620, 2001
37. Wang J, Sano A, Chen T, Huang B, Identification of Hammerstein systems without explicit parameterization of nonlinearity, International Journal of Control, 82(5): 937–952, 2010
38. Baumgartner, S., & Rugh, W. Complete identification of a class of nonlinear systems from steady state frequency response. IEEE Trans. on Circuits and Systems, 22, 753–759, 1975
39. Krzyzak, A. On nonparametric estimation of nonlinear dynamic systems by the Fourier series estimate. Signal Processing, 52, 299–321, 1996
40. Crama, P., & Schoukens, J. Initial estimates of Wiener and Hammerstein systems using multisine excitation. IEEE Transactions on Instrumentation and Measurement, 50, 1791–1795, 2001
41. Bai E.W., Frequency domain identification of Wiener models, Automatica, 39:1521–1530, 2003
42. Shanmugam K.S., Jong, M. T., Identification of nonlinear systems in frequency domain, IEEE Trans. Aerospace and Electronic Systems, AES-11(6): 1218–1225, 1975
43. Jeffreys, H. and Jeffreys, B. S. "Weierstrass's Theorem on Approximation by Polynomials" and "Extension of Weierstrass's Approximation Theory." §14.08–14.081 in Methods of Mathematical Physics, 3rd ed. Cambridge, England: Cambridge University Press, pp. 446–448, 1988.
44. Kotsios, S., Finite input/output representative of a class of Volterra polynomial systems. Automatica, 33, 257–262, 1997
45. Jing X. J., Lang Z.Q. and Billings S.A., "Frequency domain analysis for nonlinear Volterra systems with a general nonlinear output function". International Journal of Control, 81:2, 235–251, 2008
46. Li L.M., Billings S.A., Continuous time nonlinear system identification in the frequency domain, International Journal of Control, 74(11): 1052–1061, 2001
47. Levi E.C., Complex curve fitting, IRE Trans. Automatic Control, 4:37–43, 1959
48. Young P.C., The instrumental variable method: a practical approach to identification and parameter estimation, the 7th IFAC/IFORS Symposium on Identification and System Parameter Estimation, York, UK, pp 1–14, 1985
49. Silva S.D., Cogan S, Foltete E, Nonlinear identification in structural dynamics based on Wiener series and Kautz filters, Mechanical Systems and Signal Processing, 24:52–58, 2010

Chapter 3
The Parametric Characteristic of Frequency Response Functions

Abstract This chapter presents the foundational theory underlying the parametric characteristic analysis method, which serves as the theoretical cornerstone of the nonlinear Characteristic Output Spectrum (nCOS) function approach. Specifically, it addresses the formulation of accurate parametric structures for frequency response functions, enabling the explicit representation of nonlinear system behavior in terms of critical design parameters. The parametric characteristic framework is built upon a rigorously defined operator and its associated mathematical properties. This formulation facilitates a series of theoretical developments, culminating in the establishment of the nCOS function. The method has demonstrated broad applicability in various engineering domains, including vibration control, energy harvesting, and fault diagnosis.

Keywords Frequency domain analysis · Nonlinear Systems · Parametric characteristic · Frequency response functions

3.1 Introduction

3.1.1 About This Chapter

This chapter introduces a fundamental theory about the parametric characteristic analysis method which is the theoretical basis of the nCOS function method. That is, how to obtain the accurate parametric structure of frequency response functions such that an explicit nonlinear function can be achieved in terms of any critical design parameters of interest. This parametric characteristic approach includes a well-defined operator and associated properties which result in a series of the studies including the establishment of the nCOS function with applications in vibration control, energy harvesting, and fault diagnosis. More theory about this topic can also be referred to related papers as mentioned in Chap. 1.

X. Jing, *The Bio-Inspired X-Structure/Mechanism Approach for Exploring Nonlinear Benefits in Engineering*, https://doi.org/10.1007/978-981-96-5769-8_3

To clearly illustrate the relationship between system frequency response functions and model parameters that define system nonlinearities, and thereby establish a direct link from model parameters to system frequency response characteristics, a parametric characteristic analysis approach is established for Volterra systems characterized by a nonlinear differential equation (NDE). The parametric characteristics of the generalized frequency response functions (GFRFs) for the NDE model are determined, and several key properties are examined, which can clearly show which model parameters contribute to and how these parameters influence the GFRFs. Based on the parametric characteristic analysis, it is shown how the system frequency domain characteristics are connected to the system time domain model parameters, and the output frequency response function can now be explicitly determined with a detailed polynomial structure. These new findings offer significant and novel insights into the analysis and design of nonlinear systems in the frequency domain. Several examples are provided to demonstrate the results.

The first piece of this parametric characteristic method was formally published in [1] and an early version of [1] was internally circulated as a research report [2] in Sheffield University during the author studying for his PhD in 2006.

3.1.2 Research Background

Frequency domain analysis is a favored and effective methodology in engineering practices for linear systems, as it often provides intuitive insights into the system under study. However, the scenario becomes significantly more complex when dealing with nonlinear systems. Nonlinear systems are known to exhibit harmonics and intricate inter-modulations, which transfer energy across different frequencies and produce outputs at frequencies that are quite distinct from those of the input. Additionally, phenomena such as chaos and bifurcation are frequently observed in many cases [3]. These factors greatly increase the complexity of studying nonlinear systems in the frequency domain. Despite noticeable progress in this area [4–8], understanding a nonlinear term in a system, and analyzing and even designing a nonlinear system in the frequency domain, are aspects that are not fully developed in the existing literature.

The frequency domain analysis of nonlinear systems can also be carried out using the Volterra series theory [9, 10]. As pointed out in [11], any nonlinear system that is time-invariant, causal, and has fading memory can be approximated by a Volterra series of sufficiently high orders. This chapter refers to this class of nonlinear systems as Volterra systems, which indeed represents a broad range of nonlinear systems [9, 11, 12]. The existence of a Volterra series expansion for nonlinear systems was also examined in [12].

In the development of frequency domain analysis methods for Volterra systems, [13] introduced a fundamental concept known as the generalized frequency response function (GFRF), defined as the multidimensional Fourier transformation of the kernel function of the Volterra series. Based on this concept, numerous important results and techniques were developed to estimate and compute the Volterra kernels and the GFRFs of a Volterra system from input–output data over the past decades [9, 14–18].

Notably, for Volterra systems described by nonlinear integro-differential equations, a recursive algorithm for computing the GFRFs was derived in [19], providing a crucial foundation for further study on the analysis and synthesis of Volterra systems [20–23]. Despite significant achievements, many problems related to GFRF-based analysis remain unsolved.

The GFRF is a multivariate function defined in high-dimensional frequency space, and the GFRFs for a practical system model are complex-valued function polynomials in terms of model parameters. As a result, the computations of the GFRFs and the system output spectrum are typically computationally intensive [24]. Consequently, the analytical relationship between system time-domain model parameters, which may represent physical characteristics in a practical system, and these system frequency response functions, which represent the system frequency response characteristics, cannot be explicitly revealed from these results. These issues, to some extent, pose a challenge in the practical applications and understanding of GFRF-based analysis for nonlinear systems.

To clearly illustrate the analytical relationship between system frequency response functions and model parameters, which define system nonlinearities, and thereby reveal the dependency of system frequency response characteristics on system time-domain model parameters, a parametric characteristic analysis for the Generalized Frequency Response Functions (GFRFs) and output spectrum of Volterra systems described by Nonlinear Differential Equations (NDE) is conducted. This is based on some preliminary results found in [1]. For further studies on this subject by the authors, readers can refer to [24], where the understanding of the GFRFs is facilitated by a visual technique, and [25], which revealed that the NDE system output spectrum is a polynomial function of the system parameters but with no clear structures as what to be revealed here.

In this study, the parametric characteristics of the GFRFs for the NDE model are established and discussed. This reveals which model parameters contribute to and how these parameters affect the GFRFs. Some important properties of the parametric characteristics for the GFRFs are also discussed. These results provide an explicit analytical relationship between the system time-domain model parameters and the GFRFs for the NDE model. They offer a straightforward insight into how the system frequency domain characteristics are determined by the system time-domain model parameters, a relationship that was difficult to address previously in the frequency domain.

Based on these results, the nonlinear effects on system frequency response characteristics from different kinds of nonlinearities are investigated, yielding some interesting findings. It is also shown that the output frequency response function of the NDE model can be determined explicitly with a detailed polynomial structure using a numerical method, without the need for complicated computation as mentioned above. It's worth noting that the NDE model usually has a direct link between its coefficients and physical meanings. Therefore, these new results should provide a fresh approach to the frequency domain analysis and design of nonlinear systems in practice. Examples are provided to illustrate these new results.

This chapter is organized as follows. Section 3.2 presents the Generalized Frequency Response Functions (GFRFs) of the Nonlinear Differential Equation (NDE) model and discusses related issues. In Sect. 3.3, we first introduce some foundational concepts, followed by the definition of a coefficient extraction operator. Leveraging these basic concepts and the operator, we establish and thoroughly discuss the parametric characteristics of the GFRFs of the nonlinear differential equation model in this section. As will become evident, these results not only reinforce the mathematics but also significantly extend the preliminary findings for Nonlinear AutoRegressive with eXogenous inputs (NARX) models in [1]. In Sect. 3.4, we examine the nonlinear effects on system frequency response characteristics from various types of nonlinearities, based on the results in Sect. 3.3, with a particular focus on the relationship between the output spectrum of the NDE model and model parameters. A summary is provided in Sect. 3.5.

3.2 The GFRFs of the Nonlinear Differential Equation

Consider the following nonlinear differential equation (NDE)

$$\sum_{m=1}^{M}\sum_{p=0}^{m}\sum_{k_1,k_{p+q}=0}^{K} c_{p,q}(k_1,\cdots,k_{p+q})\prod_{i=1}^{p}\frac{d^{k_i}y(t)}{dt^{k_i}}\prod_{i=p+1}^{p+q}\frac{d^{k_i}u(t)}{dt^{k_i}}=0 \tag{3.1}$$

where y(t) and u(t) $\in \mathbb{R}^1$, $\left.\frac{d^k x(t)}{dt^k}\right|_{k=0} = x(t)$, $p + q = m$, $\sum_{k_1,k_{p+q}=0}^{K}(\cdot) = \sum_{k_1=0}^{K}(\cdot)\cdots\sum_{k_{p+q}=0}^{K}(\cdot)$, M is the maximum degree of nonlinearity in terms of y(t) and u(t), and K is the maximum derivative order. In this model, the parameters such as $c_{0,1}(.)$ and $c_{1,0}(.)$ are referred to as linear parameters corresponding to the linear terms in the model, i.e., $\frac{d^k y(t)}{dt^k}$ and $\frac{d^k u(t)}{dt^k}$ for $k = 0, 1, \ldots, L$, and $c_{pq}(\cdot)$ for $p + q > 1$ are referred to as nonlinear parameters corresponding to nonlinear terms in the model of the form $\prod_{i=1}^{p}\frac{d^{k_i}y(t)}{dt^{k_i}}\prod_{i=p+1}^{p+q}\frac{d^{k_i}u(t)}{dt^{k_i}}$, e.g., $y(t)^p u(t)^q$. $p + q$ is referred to as the nonlinear degree of the nonlinear parameter $c_{pq}(\cdot)$. Let

$$C(n,K) = \left(c_{pq}(k_1, \cdots, k_{p+q}) \middle| \begin{array}{c} p = 0 \cdots m,\ p+q = m, \\ 2 \le m \le n \\ k_i = 0 \cdots K,\ i = 1 \cdots p+q \end{array} \right) \tag{3.2}$$

which includes all the nonlinear parameters from degree 2 to n. Obviously, $C(M, K)$ include all the nonlinear parameters involved in model (3.1).

In this study, consider a class of nonlinear systems which are described by the NDE model (3.1) and have a convergent Volterra series expansion for its input output relationship as

$$y(t) = \sum_{n=1}^{N} \int_{-\infty}^{\infty} \cdots \int_{-\infty}^{\infty} h_n(\tau_1, \cdots, \tau_n) \prod_{i=1}^{n} u(t - \tau_i) d\tau_i \tag{3.3}$$

where N is the maximum order of the series, and $h_n(\tau_1, \cdots, \tau_n)$ is a real valued function referred to as the nth order Volterra kernel. The conditions under which a practical nonlinear system has such an expansion as (3.3) can be referred to [9, 11, 12]. The nth order generalized frequency response function (GFRF) of the nonlinear system (3.3) is defined as

$$H_n(j\omega_1, \ldots, j\omega_n) = \int_{-\infty}^{\infty} \cdots \int_{-\infty}^{\infty} h_n(\tau_1, \cdots, \tau_n) \exp(-j(\omega_1\tau_1 + \cdots + \omega_n\tau_n)) d\tau_1 \cdots d\tau_n \tag{3.4}$$

For the nonlinear differential Eq. (3.1), a recursive algorithm can be utilized to compute the nth-order GFRF as follows [19]:

$$\begin{aligned} L(n) \cdot H_n(j\omega_1, \ldots, j\omega_n) = & \sum_{k_1, k_n = 1}^{K} c_{0,n}(k_1, \ldots, k_n)(j\omega_1)^{k_1} \cdots (j\omega_n)^{k_n} \\ & + \sum_{q=1}^{n-1} \sum_{p=1}^{n-q} \sum_{k_1, k_{p+q}=0}^{K} c_{p,q}(k_1, \ldots, k_{p+q})(j\omega_{n-q+1})^{k_{p+1}} \\ & \cdots (j\omega_n)^{k_{p+q}} H_{n-q,p}(j\omega_1, \ldots, j\omega_{n-q}) \\ & + \sum_{p=2}^{n} \sum_{k_1, k_p=0}^{K} c_{p,0}(k_1, \ldots, k_p) H_{n,p}(j\omega_1, \ldots, j\omega_n) \end{aligned} \tag{3.5}$$

$$\begin{aligned} H_{n,p}(\cdot) = & \sum_{i=1}^{n-p+1} H_i(j\omega_1, \ldots, j\omega_i) \\ & H_{n-i,p-1}(j\omega_{i+1}, \ldots, j\omega_n)(j\omega_1 + \cdots + j\omega_i)^{k_p} \end{aligned} \tag{3.6}$$

$$H_{n,1}(j\omega_1,\cdots,j\omega_n)=H_n(j\omega_1,\cdots,j\omega_n)(j\omega_1+\cdots+j\omega_n)^{k_1} \tag{3.7}$$

where $L(n)=-\sum_{k_1=0}^{K}c_{1,0}(k_1)(j\omega_1+\cdots+j\omega_n)^{k_1}$. Moreover, $H_{n,p}(j\omega_1,\ldots,j\omega_n)$ in (3.6) can also be rewritten as

$$H_{n,p}(j\omega_1,\ldots,j\omega_n)=\sum_{\substack{r_1\cdots r_p=1\\ \sum r_i=n}}^{n-p+1}\prod_{i=1}^{p}H_{r_i}(j\omega_{X+1},\ldots,j\omega_{X+r_i})(j\omega_{X+1}+\cdots+j\omega_{X+r_i})^{k_i} \tag{3.8}$$

where $X=\sum_{x=1}^{i-1}r_x$. Note that a correction is made for the superscripts of $(j\omega_{n-q+1})^{k_{p+1}}\cdots(j\omega_n)^{k_{p+q}}$ in (3.5).

From Eqs. (3.4–3.8), it is evident that the Generalized Frequency Response Functions (GFRFs) are multivariate complex-valued polynomials in terms of model parameters. The computations of the nth-order GFRF are highly computationally demanding, involving numerous recursive symbolic operations. The computational complexity escalates even further for the computation of the GFRFs and system output spectrum of orders higher than 5 [24]. As a result, the relationship between the model parameters and the frequency response functions cannot be clearly demonstrated by the above recursive algorithm. In practical terms, a clear and direct analytical expression for the relationship between model parameters and system frequency response functions can significantly aid the analysis and design of nonlinear systems in the frequency domain. Therefore, the primary goal of this study is to investigate which of the nonlinear parameters in Eq. (3.2) contribute to the GFRFs and how they do so, to reveal the analytical relationship between the model parameters and the GFRFs in a clear and straightforward manner, and ultimately to provide significant insights into the analysis of the frequency domain characteristics of nonlinear systems.

3.3 Parametric Characteristics of the GFRFs

Some fundamental concepts and a useful operator are introduced at first, then the parametric characteristics for the GFRFs are established based on these concepts and the new operator.

3.3.1 Preliminaries

Definition 1. A function $h(s; x)$ is said to be separable with respect to parameter x if it can be written as $h(x; s) = g(x) \cdot f_1(s) + f_0(s)$, where $f_i(.)$ for $i = 0, 1$ are functions of variable s but independent of the parameter x.

A function $h(s; x)$ satisfying Definition 1 is referred to as x-separable function or simply separable function, where x is referred to as the interested parameter which may be a parameter to be designed for a system, and s represents the reference variable (or independent variable) of a system such as time or frequency.

Remark 1. In the definition of an x-separable function $h(s; x)$, x may be a vector including all the concerned separable parameters, and s denotes not only the independent variables of $h(.)$ but also may include all the un-separable or un-concerned parameters in $h(.)$. The detailed properties of the function $h(.)$, parameter x and s are not necessarily considered here. Note also that in Definition 1, $f_0(s)$ and $f_1(s)$ are invariant with respect to x and $g(x)$. Thus $h(s; x)$ can be regarded as a pure function of x at any a specific s. In this case, if $g(x)$ is known, and additionally the values of $h(s; x)$ and $g(x)$ under some different values of x can be obtained by certain methods (simulations or experimental tests), then the values of $f_0(s)$ and $f_1(s)$ can be achieved by the Least Square method. Thus the function $h(s; x)$ at a given point s is obtained which is an analytical function of the concerned parameter x. This provides a numerical method to determine the relationship between the concerned parameters and the corresponding function.

An x-separable function $h(s; x)$ at a given point s is denoted as $h(x)|_s$, or simply as $h(x)_s$.

Consider a parameterized series

$$\begin{aligned} H(s; x) &= g_1(x)f_1(s) + g_2(x)f_2(s) \\ &\quad + \cdots + g_n(x)f_n(s) = G \cdot F^T \end{aligned} \tag{3.9}$$

where $n > 1$, $f_i(s)$ and $g_i(x)$ for $i = 1, \ldots, n$ are all scalar functions, $F = [f_1(s), f_2(s), \ldots, f_n(s)]$, $G = [g_1(x), g_2(x), \ldots, g_n(x)]]$, x and s are both parameterized vectors including interested and non-interested parameters, respectively. The series is obviously x-separable, thus $H(x)_s$ is completely determined by the parameters in x or the values of $g_1(x), g_2(x), \ldots, g_n(x)$. Note that at a given point s, the characteristics of the series $H(s; x)$ is completely determined by G, and how the parameters in x are included in and affect $H(s; x)$ is completely demonstrated in G, too. Therefore, some important properties of the series $H(s; x)$, to a certain extent, is determined and thus can be revealed by the function vector G. With respect to any interested parameters in x, different G corresponds to different polynomial or series expansion of $H(s;\, x)$ at s and thus implies different system properties or output behaviour characteristics. This is the important motivation of this study on the parametric characteristics of frequency response functions of nonlinear systems. Hence, the vector G is referred to as the parametric characteristic vector of the series. Moreover, if the characteristic vector G is determined, then following the method mentioned in Remark 1, the function $H(x)_s$ which shows the analytical relationship between the concerned parameter x and the series is achieved, and consequently the effects on the series from each parameter in x can be studied. The function $H(x)_s$ is referred to as parametric characteristic function of the series $H(s;\, x)$. Based on the discussions above, the following result can be concluded.

Lemma 1. *$H(s;\, x)$ is a separable function with respect to the parameter x, then there must exist a parametric characteristic vector G and an appropriate function vector F, such that $H(s;\, x) = G \cdot F^T$, where the elements of G are functions of x independent of s, and the elements of F are functions of s independent of x.*

According to the definition and discussion above, it can be seen from (3.5) that each term of the nth-order GFRF is separable with respect to the nonlinear parameters in (3.2) of model (3.1). Hence, the GFRFs for model (3.1) are separable with respect to the nonlinear parameters in (3.2). In order to study the parametric characteristics of the nth-order GFRF, a useful operator is introduced for the extraction of the concerned parameters involved in a separable parameterized function series, which was firstly and simply defined in [1] and is formally redefined and discussed more here.

Let C_s be a set of some parameters which takes values in $\mathbb{C}$, let P_c be a monomial function set defined in C_s, i.e., $P_c = \left\{c_1^{r_1} c_2^{r_2} \cdots c_I^{r_I} | c_i \in C_s, r_i \in Z_0, I = |C_s|\right\}$, where $|C_s|$ is the number of the parameters in C_s, Z_0 denotes all the nonnegative integers. Let W_s be another parameter set similar to C_s but $W_s \cap C_s = \varphi$, and let P_f be a function set defined in W_s, i.e., $P_f = \{f(w_1, \cdots, w_I) | w_i \in W_s, I = |W_s|\}$. Let Ξ denote all the finite order function series with coefficients in P_c timing some functions in P_f. A series in Ξ can be written as

$$H_{CF} = s_1 f_1 + s_2 f_2 + \cdots + s_\sigma f_\sigma \in \Xi$$

where $s_i \in P_c, f_i \in P_f$ for $i = 1, \ldots, \sigma \in Z_0 \backslash \{0\}$, $C = [s_1, s_2, \ldots, s_\sigma]$, and $F = [f_1, f_2, \ldots, f_\sigma]^T$. Obviously, this series is separable with respect to the parameters in C_s and W_s. Define a **Coefficient Extraction** operator $CE : \Xi \to P_c^\sigma$, such that

$$CE(H_{CF}) = [s_1, s_2, \ldots, s_\sigma] = C \in P_c^\sigma$$

where $P_c^\sigma = \{[s_1, s_2, \ldots, s_\sigma] | s_1, \ldots, s_\sigma \in P_c\}$. This operator has the following properties:

(p1) Reduced vectorized sum "$\oplus$".

$$
\begin{aligned}
CE(H_{C_1F_1} + H_{C_2F_2}) &= CE(H_{C_1F_1}) \oplus CE(H_{C_2F_2}) \\
&= C_1 \oplus C_2 = [C_1, C_{2'}] \text{ and} \\
C_{2'} &= VEC(\overline{C}_2 - \overline{C}_1 \cap \overline{C}_2)
\end{aligned}
$$

where $\overline{C}_1 = \{C_1(i)|1 \le i \le |C_1|\}$, $\overline{C}_2 = \{C_2(i)|1 \le i \le |C_2|\}$, $VEC(.)$ is a vector consisting of all the elements in set (.). $C_{2'}$ is a vector including all the elements in C_2 except the same elements as those in C_1.

(p2) Reduced Kronecker product "$\otimes$".

$$
\begin{aligned}
CE(H_{C_1F_1} \cdot H_{C_2F_2}) &= CE(H_{C_1F_1}) \otimes CE(H_{C_2F_2}) \\
&= C_1 \otimes C_2 \\
&= VEC\left\{ c \left| \begin{array}{l} C_3 = [C_1(1)C_2, \ldots, C_1(|C_1|)C_2] \\ c = C_3(i),\ 1 \le i \le |C_3| \end{array} \right. \right\}
\end{aligned}
$$

which implies that there are no repetitive elements in $C_1 \otimes C_2$

(p3) Invariant. (a)$CE(\alpha \cdot H_{CF}) = CE(H_{CF})$, $\forall \alpha \notin C_s$; (b) $CE(H_{CF_1} + H_{CF_2}) = CE(H_{C(F_1+F_2)}) = C$.

(p4) Unitary. (a) If $\frac{\partial H_{CF}}{\partial c} = 0$ for $\forall c \in C_s$, then $CE(H_{CF}) = 1$; (b) if $H_{CF} = 0$ for $\forall c \in C_s$, then $CE(H_{CF}) = 0$. When there is a unitary 1 in $CE(H_{CF})$, there is a nonzero constant term in the corresponding series H_{CF} which has no relation with the parameters in C_s.

(p5) Inverse. $CE^{-1}(C) = H_{CF}$. This implies any a vector C consisting of the elements in P_c should correspond to at least one series in Ξ.

(p6) $CE(H_{C_1F_1}) \approx CE(H_{C_2F_2})$ if the elements of C_1 are the same as those of C_2, where "$\approx$" means equivalence. That is, both series are in fact the same result considering the order of $s_i f_i$ in the series has no effect on the value of a series H_{CF}. This further implies that the CE operator is also commutative and associative, for instance, $CE(H_{C_1F_1} + H_{C_2F_2}) = C_1 \oplus C_2 \approx CE(H_{C_2F_2} + H_{C_1F_1}) = C_2 \oplus C_1$. Hence, the results by the CE operator may be different but all may correspond to the same function series and are thus equivalent.

(p7) Separable and interested parameters only. A parameter in a series can only be extracted if the parameter is interested and the series is separable with respect to this interested parameter. Thus the operation result is different for different purposes.

Note that the CE operator above is redefined in a more mathematical and formal way, and two more properties are listed and thus emphasized as stated in (p6) and (p7), compared the original result in [1]. It can be seen that, all the operations of CE are in terms of the parameters in C_s, and the CE operator sets up a mapping from Ξ to $P_c{}^{\sigma}$. For convenience, let $\otimes_{(*)}(\cdot)$ and $\oplus_{(*)}(\cdot)$ denote the multiplication and addition by the reduced Kronecker product "$\otimes$" and vectorized sum "$\oplus$" of the terms in (.) satisfying (*), respectively; and $\otimes_{i=1}^{k} C_{p,q} = C_{p,q} \otimes \cdots \otimes C_{p,q}$ can be simply written as $C_{p,q}^k$. Moreover, define the $(p+q)$th degree nonlinear parameter vector as

$$C_{p,q} = [c_{p,q}(0, \ldots, 0), c_{p,q}(0, \ldots, 1), \ldots, c_{p,q}(\underbrace{K, \ldots, K}_{p+q=m})]$$

which includes all the nonlinear parameters of the form $c_{p,q}(.)$ in model (3.1). Note that $C_{p,q}$ can also be regarded as a set of the $(p+q)$th degree nonlinear parameters of the form $c_{p,q}(.)$. Moreover, if all the elements of CE(H_{CF}) are zero, i.e., CE(H_{CF}) = 0, then CE(H_{CF}) is also regarded as empty.

The CE operator provides a useful tool for the analysis of the parametric characteristics of separable functions. It can be shown that the nonlinear parametric characteristics of the GFRFs can be obtained by directly substituting the operations "+" and "." by "⊕" and "⊗" in the recursive algorithm (3.5–3.8), respectively, and neglecting the corresponding multiplied frequency functions. This is illustrated by the following example.

Example 1. Computation of the parametric characteristics of the 2nd order GFRF of model (3.1). It can be derived from (3.5) by applying the CE operator for the nonlinear parameters that

$$\begin{aligned} CE(H_2(\cdot)) &= CE(L(2) \cdot H_2(\cdot)) \\ &= C_{0,2} \oplus \left(\mathop{\oplus}_{q=1}^{2-1} \mathop{\oplus}_{p=1}^{2-q} (C_{p,q} \otimes CE(H_{2-q,p}(\cdot))) \right) \\ &\oplus \left(\mathop{\oplus}_{p=2}^{2} C_{p,0} \otimes CE(H_{2,p}(\cdot)) \right) \\ &= C_{0,2} \oplus \left(C_{1,1} \otimes CE(H_{1,1}(\cdot))\right) \\ &\oplus \left(C_{2,0} \otimes CE(H_{2,2}(\cdot))\right) \end{aligned}$$

It follows from Eq. (3.6) that $H_{1,1}(j\omega_1) = H_1(j\omega_1)(j\omega_1)^{k_1}$. Noting that CE($\mathrm{H}_1$(.)) = 1 since H_1(.) is independent of the nonlinear parameters in $C(M, K)$, thus CE($\mathrm{H}_{1,1}$(.)) = 1. Similarly, it can be obtained that CE($\mathrm{H}_{2,2}$(.)) = 1. Therefore, the parametric characteristic vector of the second order GFRF is

$$CE(H_2(\cdot)) = C_{0,2} \oplus C_{1,1} \oplus C_{2,0}$$

Example 1 shows that the CE operator is very effective to derive the parametric characteristic vector of a separable function series for the interested parameters. In this chapter, without speciality the CE operator is applied for all the nonlinear parameters in $C(M, K)$. If the CE operator is in particular applied for a specific nonlinear parameter c, then the parametric characteristic of the nth-order GFRF is denoted by CE(H_n(.))$_c$.

3.3.2 Parametric Characteristics

A fundamental result can be obtained for the parametric characteristic of the nth-order GFRF of model (3.1), which provides an important basis for the parametric characteristic analysis of the frequency response functions in this study.

Proposition 1. *Consider the GFRFs of the NDE model* (3.1). *There exists a complex valued function vector with appropriate dimension* $f_n(j\omega_1,\ldots,j\omega_n)$ *which is a function of* $j\omega_1,\ldots,j\omega_n$ *and the linear parameters of model* (3.1), *such that*

$$H_n(j\omega_1,\ldots,j\omega_n) = CE(H_n(j\omega_1,\ldots,j\omega_n)) \cdot f_n(j\omega_1,\ldots,j\omega_n) \tag{3.10}$$

where $CE(H_n(j\omega_1,\ldots,j\omega_n))$ *is the parametric characteristic vector of the nth-order GFRF for model* (3.1) *whose elements include and only include all the nonlinear parameters in* $C_{0,n}$ *and all the parameter monomials in* $C_{pq} \otimes C_{p_1q_1} \otimes C_{p_2q_2} \otimes \cdots \otimes C_{p_kq_k}$ *for* $0 \le k \le n-2$, *whose subscripts satisfy*

$$\begin{aligned} & p+q+\sum_{i=1}^{k}(p_i+q_i) = n+k, \\ & 2 \le p_i+q_i \le n-k, \\ & 2 \le p+q \le n-k \text{ and} \\ & 1 \le p \le n-k \end{aligned} \tag{3.11}$$

Proof. Equation (3.10) is obvious from Lemma 1 and the discussions above. It can be derived by applying the CE operator to Eqs. (3.4–3.7) that

$$\begin{aligned} CE(H_n(j\omega_1,\ldots,j\omega_n)) = {} & C_{0,n} \oplus \left(\mathop{\oplus}_{q=1}^{n-1} \mathop{\oplus}_{p=1}^{n-q} C_{p,q} \otimes CE(H_{n-q,p}(\cdot)) \right) \\ & \oplus \left(\mathop{\oplus}_{p=2}^{n} C_{p,0} \otimes CE(H_{n,p}(\cdot)) \right) \end{aligned} \tag{3.12a}$$

$$\begin{aligned} CE(H_{n,p}(\cdot)) &= \mathop{\oplus}_{i=1}^{n-p+1} CE(H_i(\cdot)) \otimes CE(H_{n-i,p-1}(\cdot)) \quad \text{or} \\ CE(H_{n,p}(\cdot)) &= \mathop{\oplus}_{\substack{r_1 \cdots r_p = 1 \\ \sum r_i = n}}^{n-p+1} \mathop{\otimes}_{i=1}^{p} CE(H_{r_i}(\cdot)) \end{aligned} \tag{3.12b}$$

$$CE(H_{n,1}(\cdot)) = CE(H_n(\cdot))) \tag{3.12c}$$

Obviously, $C_{0,n}$ is the first term in Eq. (3.12a). For clarity, consider a simpler case that there is only output nonlinearities in (3.12a), then (3.12a) is reduced to only the last term of Eq. (3.12a), i.e., $\oplus_{p=2}^{n} C_{p,0} \otimes CE(H_{n,p}(\cdot)) = \oplus_{p=2}^{n} C_{p,0} \otimes$

$\bigoplus_{\substack{r_1 \cdots r_p = 1 \\ \sum r_i = n}}^{n-p+1} \bigotimes_{i=1}^{p} CE(H_{r_i}(\cdot))$. At this stage, the proof can be finished by following the proof for Proposition 2 in [1]. □

Remark 2. Although the main result in Proposition 1 has been derived in [1] for the NARX model, it is systematically established and extended here for the NDE model (3.1) and a proof outline is given. It shall be noted that, the condition described by (3.11) in Proposition 1 provides a sufficient and necessary condition on what nonlinear parameters of model (3.1) can appear in the *n*th-order GFRF and how they do. Moreover, in Proposition 1, $f_n(j\omega_1, \ldots, j\omega_n)$ is not a function of $CE(H_n(j\omega_1, \ldots, j\omega_n))$ and is invariant at a specific point $(\omega_1, \ldots, \omega_n)$ once the linear parameters of model (3.1) are determined. Proposition 1 provides an explicitly analytical expression for the *n*th-order GFRF which reveals a straightforward relationship between the nonlinear parameters of model (3.1) and the system GFRFs, and is an explicit function of the nonlinear parameters at any a specific frequency point $(\omega_1, \ldots, \omega_n)$. Equation (3.10) is referred to as the parametric characteristic function of the *n*th-order GFRF, which is denoted by $H_n(C(n, K))_{(\omega_1, \ldots, \omega_n)}$.

Remark 3. As mentioned above, the CE operator sets up a mapping from Ξ to P_c^σ. When applying the CE operator to the GFRFs of model (3.1), $C_s = C(M, K)$, $W_s = \{\omega_1, \ldots, \omega_N\} \cup \{c_{1,0}(k_1), c_{0,1}(k_1) | 0 \leq k_1 \leq K\}$, $P_c = \left\{c_1^{r_1} c_2^{r_2} \cdots c_I^{r_I} | c_i \in C(M, K), r_i \in Z_0, I = |C(M, K)|\right\}$ and $\Xi = \{H_n(\cdot) | 1 \leq n \leq N\}$. Obviously, compared with the result in [1], the proposition is established in a more rigorous way and also provides a clear insight into the CE operator and the nature of the parametric characteristics.

Based on Proposition 1, the following properties of $CE(H_n(j\omega_1, \ldots, j\omega_n))$ for the NDE model (3.1) can be obtained. These further demonstrate the complicated relationship between the nth-order GFRF and model parameters, and also provide a way to understand the parametric characteristics of the GFRFs. The proofs for these properties are given in Appendix.

Definition 2. If a nonlinear parameter monomial $\prod_{i=1}^{k} c_{p_i q_i}^{j_i}(\cdot)(\mathrm{k} > 0, j_i \geq 0)$ is an element of $CE(H_n(j\omega_1, \ldots, j\omega_n))$, then it has an independent contribution to $H_n(j\omega_1, \ldots, j\omega_n)$, and is said to be a complete monomial of order *n* (simply as *n*-order complete); otherwise, if it is part of an *n*-order complete monomial, then it is referred to as *n*-order incomplete.

Obviously, all the elements in $CE(H_n(j\omega_1, \ldots, j\omega_n))$ are *n*-order complete.

Property 1. *The largest nonlinear degree of the nonlinear parameters appearing in* $CE(H_n(j\omega_1, \ldots, j\omega_n))$ *is n corresponding to nonlinear parameters* $c_{p,q}(.)$ *with* $p + q = n$, *and the n-degree nonlinear parameters of form* $c_{p,q}(.)$ $(p + q = n)$ *are all n-order complete.*

Property 2. $c_{p,q}(.)$ *is j-order incomplete for* $j > p + q$. *That is, for a nonlinear parameter* $c_{p,q}(.)$, *it will appear in all the GFRFs of order larger than* $p + q$.

Properties 1 and 2 show that only the nonlinear parameters of degree from 2 to n have contribution to $CE(H_n(j\omega_1, \ldots, j\omega_n))$, and the n-degree nonlinear parameters contribute to all the GFRFs of order $\geq n$.

Property 3. *If* $2 \leq p_i + q_i$, $1 \leq k$ *and there is at least one* p_i *satisfying* $1 \leq p_i$ *except for* $k = 1$, *then* $c_{p_1q_1}(\cdot)c_{p_2q_2}(\cdot)\cdots c_{p_kq_k}(\cdot)$ *is Z-order complete, where* $Z = \sum_{i=1}^{k}(p_i + q_i) - k + 1$. *Moreover,* $\prod_{i=1}^{k} c_{p_iq_i}(\cdot)$ *are j-order incomplete for* $j > Z$, *and have no effect on the GFRFs of order less than Z.*

Given any a monomial $c_{p_1q_1}(\cdot)c_{p_2q_2}(\cdot)\cdots c_{p_kq_k}(\cdot)$, it can be easily determined from Property 3 that, to which order GFRF the monomial contributes independently. For instance, a nonlinear parameter $c_{3,2}(.)$, which corresponds to the nonlinear term $\prod_{i=1}^{3}\frac{d^{k_i}y(t)}{dt^{k_i}}\prod_{i=4}^{5}\frac{d^{k_i}u(t)}{dt^{k_i}}$. It follows from Property 3 that Z = (3 + 2) − 1 + 1 = 5. Thus this nonlinear term has an independent contribution in the 5th order GFRF $H_5(.)$ and affects all the GFRFs of order larger than 5. Moreover, it has no effect on the GFRFs less than the 5th order.

Property 4. *If* $1 \leq r_i$ *and* $1 \leq k$, *then the elements of* $CE(\prod_{i=1}^{k} H_{r_i}(\cdot))$ *are all Z-order complete, where* $Z = \sum_{i=1}^{k} r_i - k + 1$, *and are all j-order incomplete for* $j > Z$, *and have no effect on the GFRFs of order less than Z. Similarly, the elements of* $\prod_{i=1}^{k_1} c_{p_iq_i}(\cdot) \otimes CE(\prod_{i=1}^{k_2} H_{r_i}(\cdot))$ *are all Z-order complete, where* $Z = \sum_{i=1}^{k_1}(p_i + q_i) + \sum_{i=1}^{k_2} r_i - k_1 - k_2 + 1$, *and are all j-order incomplete for* $j > Z$, *and have no effect on the GFRFs of order less than Z.*

Obviously, this property is an extension of Property 3, which shows that some computation by "⊗" between some parameters and the parametric characteristics of some different order GFRFs may result in the same parametric characteristic.

Property 5. $CE(H_{n,p}(\cdot)) = CE\Big(H_{n-p+1}(\cdot)\Big)$.

This property, together with Property 4, provides a simplified way to the recursive computation of the parametric characteristic of the nth-order GFRF in Eqs. (3.12a–c) as follows.

Corollary 1. *The parametric characteristic of the nth-order GFRF for model* (3.1) *can be recursively determined as*

$$\begin{aligned} &CE(H_n(j\omega_1, \ldots, j\omega_n)) \\ &= C_{0,n} \oplus \bigoplus_{q=1}^{n-1}\left\{C_{n-q,q} \oplus \left(\bigoplus_{p=1}^{n-q-1} C_{p,q} \otimes \chi_C\left(n, p, q, \left\lfloor \tfrac{n-q}{2} \right\rfloor\right)\right)\right\} \\ &\oplus \left\{C_{n,0} \oplus \left(\bigoplus_{p=2}^{n-1} C_{p,0} \otimes \chi_C\left(n, p, 0, \left\lfloor \tfrac{n+1}{2} \right\rfloor\right)\right)\right\} \end{aligned} \tag{3.13}$$

where $\lfloor\cdot\rfloor$ *is to take the integer part,* $\chi_C(n, p, q, \aleph) = \begin{cases} CE(H_{n-p-q+1}(\cdot)) & p \leq \aleph \\ C_{0,n-p-q+1} & p > \aleph \end{cases}$, *and* $\aleph$ *is a positive integer.*

Proof. Using Property 5, (3.12a) can be written as ($n > 1$)

$$CE(H_n(j\omega_1, \ldots, j\omega_n)) = C_{0,n} \oplus \left(\mathop{\oplus}_{q=1}^{n-1} \mathop{\oplus}_{p=1}^{n-q} C_{p,q} \otimes CE(H_{n-q-p+1}(\cdot)) \right)$$

$$\oplus \left(\mathop{\oplus}_{p=2}^{n} C_{p,0} \otimes CE(H_{n-p+1}(\cdot)) \right) \tag{3.14}$$

Note from Property 4 that some computations in the second and third part of the last equation are repetitive. For example, the monomials in $C_{n-2,1} \otimes CE(H_{n-n+2-1+1}(\cdot)) = C_{n-2,1} \otimes CE(H_2(\cdot))(\mathrm{n} > 2)$ are included in $C_{1,1} \otimes CE(H_{n-1}(\cdot)) \cup C_{2,0} \otimes CE(H_{n-1}(\cdot))$, except the monomials in $C_{n-2,1} \cdot C_{0,2}$. For this reason, (3.14) can be further written as

$$\begin{aligned}
&CE(H_n(j\omega_1, \ldots, j\omega_n)) \\
&= C_{0,n} \oplus \mathop{\oplus}_{q=1}^{n-1} \left\{ C_{n-q,q} \oplus \left(\mathop{\oplus}_{p=1}^{\lfloor (n-q)/2 \rfloor} C_{p,q} \otimes CE(H_{n-q-p+1}(\cdot)) \right) \right. \\
&\left. \oplus \left(\mathop{\oplus}_{p=\lfloor (n-q)/2 \rfloor + 1}^{n-q-1} C_{p,q} \otimes C_{0,n-q-p+1}) \right) \right\} \\
&\oplus \left\{ C_{n,0} \oplus \left(\mathop{\oplus}_{p=2}^{\lfloor (n+1)/2 \rfloor} C_{p,0} \otimes CE(H_{n-p+1}(\cdot)) \right) \right. \\
&\left. \oplus \left(\mathop{\oplus}_{p=\lfloor (n+1)/2 \rfloor + 1}^{n-1} C_{p,0} \otimes C_{0,n-p+1} \right) \right\}
\end{aligned}$$

This produces Eq. (3.13). The proof is completed. □

Remark 4. Corollary 1 provides an alternative and recursive way to determine the parametric characteristic of the nth-order GFRF. If there are only some nonlinear parameters in (3.2) of interest, then Eq. (3.13) and all the results above can still be used by only regarding the other parameters to be 1 if they are nonzero or zero if they are zero. Therefore, whatever nonlinear parameters (for instance x) are concerned, the parametric characteristic function with respect to x denoted by $H_n(x)_{(\omega_1,\ldots,\omega_n;C(n,K)\backslash x)}$ and the parametric characteristic $CE(H_n(j\omega_1, \ldots, j\omega_n))$ can both be derived by following the same method established above.

The parametric characteristic analysis discussed in this section can do more than just guide the computation and analysis of the Generalized Frequency Response Functions (GFRFs). It can also show how the parameters of interest influence the

GFRFs, thereby offering valuable insights for system analysis without the need for complex computations. The subsequent example serves as a straightforward demonstration of this concept.

Example 2. Consider the parametric characteristics of the following two cases:

Case 1: Suppose there is only input nonlinearity $C_{0,3} \neq 0$, and all the other nonlinear parameters are zero in model (3.1). Then the parametric characteristics of the nth-order GFRF can be computed as

If n < 3, it follows from Property 1 that $CE(H_n(j\omega_1,...,j\omega_n)) = 0$.
If n = 3, also from Property 1, the parameters in $C_{0,3}$ are all 3-order complete. Thus $CE(H_3(j\omega_1,...,j\omega_3)) = C_{0,3}$.
If n > 3, from Property 2, $C_{0,3}$ should be *n*-order incomplete in this case. However, from the Definition 2, a complete monomial should have at least one $p \geq 1$. Since there are no other nonzero nonlinear parameters, $CE(H_n(j\omega_1,...,j\omega_n)) = 0$ for this case.

Therefore, $CE(H_n(j\omega_1,...,j\omega_n)) = 0$ for n $\neq$ 1 and n $\neq$ 3 in Case 1. That is, only $H_1(j\omega)$ and $H_3(j\omega_1,...,j\omega_3)$ are nonzero in this case. Obviously, the computation of the parametric characteristics can provide guidance to the computation and analysis of the GFRFs as demonstrated in this case study.

Case 2: Suppose only $C_{0,3} \neq 0$ and $C_{2,0} \neq 0$, and all the other nonlinear parameters are zero. Then the parametric characteristics of the GFRFs can be simply determined as

$$
\begin{aligned}
&CE(H_1(j\omega_1)) = 1,\\
&CE(H_2(j\omega_1, j\omega_2)) = C_{2,0},\\
&CE(H_3(j\omega_1,...,j\omega_3)) = C_{2,0}^2 \oplus C_{0,3}\\
&CE(H_4(j\omega_1,...,j\omega_4)) = C_{2,0}^3 \oplus C_{0,3} \otimes C_{2,0},\\
&CE(H_5(j\omega_1,...,j\omega_5)) = C_{2,0}^4 \oplus C_{0,3} \otimes C_{2,0}^2\\
&CE(H_6(j\omega_1,...,j\omega_6)) = C_{2,0}^6 \oplus C_{0,3} \otimes C_{2,0}^3 \oplus C_{0,3}^2 \otimes C_{2,0}
\end{aligned}
$$

Especially, if only $C_{0,3}$ is interested for analysis, then $C_{2,0}$ can be regarded as a constant 1. In this case, the parametric characteristics of the GFRFs can be obtained as

$$
\begin{aligned}
&CE(H_1(j\omega_1)) = CE(H_2(j\omega_1, j\omega_2)) = 1,\\
&CE(H_3(j\omega_1,...,j\omega_3)) = C_{0,3}\\
&CE(H_4(j\omega_1,...,j\omega_4)) = C_{0,3},\\
&CE(H_5(j\omega_1,...,j\omega_5)) = C_{0,3},\\
&CE(H_6(j\omega_1,...,j\omega_6)) = C_{0,3} \oplus C_{0,3}^2
\end{aligned}
$$

Note that different parametric characteristics of the GFRFs correspond to different polynomial functions with respect to the interested parameters. Therefore, it can be known how the interested parameters affect the GFRFs and thus some useful techniques or information can be derived for the system analysis without directly analytical computation of the GFRFs as in (3.5–3.8). For example, from the parametric characteristics in Case 2, it can be seen that the sensitivity of the GFRFs for $n < 6$ with respect to $C_{0,3}$ is a constant when $C_{2,0}$ and the linear parameters are constant. This may imply that in order to make the system less sensitive to the input nonlinearity with coefficient $C_{0,3}$, it needs only to adjust the parameters in $C_{2,0}$ and the linear parameters of model (3.1) to reduce the corresponding constants in Case 2 under certain conditions.

As illustrated in Example 2, the parametric characteristic and its properties, developed for the nth-order Generalized Frequency Response Function (GFRF) in this section, can be utilized to uncover how the nonlinear parameters influence each order GFRF in its polynomial form. It also shows how the frequency response functions of model (3.1) are constructed and predominantly influenced by the model parameters that define nonlinearities. More detailed results will be discussed in the subsequent section.

3.4 Parametric Characteristics Based Analysis

The nonlinearities in model (3.1) can not only be classified by its nonlinear degree, but also be grouped into three kinds of nonlinearity as: (a) pure input nonlinearities corresponding to the nonlinear parameters $c_{0,n}(\cdot)$, which lead to the first term in the parametric characteristics of Eq. (3.13); (b) pure output nonlinearities corresponding to the nonlinear parameters $c_{n,0}(\cdot)$, which lead to the last term of Eq. (3.13); (c) input–output cross nonlinearities corresponding to the nonlinear parameters $c_{p,q}(\cdot)$, which lead to the second term of (3.13). It is known that different nonlinearity has different effect on system dynamics. The frequency characteristics of the frequency response functions and the nonlinear effects on the system output behaviour can be revealed, to a certain extent, by the parametric characteristic analysis of the frequency response functions. In this section, the parametric characteristics based analysis is investigated and discussed in order to provide a significant insight into how the different model parameters work in and make an effect on the frequency response functions for model (3.1), and therefore affect the system frequency characteristics.

3.4.1 *Nonlinear Effect on the GFRFs from Different Nonlinear Parameters*

The following results demonstrate how different nonlinearities affect the polynomial structure or polynomial expansion of the GFRFs in (3.10), which can therefore provide an insight into the different nonlinear effect on the GFRFs. In what follows, $k+1$ for a monomial in $C_{p,q} \otimes C_{p_1,q_1} \otimes C_{p_2,q_2} \otimes \cdots \otimes C_{p_k,q_k}$ is referred to as the power of the monomial.

A. **Pure Input Nonlinearities**

As mentioned, this kind of nonlinearities correspond to the nonlinear parameters of the form $c_{0,q}(\cdot)$ with $q > 1$. Then from Property 1

$$CE(H_n(j\omega_1,\ldots,j\omega_n))_{C_{0,q}} = C_{0,q} \;\text{ for }\; n = q \tag{3.15a}$$

$$CE(H_n(j\omega_1,\ldots,j\omega_n))_{C_{0,q}} = 1 \;\text{ for }\; n < q \tag{3.15b}$$

For $n > q$, according to Proposition 1, $c_{0,q}(.)^{\rho}$ for any $\rho > 0$ cannot be an independent entry in $CE(H_n(j\omega_1,\ldots,j\omega_n))_{c_{0,q}(\cdot)}$. The largest power ρ can only appear in the monomial $c_{0,q}(.)^{\rho} c_{p\prime,q\prime}(.)$, where $c_{p\prime,q\prime}(.)$ is nonzero, satisfies $p\prime \ge 1$ and $p\prime + q\prime \ge 2$ and has the smallest $p\prime + q\prime$. In this case, ρ can be computed from Property 3 as

$$\rho(n,0,q) = \tfrac{n-p'-q'}{q-1}$$

For example, if $p\prime + q\prime = 2$, then

$$\rho(n,0,q) = \begin{cases} \left\lfloor \frac{n-1}{q-1} \right\rfloor & \text{if } \frac{n-1}{q-1} \text{ is not an integer} \\ \frac{n-q}{q-1} & \text{else} \end{cases}$$

Therefore, for $n > q$,

$$CE(H_n(j\omega_1,\ldots,j\omega_n))_{C_{0,q}} = \begin{bmatrix} 1 & C_{0,q} & C_{0,q}^2 & \cdots & C_{0,q}^{\rho(n,0,q)} \end{bmatrix} \tag{3.15c}$$

In particular, when all the other nonlinear parameters are zero except for $C_{0,q}$, then ($n > 1$)

$$CE(H_n(j\omega_1,\ldots,j\omega_n))_{C_{0,q}} = \begin{cases} C_{0,q} & \text{if } q = n \\ 0 & \text{else} \end{cases} \tag{3.15d}$$

It can be further verified that the parametric characteristic $CE(H_n(j\omega_1,\ldots,j\omega_n))_{C_{0,q}}$ is the same as (3.15d) even when only all the other kinds of nonlinear parameters are zero except for the input nonlinearity. From the parametric characteristic analysis of the nth-order GFRF for the input nonlinearity, it can be concluded that,

(A1) The parametric characteristic function with respect to the input nonlinearity for the nth-order GFRF is a polynomial of the largest degree $\rho(n, 0, q)$, i.e.,

$$H_n(C_{0,q})_{(j\omega_1,\ldots,j\omega_n;C(n,K)\backslash C_{0,q})} = \left[1\ C_{0,q}\ C_{0,q}^2\ \cdots\ C_{0,q}^{\rho(n,0,q)} \right] \cdot f_n(j\omega_1, \ldots, j\omega_n; C(n, K)\backslash C_{0,q})$$

where $f_n(j\omega_1, \ldots, j\omega_n; C(n, K)\backslash C_{0,q})$ is an appropriate function vector.

(A2) The largest power for the input nonlinearity of an independent contribution in $CE(H_n(j\omega_1, \ldots, j\omega_n))$ is 1, which corresponds to the nonlinear parameters in $C_{0,n}$.

(A3) When there are no other kinds of nonlinearities, i.e., output nonlinearity and input–output cross nonlinearity, it can be seen from (3.15d) that the input nonlinearities have no auto-crossing effects on system dynamics. That is, each degree of the input nonlinearities has an independent contribution to the corresponding order GFRF and the largest power of a complete monomial from input nonlinearities is only 1, i.e., the nth-order GFRF is simply $H_n(j\omega_1, \ldots, j\omega_n) = C_{0,q} \cdot f_n(j\omega_1, \ldots, j\omega_n)$ from Proposition 1. Obviously, if $C_{0,n} = 0$, there will be no contribution from the input nonlinearities in the nth-order GFRF. It will be seen that these demonstrate a quite different property for the input nonlinearity from the other kinds of nonlinearities.

It is recognized that a challenge in analyzing Volterra systems is that the Volterra kernel functions in the time domain typically interact with each order due to the intersecting nonlinear effects from various nonlinearities. This is also true for the Generalized Frequency Response Functions (GFRFs) in the frequency domain. However, as discussed earlier, this challenge does not apply when there are only input nonlinearities, such as in the class of Volterra systems studied in [26]. The parametric characteristic analysis for the input nonlinearities can shed light on the selection of different parameters for the energy transfer filter design in [27], since different nonlinearities may correspond to different output frequencies [22]. Determining the relationship between output frequencies and input nonlinearities is a crucial step in the filter design in [27]. A more detailed study on this topic is currently underway, with some aspects discussed in Sect. 3.4.2.2.

B. **Pure Output Nonlinearities**

This kind of nonlinearities correspond to the nonlinear parameters of the form $c_{p,0}(.)$ with $p > 1$. From Property 1,

$$CE(H_n(j\omega_1, \ldots, j\omega_n))_{C_{p,0}} = C_{p,0} \text{ for } n = p \tag{3.16a}$$

$$CE(H_n(j\omega_1, \ldots, j\omega_n))_{C_{p,0}} = 1 \text{ for } n < p \tag{3.16b}$$

These are similar to the input nonlinearity. If $n > p$, then from Properties 1–3 $C_{p,0}$ will contribute to all the GFRFs of order lager than p. From Property 3, $c_{p,0}(.)^{\rho}$ for $\rho > 0$ is a complete monomial for the Zth-order GFRFs where $Z = (p-1)\rho + 1$. For the nth-order GFRF with $n > p$, the largest power ρ can be computed from Property 3 as

$$\rho(n,p,0) = \left\lfloor \frac{n-1}{p-1} \right\rfloor$$

Thus, for $n > p$,

$$CE(H_n(j\omega_1,\ldots,j\omega_n))_{C_{p,0}} = \left[1 \ C_{p,0} \ C_{p,0}^2 \ \cdots \ C_{p,0}^{\rho(n,p,0)} \right] \tag{3.16c}$$

Consider the particular case that all the other nonlinear parameters are zero except for the parameters in $C_{\mathrm{p,0}}$, then for $n > 1$

$$CE(H_n(j\omega_1,\ldots,j\omega_n))_{C_{p,0}} = \begin{cases} 0 & \text{if p > n or } \frac{\text{n}-1}{\text{p}-1} \text{ is not an integer} \\ C_{p,0}^{\rho(n,p,0)} & \text{else} \end{cases} \tag{3.16d}$$

However, when only all the other kinds of nonlinear parameters are zero except for the output nonlinearity, the parametric characteristic $CE(H_n(j\omega_1,\ldots,j\omega_n))_{C_{p,0}}$ for $n > p$ is the same as (3.16c). From the parametric characteristic analysis of the nth-order GFRF for the pure output nonlinearity, it can be concluded that,

(B1) The parametric characteristic function with respect to the output nonlinearity for the nth-order GFRF is a polynomial of the largest degree $\rho(n,p,0)$, i.e.,

$$H_n(C_{p,0})_{(j\omega_1,\ldots,j\omega_n;C(n,K)\backslash C_{p,0})} = \left[1 \ C_{p,0} \ C_{p,0}^2 \ \cdots \ C_{p,0}^{\rho(n,p,0)} \right] \cdot f_n(j\omega_1,\ldots,j\omega_n;\, C(n,K)\backslash C_{p,0})$$

where $f_n(j\omega_1,\ldots,j\omega_n;\, C(n,K)\backslash C_{p,0})$ is an appropriate function vector. Note that $\rho(n,p,0) \geq \rho(n,0,q)$, which may imply that for the same nonlinear degree, output nonlinearity has a larger effect on the system than input nonlinearity.

(B2) The largest power for the output nonlinear parameter $C_{p,0}$ of an independent contribution in $CE(H_n(j\omega_1,\ldots,j\omega_n))$ is $\rho(n,p,0)$, which corresponds to the n-order complete monomial $C_{p,0}^{\rho(n,p,0)}$. However, the largest power for the output nonlinearity of a complete monomial in $CE(H_n(j\omega_1,\ldots,j\omega_n))$ is k, corresponding to the monomial $C_{p_1,0} \otimes C_{p_2,0} \otimes \cdots \otimes C_{p_k,0}$, where $k = p_1 + \cdots + p_k + 1 - n$. This is quite different with the input nonlinearity whose power is 1 in this case.

(B3) When there are no other kinds of nonlinearities, i.e., input nonlinearity and input–output cross nonlinearity, it can be seen from (3.16c) that the output nonlinearities have auto-crossing nonlinear effects on system dynamics. That is, different degree of the output nonlinearities can form a complete monomial in the nth-order GFRF and the largest power of this kind of complete monomials from output nonlinearities is k as mentioned in (B2). Obviously, if the n-degree nonlinear parameter $C_{n,0} = 0$, there are still contributions from the output nonlinearities in the nth-order GFRF if there are some other nonzero output nonlinear parameters of nonlinear degree less n. These may imply that output nonlinearity has more complicated and larger effect on the system than input nonlinearity of the same order, which demonstrates a very different property compared with the input nonlinearity as mentioned in (A3).

(B4) It can be seen from (3.16c–d) that, when p is smaller, $C_{p,0}$ will contribute independently to more orders of the GFRFs whose orders are $(p-1)i+1$ for $i = 1, 2, 3, \ldots$; and if p is larger, $C_{p,0}$ can only affect the GFRFs of order higher than p. It known that for a Volterra system, the system nonlinear dynamics is usually dominated by the first several order GFRFs [11]. This implies that the nonlinear terms with coefficient $C_{p,0}$ of smaller nonlinear degree, e.g., 2 and 3, take much greater roles in the GFRFs than other pure output nonlinearities. This property is significant for the design of nonlinear feedback controller design, where the desired kind and degree of nonlinearities should be determined for some control objectives [1, 28]. For some detailed results in this topic, refer to [29].

C. **Input–output Cross Nonlinearities**

This kind of nonlinearities correspond to the nonlinear parameters of the form $c_{p,q}(.)$ with $p \geq 1$ and $q \geq 1$. It can be verified that the parametric characteristics of the GFRFs with respect to this kind of nonlinearities are very similar to those for the pure output nonlinearities as shown in B, and the conclusions held for the output nonlinearity still hold for the input–output cross nonlinearity. Thus the detailed discussions are omitted here. For a summary, the following parametric characteristics hold for both of these two kinds of nonlinearities

$$CE(H_n(j\omega_1, \ldots, j\omega_n))_{C_{p,q}} = \begin{cases} 1 & \text{if } n < p+q \\ \left[1 \; C_{p,q} \; C_{p,q}^2 \; \cdots \; C_{p,q}^{\rho(n,p,q)} \right] & \text{else} \end{cases} \tag{3.17}$$

where, n > 1, $\rho(n,p,q) = \left\lfloor \frac{n-1}{p+q-1} \right\rfloor$, $p \geq 1$ and $p+q \geq 2$.

A straightforward difference between the input–output cross nonlinearity and the pure output nonlinearity may be that the output nonlinearity can be more easily realized by nonlinear state or output feedback in a practical design. A simple comparison is summarized in Table 3.1.

Remark 5. Based on the parametric characteristic of the nth-order GFRF with respect to nonlinear parameters in $C_{p,q}$, the sensitivity of the GFRFs with respect to these nonlinear parameters can also be studied. From Proposition 1, the sensitivity of $H_n(j\omega_1, \cdots, j\omega_n)$ with respect to a specific nonlinear parameter c can be computed as

Table 3.1 Comparisons between different nonlinearities for $H_n(.)$

For $H_n(\cdot)$	Input nonlinearities	Output nonlinearities	Input–Output nonlinearities
The parametric characteristic function with respect to different nonlinearity	$H_n(C_{0,q})_{(j\omega_1,\cdots,j\omega_n;C(n,K)\backslash C_{0,q})} = \left[1\ C_{0,q}\ C_{0,q}^2\ \cdots\ C_{0,q}^{\rho(n,0,q)}\right]$ for $\cdot f_n(j\omega_1,\cdots,j\omega_n;C(n,K)\backslash C_{0,q})$ $n \geq q$	$H_n(C_{p,0})_{(j\omega_1,\cdots,j\omega_n;C(n,K)\backslash C_{p,0})} = \left[1\ C_{p,0}\ C_{p,0}^2\ \cdots\ C_{p,0}^{\rho(n,p,0)}\right]$ for $\cdot f_n(j\omega_1,\cdots,j\omega_n;C(n,K)\backslash C_{p,0})$ $n \geq p$	$CE(H_n(j\omega_1,\cdots,j\omega_n))_{C_{p,q}} = \begin{cases} 1 & \text{if } n < p+q \\ \left[1\ C_{p,q}\ C_{p,q}^2\ \cdots\ C_{p,q}^{\rho(n,p,q)}\right] & \text{else} \end{cases}$
The largest power for a specific nonlinear parameter in an independent contribution to $H_n(.)$	1 (not $\rho(n, 0, q)$)	$\rho(n,p,0) = \left\lfloor \frac{n-1}{p-1} \right\rfloor$	$\rho(n,p,q) = \left\lfloor \frac{n-1}{p+q-1} \right\rfloor$
The largest power for a pure nonlinearity of a complete monomial in $CE(H_n(.))$	1, corresponding to the monomial $C_{0,n}$	$k = p_1 + \cdots + p_k + 1 - n$ corresponding to the monomial $C_{p_1,0}C_{p_2,0}\cdots C_{p_k,0}$	$k = p_1 + q_1 \cdots + p_k + q_k + 1 - n$, corresponding to the monomial $C_{p_1,q_1}C_{p_2,q_2}\cdots C_{p_k,q_k}$
When there are no other kinds of nonlinearities	No auto-crossing effects on system dynamics; have straightforward relationship with output frequencies; and parameter sensitivity is zero or constant	Have complicated auto-crossing nonlinear effects; can be realized in feedback control; and lower degree parameters have more contributions to $H_n(.)$	Have complicated auto-crossing nonlinear effects; and lower degree parameters have more contributions to $H_n(.)$

$$\frac{\partial H_n(c)_{(\omega_1,\ldots,\omega_n;C(M,K,n)\backslash c)}}{\partial c} = \frac{\partial H_n(j\omega_1,\ldots,j\omega_n)}{\partial c} = \frac{\partial CE(H_n(j\omega_1,\ldots,j\omega_n))}{\partial c} \cdot f_n(j\omega_1,\ldots,j\omega_n) \tag{3.18}$$

Thus, the sensitivity of the nth-order GFRF with respect to any a nonlinear parameter $c = c_{p,q}(\cdot)$ with $p \geq 1$ and $p + q \geq 2$ can be obtained from (3.17) as:

$$\frac{\partial H_n(c)_{(\omega_1,\ldots,\omega_n;C(K,n)\backslash c)}}{\partial c} = \left[0\ 1\ 2c \cdots \rho(n,p,q)c^{\rho(n,p,q)-1}\right] \cdot \bar{f}_n(j\omega_1,\ldots,j\omega_n;C(K,n)\backslash c) \tag{3.19}$$

where $\bar{f}_n(j\omega_1,\ldots,j\omega_n;C(K,n)\backslash c)$ is an appropriate function vector defined in Proposition 1. Clearly, the sensitivity towards a specific parameter remains an analytical polynomial function of the nonlinear parameter. From the parametric characteristics in Eqs. (3.15–3.17), it can be inferred that the sensitivity of the nth-order Generalized Frequency Response Function (GFRF) with respect to an input nonlinear parameter must be either zero or constant when there are no other types of nonlinearities. However, this can only occur for the output nonlinear parameters and input–output cross nonlinear parameters if the nonlinear degree of the parameter of interest is n. Otherwise, the sensitivity function for a non-input nonlinear parameter remains an analytical polynomial function of the parameter of interest and some other nonzero parameters.

An application of these results will be seen in Sect. 3.4.2.2.

3.4.2 Parametric Characteristic Analysis of System Output Spectrum

Based on the parametric characteristics of the GFRFs discussed above, the system output spectrum and related homogeneous functions of model (3.1) can be analyzed.

3.4.2.1 Parametric Characteristic of System Output Frequency Response Function

The system output spectrum of model (3.1) can be described as [21]:

$$Y(j\omega) = \sum_{n=1}^{N} Y_n(j\omega) \tag{3.20}$$

when subject to a general input u(t),

$$Y_n(j\omega) = \frac{1}{\sqrt{n}(2\pi)^{n-1}} \int\limits_{\omega_1+\cdots+\omega_n=\omega} H_n(j\omega_1, \ldots, j\omega_n) \prod_{i=1}^{n} U(j\omega_i) d\sigma_\omega \tag{3.21}$$

When the input is a specific multi-tone function described by

$$u(t) = \sum_{i=1}^{K} |F_i| \cos(\omega_i t + \angle F_i) \tag{3.22}$$

then

$$Y_n(j\omega) = \frac{1}{2^n} \sum_{\omega_{k_1}+\cdots+\omega_{k_n}=\omega} H_n(j\omega_{k_1}, \ldots, j\omega_{k_n}) F(\omega_{k_1}) \cdots F(\omega_{k_n}) \tag{3.23}$$

where $F(\omega) = \begin{cases} |F_i| e^{j\angle F_i} & \text{if } \omega \in \{\omega_k, k = \pm 1, \ldots, \pm K\} \\ 0 & \text{else} \end{cases}$.

Definition 3. A function $y(h;s)$ is homogeneous of degree d with respect to h if $y(ch;s) = c^d y(h;s)$, where c is a constant, s denotes the independent variables of $y(.)$, and h may be a parameter or a function of certain variables and parameters.

The detailed properties of the functions and variables in Definition 3 are not necessarily considered here as Definition 1. The definition of the homogeneous function can also be referred to [9]. From Definition 3, it can be verified that Eqs. (3.21) and (3.23) are both 1-degree homogeneous with respect to the nth-order GFRF $H_n(\cdot)$. With this definition, the following lemma is obvious.

Lemma 2. *If $y(h;s_1)$ is a homogeneous function of degree d, and $h(.)$ is a separable function with respect to parameter x whose parametric characteristic function can be written as $h(x) = g(x)f(s_2)$, then $y(h;s_1)$ is a separable function with respect to x and its parametric characteristic function can be written as $y(x)_s = g(x))^{[d]} f_y(f(s_2);s_1)$, where s_1 denotes the un-separable and un-interested parameters and variables of $h(.)$, s_2 denotes some variables of $y(.)$,$f_y(f(s_2);s_1)$ is an appropriate function vector, and $g(x)^{[d]}$ is the d times reduced kronecker product of $g(x)$.*

From Proposition 1, Lemma 2 and Eqs. (3.20–3.21), the following result can be obtained for a homogeneous function $Y(H_n(.);s)$ of degree d, where $H_n(.)$ is the nth-order GFRF of Eq. (3.1).

Proposition 2. *$Y_n(H_n(j\omega_1, ..., j\omega_n);\ \omega_1, \ldots, \omega_n)$ is a homogeneous function of degree d with respect to the nth-order GFRF $H_n(j\omega_1, ..., j\omega_n)$ of Eq.* (3.1). *Then $Y_n(H_n(j\omega_1, ..., j\omega_n);\omega_1, \ldots, \omega_n)$ is a separable function with respect to the nonlinear parameters in* (3.2), *whose parametric characteristic function can be described by*

$$Y_n(C(M, K)) = \mathrm{CE}(H_n(j\omega_1, \ldots, j\omega_n)))^{[d]} \tilde{Y}_n(f_n(j\omega_1, \ldots, j\omega_n); \omega_1, \ldots, \omega_n) \quad (3.24)$$

The sensitivity of the homogeneous function with respect to a specific parameter c is

$$\frac{\partial Y_n(C(M, K))_{\omega_1,\ldots,\omega_n}}{\partial c} = \frac{\partial CE(H_n(j\omega_1, \ldots, j\omega_n))^{[d]}}{\partial c} \cdot Y_n(f_n(j\omega_1, \ldots, j\omega_n); \omega_1, \ldots, \omega_n) \quad (3.25)$$

where $\tilde{Y}_n(f_n(j\omega_1, \ldots, j\omega_n);\omega_1, \ldots, \omega_n)$ is an appropriate function vector, and when $d = 1$

$$\tilde{Y}_n(f_n(j\omega_1, \ldots, j\omega_n); \omega_1, \ldots, \omega_n) = Y_n(f_n(j\omega_1, \ldots, j\omega_n); \omega_1, \ldots, \omega_n) \quad (3.26)$$

Proof. The results are straightforward. □

The following result is straightforward from Proposition 2 for the output spectrum of model (3.1).

Corollary 2. *The output spectrum* $Y(j\omega)$ *in* (3.20) *for model* (3.1) *is separable with respect to the nonlinear parameters in* (3.2)*, whose parametric characteristic function can be described by*

$$Y(C(M, K))_\omega = \sum_{n=1}^{N} CE(H_n(\cdot)) \cdot Y_n(f_n(\cdot); j\omega) \quad (3.27)$$

The sensitivity of the output frequency response with respect to a specific parameter c is

$$\frac{\partial Y(j\omega)}{\partial c} = \sum_{n=1}^{N} \frac{\partial CE(H_n(\cdot))}{\partial c} \cdot Y_n(f_n(\cdot); j\omega) \quad (3.28)$$

where, if the input is a general function, then $\omega = \omega_1 + \cdots + \omega_n$,

$$\begin{aligned} Y_n(f_n(\cdot); j\omega) &= Y_n(f_n(j\omega_1, \ldots, j\omega_n); j\omega) \\ &= \frac{1}{\sqrt{n}(2\pi)^{n-1}} \int_{\omega_1+\cdots+\omega_n=\omega} f_n(j\omega_1, \ldots, j\omega_n) \prod_{i=1}^{n} U(j\omega_i) d\sigma_\omega \end{aligned} \quad (3.29)$$

if the input is the multi-tone function given in (3.22), then $\omega = \omega_{k_1} + \cdots + \omega_{k_n}$,

$$
\begin{aligned}
Y_n(f_n(\cdot); j\omega) &= Y_n(f_n(j\omega_{k_1}, \ldots, j\omega_{k_n}); j\omega) \\
&= \frac{1}{2^n} \sum_{\omega_{k_1}+\cdots+\omega_{k_n}=\omega} f_n(j\omega_{k_1}, \ldots, j\omega_{k_n}) F(\omega_{k_1}) \cdots F(\omega_{k_n})
\end{aligned} \tag{3.30}
$$

From these results, it is noted that the system output spectrum can also be expressed by a polynomial function of the nonlinear parameters in $C(M, K)$ based on the parametric characteristics of the GFRFs, and the detailed structure of this polynomial function with respect to any interested parameters is completely determined by its parametric characteristic. Therefore, how the nonlinear parameters affect the system output spectrum can be studied through the parametric characteristic analysis.

Remark 6. Note that $CE(H_n(\cdot))$ can be derived from the system model parameters according to the results developed in Sect. 3.3. Given a specific system described by model (3.1), $Y(C(M, K)_\omega$ can be achieved by the FFT of the time domain output data from simulations or experiments at frequency ω. Therefore, $Y_n(f_n(\cdot); j\omega)$ for $n = 1, \ldots, N$ can be obtained by the Least Square method as mentioned in Remark 1. And then $Y_n(C(n, K))_{\omega=\omega_1+\cdots+\omega_n} = CE(H_n(\cdot)) \cdot Y_n(f_n(\cdot); j\omega)$ for $n = 1, \ldots, N$ and the sensitivity Eqs. (3.25, 3.27) can all be obtained readily. This provides a numerical method to compute the output spectrum and its each order component which are now determined as analytical polynomial functions of any interested nonlinear parameters. Thus the analysis and design of the output performance of nonlinear systems can now be conducted in terms of these interested model parameters. Compared with the direct computation by using (3.5–3.8) and (3.20–3.23), the computational complexity is reduced. And compared with the results in [25], the parametric characteristic analysis of this study provides an explicitly analytical expression for the relationship between system output spectrum and model parameters with detailed polynomial structure up to any high order and each order output spectrum component can also be determined definitely. Moreover, let

$$
G_n(C(M, K, n)_{\omega=\omega_1+\cdots+\omega_n} = \frac{CE(H_n(\cdot)) \cdot Y_n(f_n(\cdot); j\omega)}{\frac{1}{\sqrt{n}(2\pi)^{n-1}} \int_{\omega_1+\cdots+\omega_n=\omega} \prod_{i=1}^{n} U(j\omega_i) d\sigma_\omega} \tag{3.31}
$$

This is the parametric characteristic function of the nth-order nonlinear output frequency response function defined in [23], which can be used for the fault diagnosis of engineering systems and structures.

Corresponding to Proposition 2 and Corollary 2, the following results for some special cases can be established immediately. Let

$$
\delta(p) = \begin{cases} 1 \text{ if } p = 0 \\ 0 \text{ else} \end{cases}, \quad \text{and } pos(x) = \begin{cases} 1 \text{ if } x > 0 \\ 0 \text{ else} \end{cases} \tag{3.32}
$$

Corollary 3. *Consider only the nonlinear parameter $C_{p,q} = c$. The parametric characteristic vector of the output spectrum in* (3.20) *with respect to the parameter c can* *be written as*

$$CE(Y(j\omega)) = \mathop{\oplus}_{n=1}^{N} CE(H_n(\cdot))$$
$$= \left[1\ c\ c^2 \cdots c^{\left\lfloor \frac{N-1}{p+q-1} \right\rfloor - \delta(p)\cdot pos(N-q)\cdot\delta\left(\frac{N-1}{p+q-1} - \left\lfloor \frac{N-1}{p+q-1} \right\rfloor\right)}\right] \tag{3.33}$$

Then there exists a complex valued function vector $F(j\omega_1, \cdots, j\omega_n; C(M, K)\backslash c)$ with appropriate dimension such that

$$Y(c)_{\omega; C(M,K)\backslash c} = CE(Y(j\omega)) \cdot F(j\omega_1, \ldots, j\omega_n; C(M, K)\backslash c) \tag{3.34}$$

If all the other nonlinear parameters are zero except that $C_{p,q} = c$ is nonzero (p + q > 1). Then the parametric characteristic vector of the output spectrum in (3.20) with respect to the parameter c is: if $p = 0$

$$CE(Y(j\omega)) = 1 \oplus CE(H_q(\cdot)) \cdot (1 - pos(q - N))$$
$$= [1\ c \cdot (1 - pos(q - N))] \tag{3.35}$$

else

$$CE(Y(j\omega)) = \mathop{\oplus}_{i=0}^{\left\lfloor N-1/p+q-1 \right\rfloor} CE(H_{(p+q-1)i+1}(\cdot))$$
$$= \left[1\ c\ c^2 \cdots c^{\left\lfloor N-1/p+q-1 \right\rfloor}\right] \tag{3.36}$$

Proof. See Appendix. □

From Corollary 2–3, it can be seen that different nonlinearities will result in a quite different polynomial structure for the output spectrum, and thus greatly affect the system output frequency response. By using the results established in this chapter, the effect from different nonlinearities on system output frequency characteristics can now be studied readily.

Example 3. To illustrate the results, consider a nonlinear system,

$$a_1\ddot{x} = -a_2 x - a_3\dot{x} - c_1\dot{x}^3 - c_2\dot{x}^2 x - c_3 x^3 + bu(t) \tag{3.37}$$

which is a simple case of model (3.1) with $M = 3, K = 2, c_{10}(2) = a_1, c_{10}(1) = a_3$, $C_{10}(0) = a_2$, $c_{30}(111) = c_1$, $c_{30}(110) = c_2$, $c_{30}(000) = c_3$, $c_{01}(0) = -b$, all other parameters are zero. The GFRFs for system (3.37) can be computed according to Eqs. (3.5–3.8). In the following, the parametric characteristics of the GFRFs for system (3.37) are demonstrated.

When all the other nonlinear parameters are zero except $C_{p,q}$, it can be summarized from Sect. 3.4.1 that the parametric characteristic of the nth-order GFRF with respect to $C_{p,q}$ is

$$
\begin{aligned}
CE(H_n(j\omega_1,\ldots,j\omega_n)) = {} & C_{p,q}^{\overline{\frac{n-1}{p+q-1}}} \\
& \cdot \delta\left(\frac{n-1}{p+q-1} - \left\lfloor \frac{n-1}{p+q-1} \right\rfloor\right) \\
& \cdot (1 - \delta(p)pos(n-q))
\end{aligned}
\tag{3.38}
$$

For system (3.37), note that a_1, a_2, a_3 and b are all linear parameters, and the nonzero nonlinear parameters are $C_{30} = \begin{bmatrix} c_{30}(000) & c_{30}(110) & c_{30}(111) \end{bmatrix} = \begin{bmatrix} c_3 & c_2 & c_1 \end{bmatrix}$. Hence,

$$
\begin{aligned}
& CE(H_n(j\omega_1,\ldots,j\omega_n)) = C_{3,0}^i \\
& = \begin{bmatrix} c_3 & c_2 & c_1 \end{bmatrix}^i \text{ for} \\
& \mathrm{n} = 2\mathrm{i}+1,\ \mathrm{i} = 1, 2, 3, \ldots, \\
& \text{else } CE(H_n(j\omega_1,\ldots,j\omega_n)) = 0
\end{aligned}
\tag{3.39}
$$

It is easy to compute from (3.39) as follows:

For n = 3, $CE(H_3(j\omega_1,\ldots,j\omega_3)) = \begin{bmatrix} c_3 & c_2 & c_1 \end{bmatrix}$;

For n = 5, $CE(H_5(j\omega_1,\ldots,j\omega_5)) = \begin{bmatrix} c_3 & c_2 & c_1 \end{bmatrix}^2 = \begin{bmatrix} c_3 & c_2 & c_1 \end{bmatrix} \otimes \begin{bmatrix} c_3 & c_2 & c_1 \end{bmatrix} = [c_3^2, c_3c_2, c_3c_1, c_2^2, c_2c_1, c_1^2]$;

For n = 7, $CE(H_7(j\omega_1,\ldots,j\omega_7)) = \begin{bmatrix} c_3 & c_2 & c_1 \end{bmatrix}^3 = \begin{bmatrix} c_3 & c_2 & c_1 \end{bmatrix} \otimes \begin{bmatrix} c_3 & c_2 & c_1 \end{bmatrix} \otimes \begin{bmatrix} c_3 & c_2 & c_1 \end{bmatrix} = [c_3^3, c_3^2c_2, c_3^2c_1, c_3c_2^2, c_3c_2c_1, c_3c_1^2, c_2^3, c_2^2c_1, c_2c_1^2, c_1^3]$.

From Proposition 1, there must exist a complex valued function vector $\overline{f}_n(j\omega_1,\cdots,j\omega_n)$ with appropriate dimension, such that for n = 2i + 1, i = 1,2,3,…,

$$
H_n(c_1, c_2, c_3)_{(\omega_1,\ldots,\omega_n)} = \begin{bmatrix} c_3 & c_2 & c_1 \end{bmatrix}^i \cdot \overline{f}_n(j\omega_1,\ldots,j\omega_n)
\tag{3.40}
$$

else $H_n(c_1, c_2, c_3)_{(\omega_1,\ldots,\omega_n)} = 0$.

When there is only one parameter for example c_1 is interested for analysis, the parametric characteristic can be obtained by simply letting $C_{3,0} = c_1$ in (3.39), i.e., the parametric characteristic vector is: for n = 2i + 1 and i = 1, 2, 3, …

$$
CE(H_n(j\omega_1,\ldots,j\omega_n)) = [\,1\ c_1\ c_1^2 \cdots c_1^i\,]
\tag{3.41}
$$

else,

$$CE(H_n(j\omega_1, \ldots, j\omega_n)) = 0 \tag{3.42}$$

Thus the parametric characteristic function with respect to the parameter c_1 is: for n = 2i + 1 and i = 1, 2, 3, …

$$H_n(c_1)_{(\omega_1,\ldots,\omega_n;c_2,c_3)} = [\,1\ c_1\ c_1^2 \cdots c_1^i\,] \cdot \overline{f}_n(j\omega_1, \ldots, j\omega_n; c_2, c_3) \tag{3.43}$$

else,

$$H_n(c_1)_{(\omega_1,\ldots,\omega_n;c_2,c_3)} = 0 \tag{3.44}$$

where, $\overline{f}_n(j\omega_1, \ldots, j\omega_n; c_2, c_3)$ is a complex valued function vector with appropriate dimension. The sensitivity of the nth-order GFRFs for n = 2i + 1 and i = 1, 2, 3, … with respect to the parameter c_1 can also be obtained as

$$\frac{\partial H_n(c_1)_{(\omega_1,\ldots,\omega_n;c_2,c_3)}}{\partial c_1} = \left[0\ 1\ 2c_1 \cdots ic^{i-1}\right] \cdot \overline{f}_n(j\omega_1, \ldots, j\omega_n; c_2, c_3) \tag{3.45}$$

Consider the output spectrum of system (3.37). From Corollary 3,

$$CE(X(j\omega)) = \bigoplus_{i=0}^{\lfloor N-1/2 \rfloor} CE(H_{2i+1}(\cdot)) = \bigoplus_{i=0}^{\lfloor N-1/2 \rfloor} C_{30}^i \tag{3.46}$$

Suppose the output function of interest is

$$y = a_2x + a_3\dot{x} - c_1\dot{x}^3 - c_2\dot{x}^2x - c_3x^3 \tag{3.47}$$

It was shown in [29] that

$$CE(Y(j\omega)) = CE(X(j\omega)) \tag{3.48}$$

Then from Corollary 3, the parametric characteristic function for the output frequency response $Y(j\omega)$ of system (3.27) with respect to nonlinear parameters c_1, c_2 and c_3 is

$$\begin{aligned} Y(c_1, c_2, c_3)_\omega &= \sum_{i=0}^{\lfloor N-1/2 \rfloor} C_{30}^i \cdot Y_i(f_i(\cdot); j\omega) \\ &= \left(\bigoplus_{i=0}^{\lfloor N-1/2 \rfloor} C_{30}^i \right) \end{aligned}$$

$$\cdot \left[Y_0(f_0(\cdot); j\omega)\ Y_1(f_1(\cdot); j\omega)^T \cdots Y_{\lfloor N-1/2 \rfloor}(f_{\lfloor N-1/2 \rfloor}(\cdot); j\omega)^T \right]^T \tag{3.49}$$

For convenience, consider a much simpler case. Let $c_2 = c_3 = 0$, then $C_{30} = c_{30}(111) = c_1$. Therefore the parametric characteristic function in this simple case is

$$\begin{aligned} Y(c_1)_\omega &= Y_0(f_0(\cdot); j\omega) + c_1 \cdot Y_1(f_1(\cdot); j\omega) \\ &\quad + \cdots + c_1^{\lfloor N-1/2 \rfloor} \cdot Y_{\lfloor N-1/2 \rfloor}(f_{\lfloor N-1/2 \rfloor}(\cdot); j\omega) \\ &= \left[1\ c_1 \cdots c_1^{\lfloor N-1/2 \rfloor} \right] \\ &\quad \cdot \left[Y_0(f_0(\cdot); j\omega)\ Y_1(f_1(\cdot); j\omega) \cdots Y_{\lfloor N-1/2 \rfloor}(f_{\lfloor N-1/2 \rfloor}(\cdot); j\omega) \right]^T \end{aligned} \tag{3.50}$$

As mentioned in Remark 1, $\left[Y_0(f_0(\cdot); j\omega)\ Y_1(f_1(\cdot); j\omega) \cdots Y_{\lfloor N-1/2 \rfloor}(f_{\lfloor N-1/2 \rfloor}(\cdot); j\omega) \right]^T$ can be computed by a numerical method for a specific input u(t) and at a specific frequency ω. The idea [25] is to obtain $\lfloor N-1/2 \rfloor + 1$ system output frequency responses from $\lfloor N-1/2 \rfloor + 1$ simulations or experimental tests on the system (3.37) under $\lfloor N-1/2 \rfloor + 1$ different values of the nonlinear parameter c_1 and the same input u(t), then yielding

$$\begin{bmatrix} Y(j\omega)_0 \\ Y(j\omega)_1 \\ \vdots \\ Y(j\omega)_{\lfloor N-1/2 \rfloor} \end{bmatrix} = \begin{bmatrix} 1 & c_1(0) & \cdots & c_1(0)^{\lfloor N-1/2 \rfloor} \\ 1 & c_1(1) & \cdots & c_1(1) \\ 1 & \vdots & \ddots & \vdots \\ 1 & c_1(\lfloor N-1/2 \rfloor) & \cdots & c_1(\lfloor N-1/2 \rfloor)^{\lfloor N-1/2 \rfloor} \end{bmatrix} \cdot \begin{bmatrix} Y_0(f_0(\cdot); j\omega) \\ Y_1(f_1(\cdot); j\omega) \\ \vdots \\ Y_{\lfloor N-1/2 \rfloor}\left(f_{\lfloor N-1/2 \rfloor}(\cdot); j\omega\right) \end{bmatrix} \tag{3.51}$$

Hence,

$$\begin{bmatrix} Y_0(f_0(\cdot); j\omega) \\ Y_1(f_1(\cdot); j\omega) \\ \vdots \\ Y_{\lfloor N-1/2 \rfloor}\left(f_{\lfloor N-1/2 \rfloor}(\cdot); j\omega\right) \end{bmatrix} = \begin{bmatrix} 1 & c_1(0) & \cdots & c_1(0)^{\lfloor N-1/2 \rfloor} \\ 1 & c_1(1) & \cdots & c_1(1) \\ 1 & \vdots & \ddots & \vdots \\ 1 & c_1\left(\lfloor N-1/2 \rfloor\right) & \cdots & c_1\left(\lfloor N-1/2 \rfloor\right)^{\lfloor N-1/2 \rfloor} \end{bmatrix}^{-1} \cdot \begin{bmatrix} Y(j\omega)_0 \\ Y(j\omega)_1 \\ \vdots \\ Y(j\omega)_{\lfloor N-1/2 \rfloor} \end{bmatrix} \tag{3.52}$$

Then Eq. (3.50) is determined explicitly, which is an analytical function of the nonlinear parameter c_1. The system output frequency response can therefore be analyzed and optimized in terms of the nonlinear parameters. And also from (3.50), the sensitivity of the system output frequency response with respect to the nonlinear parameter, and the nonlinear output frequency response function defined in (3.31) can both be studied readily. For more complicated cases, the similar process above can be followed to conduct the desired analysis and design in terms of multiple nonlinear parameters for model (3.1). Obviously, the parametric characteristic analysis provides a useful technique and significant step for these kinds of system analysis. For an application of this method, please refer to [29]. Compared with [25] where the polynomial structure of output spectrum is unknown, since the detailed polynomial structure for the output spectrum up to any high order can explicitly be determined, it is much easier to carry out the numerical method as above when multiple parameters are considered.

3.4.2.2 Nonlinear Effect from Different Nonlinearities on System Output Spectrum

Different nonlinearities can lead to varying output spectrum characteristics for a system. It's recognized that nonlinear systems exhibit a wider range of output frequencies than the driving frequency [14, 22], illustrating an energy transfer phenomenon [23, 27]. Predicting when and where an output will occur is typically challenging [26]. To achieve a desired output spectrum at specific frequencies, the system should be appropriately designed to incorporate or exclude certain nonlinearities [27]. In this subsection, the study explores how different nonlinear parameters influence the system output frequencies when subjected to a harmonic input. This further underscores the utility of the parametric characteristic analysis conducted in this study.

Consider model (3.1) subject to a harmonic input

$$u(t) = F_d \sin(\Omega t) \tag{3.53}$$

In this case, $F(\omega_{k_1})$ in the nth-order output spectrum (3.23) is

$$F(\omega_{k_l}) = -jk_l F_d, \quad \text{for } k_l = \pm 1, \\ \omega_{k_l} = k_l \Omega, \quad \text{and } l = 1, \ldots, n \tag{3.54}$$

Thus Eq. (3.23) can be written as

$$Y_n(j\omega) = \frac{1}{2^n} \sum_{\omega_{k_1}+\cdots+\omega_{k_n}=\omega} H_n(j\omega_{k_1}, \ldots, j\omega_{k_n}) \cdot F_d^n \cdot (k_1 \cdots k_n) \cdot (-j)^n \\ = \frac{C_F(n)}{2^n} \sum_{\omega_{k_1}+\cdots+\omega_{k_n}=\omega} H_n(j\omega_{k_1}, \ldots, j\omega_{k_n}) \tag{3.55}$$

where

$$C_F(n) = F_d^n \cdot (k_1 \cdots k_n) \cdot (-j)^n \tag{3.56}$$

From (3.55), it can be seen that the output frequency range for the nth-order output spectrum in this case is completely determined by $\omega_{k_1} + \cdots + \omega_{k_n} = \omega$ and the nth-order GFRF. Note that $\omega_{k_l} = \pm\Omega$ for $l = 1, \ldots, n$. Thus the output frequency range for the nth-order output spectrum can only possibly be $\{k\Omega | k = 0, 1, 2, \ldots\}$, which can be written as

$$W_n = \{\omega \big| \omega = (\omega_{k_1} + \cdots + \omega_{k_n}) \cdot (1 - \delta(H_n(\cdot))), \ \omega_{k_l} = \pm\Omega, \ 1 \le l \le n\} \tag{3.57}$$

Therefore the output frequency range for the system output spectrum is

$$W_O = \bigcup_{1 \le n \le N} W_n \tag{3.58}$$

Obviously, when certain orders of the GFRFs are zero, it will lead to no output spectrum at certain frequencies. From Proposition 1, Eq. (3.57) can also be written as

$$W_n = \{\omega \big| \omega = (\omega_{k_1} + \cdots + \omega_{k_n}) \cdot (1 - \delta(CE(H_n(\cdot)))), \ \omega_{k_l} = \pm\Omega, \ 1 \le l \le n\} \tag{3.59}$$

This demonstrates the effects of nonlinear parameters on the system output frequency characteristics. From Eq. (3.59), the following lemma is straightforward. In what follows, $\omega = 2l\Omega$ for $l = 0, 1, 2, \ldots$ are called even frequencies, and $\omega = (2l + 1)\Omega$ for $l = 0, 1, 2, \ldots$ are called odd frequencies.

Lemma 3. *Consider the output frequencies of model* (3.1) *when subject to a harmonic input* (3.53). *W_n can only include even frequencies, when n is an even number, and W_n only includes odd frequencies, when n is an odd number.*

Consider the effects of different nonlinear parameters in the following cases.

(A) **Pure Input Nonlinearities**

When there are only input nonlinearity in model (3.1) and suppose $C_{0,n} \neq 0$, then from Sect. 3.4.1(A), $CE(H_n(\cdot)) = C_{0,n}$. Equation (3.59) can be written as

$$W_n = \{\omega | \omega = (\omega_{k_1} + \cdots + \omega_{k_n}) \cdot (1 - \delta(C_{0,n})),\ \omega_{k_l} = \pm\Omega,\ 1 \leq l \leq n\} \quad (3.60)$$

which further yields

$$W_n = \begin{cases} \{2l\Omega \cdot (1 - \delta(C_{0,n})) \,|0 \leq l \leq n/2\,, n \geq 2\} & \text{if n is an even integer} \\ \{(2l+1)\Omega \cdot (1 - \delta(C_{0,n})) \,|0 \leq l \leq (n-1)/2\,, n \geq 3\} & \text{if n is an odd integer} \end{cases} \quad (3.61)$$

Obviously $W_1 = \{\Omega\}$, which represents the linear output frequency. Equation (3.61) shows that for any a pure input nonlinear term of nonlinear degree n, it will introduce a finite output frequency range from 0 to $n\Omega$. If the nonlinear degree n is an even integer, the introduced output spectrum will appear at even frequencies, and if the nonlinear degree n is an odd integer, it will appear at odd frequencies.

Example 4. To verify the result for pure input nonlinearity, a simulation result is provided for system (3.37) with $a_1 = 1$, $a_2 = 1$, $a_3 = 0.5$, $c_1 = c_2 = c_3 = 0$, $b = u(t)^4$ or $u(t)^5$. See Fig. 3.1.

(B) **Output Nonlinearity and Input–output Cross Nonlinearity**

When all the other nonlinear parameters are zero except for $C_{p,q}$, then from (3.16d) and Sect. 3.4.1(C), the parametric characteristic of the nth-order GFRF can be summarized as for $n > 1$

$$\begin{aligned} CE(H_n(j\omega_1, \ldots, j\omega_n))_{C_{p,q}} &= C_{p,q}^{\overline{\frac{n-1}{p+q-1}}} \cdot step(n-p-q) \\ &\quad \cdot \delta\left(\frac{n-1}{p+q-1} - \left\lfloor \frac{n-1}{p+q-1} \right\rfloor\right) \end{aligned} \quad (3.62)$$

where

$$step(x) = \begin{cases} 1 \text{ if } x \geq 0 \\ 0 \text{ else} \end{cases} \quad (3.63)$$

Equation (3.62) demonstrates the independent contribution from the parameters in $C_{p,q}$ to the nth-order GFRF. Only when $\frac{n-1}{p+q-1}$ is an integer and $n \geq p+q$, then $CE(H_n(j\omega_1, \ldots, j\omega_n))_{C_{p,q}}$ is nonzero. In this case, it can be obtained from Eq. (3.59) as follows.

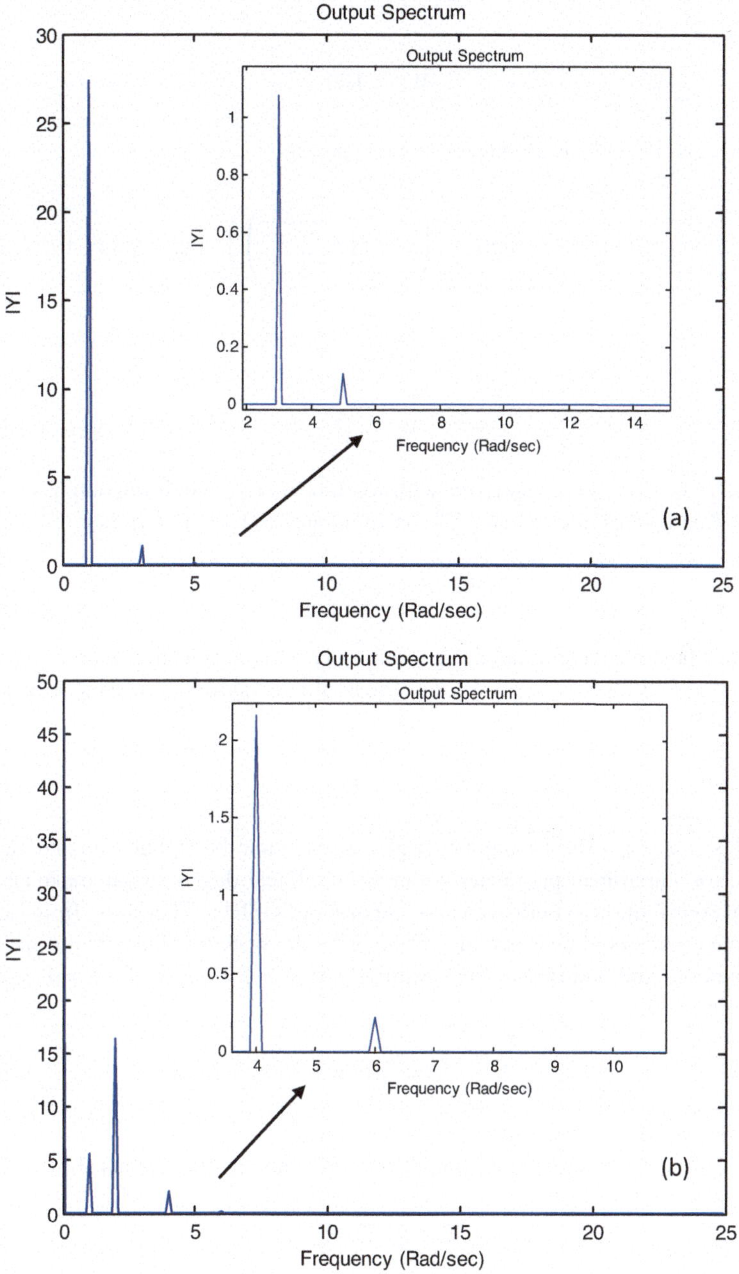

Fig. 3.1 Output spectrum of a nonlinear system with only one input nonlinear term having coefficient $c_{0,5}(.)$ (**a**) or $c_{0,6}(.)$ (**b**), subject to the input $u(t) = 10\sin(t)$

for $n = 1$,

$$W_1 = \{\Omega\} \tag{3.64}$$

for $n > 1$, and if $\frac{n-1}{p+q-1}$ is an integer and $n \geq p+q$,

$$W_n = \left\{ \omega \middle| \omega = (\omega_{k_1} + \cdots + \omega_{k_n}) \cdot \left(1 - \delta\left(C_{p,q}^{\frac{n-1}{p+q-1}}\right)\right),\ \omega_{k_l} = \pm\Omega,\ 1 \leq l \leq n \right\} \tag{3.65}$$

else

$$W_n = \{\} \tag{3.66}$$

From (3.65), the nonlinear terms with coefficients $C_{\mathrm{p,q}}$ will bring output spectrum at some frequencies only when $\frac{n-1}{p+q-1}$ is an integer and $n \geq p+q$. Let

$$\overline{\rho} = \frac{n-1}{p+q-1} \tag{3.67}$$

which is a positive integer and can go to infinity when n goes to infinity.

If $p + q = 2l$ for $l = 1, 2, 3, \ldots$, i.e., the nonlinear degree is an even integer, then from (3.67)

$$n = (2l - 1)\overline{\rho} + 1 \tag{3.68}$$

In this case, Eq. (3.68) can not only give an even number but also an odd number. That is, for a nonlinear parameter with even nonlinear degree, it can make an independent contribution to both even and odd order of GFRFs. Therefore, from Lemma 1 an even degree nonlinear parameter will drive the system to have output spectrum at all the even and odd frequencies. Similarly, if $p + q = 2l + 1$ for $l = 1, 2, 3, \ldots$, then from (3.67)

$$n = 2l\overline{\rho} + 1 \tag{3.69}$$

In this case, Eq. (3.69) must give an odd integer.

The following conclusion is straightforward from the discussions above.

Proposition 3. *Consider the output frequencies of model* (3.1) *when subject to a harmonic input* (3.53)*. For any nonlinear term with coefficient* $c_{p,q}(.)$*, where* $p + q > 1$ *and* $p > 0$*, if the nonlinear degree* $p + q$ *is an odd integer, it will bring super-harmonics to the system output spectrum only at these frequencies which are odd integer multiples of the input frequency; if* $p + q$ *is an even integer, it will introduce super-harmonics at all frequencies which are nonnegative integer multiples of the input frequency.*

These results can be verified by a simple example in simulation below.

Example 5. Consider system (3.37) subject to the harmonic input $u(t) = 5\sin(t)$ with $a_1 = 1$, $a_2 = 1$, $a_3 = 0.5$, $c_3 = 0$, $b = 1$ and two cases: $c_1 = c_{3,0}(111)$, $c_2 = 0$ and $c_1 = 0$, $c_2 = c_{6,0}(000, 111)x^2\dot{x}$. See Fig. 3.2, where the output frequency responses of the two cases are given. It can be seen that the harmonics happen only at odd frequencies when subject to the third degree nonlinearity; however, there are harmonics at all the integer frequencies when subject to the sixth degree nonlinearity.

3.5 Conclusions

Essential findings for the parametric characteristics of the Generalized Frequency Response Functions (GFRFs) for Volterra systems, as described by a Nonlinear Differential Equation (NDE) model, have been established. This is based on a novel coefficient extraction operator for functions or series that are separable by parameters. This study represents a significant and beneficial development of the authors' previous results. The parametric characteristics of the GFRFs can explicitly illustrate the relationship between model parameters and system frequency response functions, and clearly reveal the system frequency characteristics' dependence on the system's nonlinear parameters. This provides valuable insights into the analysis and design of nonlinear systems in the frequency domain. By utilizing the parametric characteristic analysis, the nonlinear frequency domain behavior of a system can be studied in terms of the time-domain model parameters that define system nonlinearities. This reveals the dependence of the frequency response functions of nonlinear systems on model parameters. This also offers a new approach to the frequency domain analysis and design of nonlinear systems in the context of Volterra series expansion.

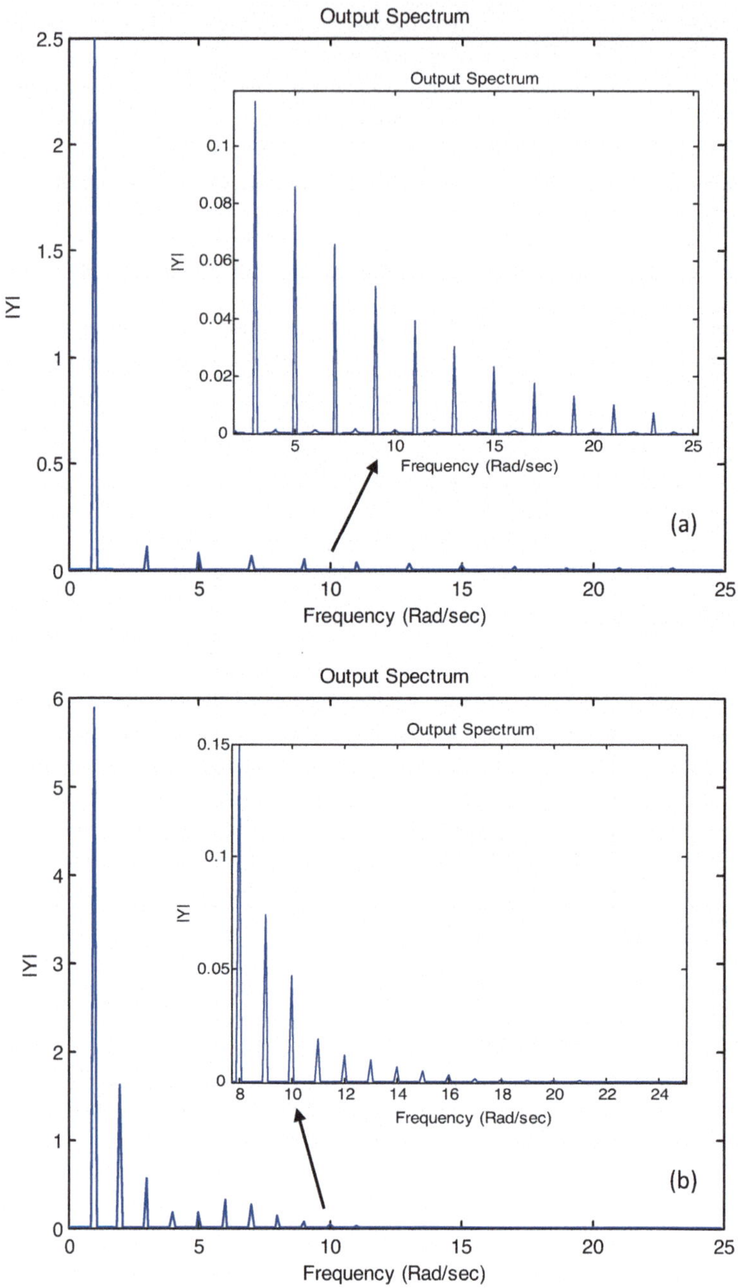

Fig. 3.2 Output spectrum of a nonlinear system with only one output nonlinear term having coefficient $c_{3,0}(.)$ (**a**) or $c_{6,0}(.)$ (**b**)

Acknowledgements The main results of this chapter can also be referred to the published work in [30].

Appendix: Proofs

Proof of Property 1

In (3.11) when $p+q=n$, then $p+q+\sum_{i=1}^{k}(p_i+q_i)=n+\sum_{i=1}^{k}(p_i+q_i)=n+k$, which further yields $\sum_{i=1}^{k}(p_i+q_i)=k$. Note that $2\le p_i+q_i\le n-k$ and $0\le k\le n-2$, thus $k=p_i=q_i=0$. Therefore, the property is proved. □

Proof of Property 2

This property can be seen from the recursive Eqs. (3.12a–c) and can also be proved from Proposition 1. Suppose $c_{p,q}(.)$ does not appear in $H_n(j\omega_1,\ldots,j\omega_n)$, where n > p + q. Consider a monomial $c_{p,q}(.)c_{2,0}^{k}(.)$ with k = *n-p-q*. It can be verified from Proposition 1 that $c_{p,q}(.)c_{2,0}^{n-p-q}(.)$ is n-order complete. This results in a contradiction. □

Proof of Property 3

Although a part of this property has given in [3], it is summarized into a complete one and the proof is given here. From Proposition 1, $CE(H_Z(\cdot))$ includes all non-repetitive monomial functions of the nonlinear parameters in Eq. (3.1) of the form $C_{pq}\otimes C_{p_1q_1}\otimes C_{p_2q_2}\otimes\cdots\otimes C_{p_kq_k}$, where the subscripts satisfy $p+q+\sum_{i=1}^{k\prime}(p_i+q_i)=Z+k\prime$, $2\le p_i+q_i\le Z-k\prime$, $0\le k\prime\le Z-2$, $2\le p+q\le Z-k\prime$, and noting $1\le p\le Z-k\prime$, thus $\mathop{\otimes}\limits_{i=1}^{k} C_{p_iq_i}$ is included in $CE(H_Z(\cdot))$. Moreover, substitute k by $k+x$, where x > 0 is an integer, then $Z\prime=\sum_{i=1}^{k+x}(p_i+q_i)-k-x+1$, which further yields $Z\prime-Z=\sum_{i=1}^{x}(p_i+q_i)-x$. Note that $2\le p_i+q_i$, thus $Z\prime-Z\ge\sum_{i=1}^{x}2-x=x$. Therefore, $\mathop{\otimes}\limits_{i=1}^{k} C_{p_iq_i}$ must appear in $CE\big(H_j(j\omega_1,\ldots,j\omega_j)\big)$ for $j>Z$ and not in the GFRFs of order less than j. This completes the proof. □

Proof of Property 4

From Proposition 1, any element $c_{p_1,q_1}(\cdot)c_{p_2,q_2}(\cdot)\cdots c_{p_{k_{r_i}},q_{k_{r_i}}}(\cdot)$ in CE($H_{r_i}(\cdot)$) with $r_i > 1$ must satisfy $r_i = \sum_{i=1}^{k_{r_i}}(p_i + q_i) - k_{r_i} + 1$. Note that if $r_i = 1$, then CE($H_{r_i}(\cdot)$) $= 1$. In this case, suppose $(p_i + q_i) = 1$ for consistence. Therefore,

$$\sum_{i=1}^{k} r_i - k + 1 = \left(\sum_{i=1}^{k}\sum_{j=1}^{k_{r_i}}(p_j + q_j) - \sum_{i=1}^{k} k_{r_i} + k\right) - k + 1$$
$$= \sum_{i=1}^{k}\sum_{j=1}^{k_{r_i}}(p_j + q_j) - \sum_{i=1}^{k} k_{r_i} + 1 = Z.$$

This proves the first part of this property. The second part follows from the first part and Property 3. □

Proof of Property 5

This was proved in Proposition 3 of [3], but here presents a more concise proof based on the properties above. Applying the CE operator to Eq. (3.8), it can be obtained that

$$CE\big(H_{n,p}(j\omega_1,\ldots,j\omega_n)\big) = \mathop{\oplus}_{\substack{r_1\cdots r_p=1\\ \sum r_i=n}}^{n-p+1} \mathop{\otimes}_{i=1}^{p} CE\big(H_{r_i}(\cdot)\big)$$
$$= CE\big(H_{n-p+1}(\cdot)\big) \oplus \left(\mathop{\oplus}_{\substack{r_1\cdots r_p=1\\ \sum r_i=n}}^{n-p} \mathop{\otimes}_{i=1}^{p} CE\big(H_{r_i}(\cdot)\big)\right)$$

From Property 4, it follows that all the elements in $\mathop{\oplus}_{\substack{r_1\cdots r_p=1\\ \sum r_i=n}}^{n-p} \mathop{\otimes}_{i=1}^{p} CE\big(H_{r_i}(\cdot)\big)$ should be Z-order complete, where $Z = \sum_{i=1}^{p} r_i - p + 1 = n - p + 1$. This completes the proof. □

Proof of Corollary 3

Equation (3.33) is summarized from Eqs. (3.15–3.17), and when all the other parameters are zero except $c = c_{p,q}(.)$, the following equation can also be summarized from (3.15–3.17)

$$CE(H_n(j\omega_1,\ldots,j\omega_n)) = c^{\frac{n-1}{p+q-1}} \cdot \delta\left(\frac{n-1}{p+q-1} - \left\lfloor\frac{n-1}{p+q-1}\right\rfloor\right)$$
$$\cdot (1-\delta(p)pos(n-q))$$

Therefore, it can be shown that

$$\begin{aligned} CE(Y(j\omega)) &= \bigoplus_{i=0}^{\left\lfloor N-1/_{p+q-1}\right\rfloor} CE(H_{(p+q-1)i+1}(\cdot)) \\ &= \bigoplus_{i=0}^{\left\lfloor N-1/_{p+q-1}\right\rfloor} c^i \cdot \delta(i-\lfloor i\rfloor)\cdot(1-\delta(p)pos((p+q-1)i+1-q)) \\ &= \bigoplus_{i=0}^{\left\lfloor N-1/_{p+q-1}\right\rfloor} c^i \cdot (1-\delta(p)pos((p+q-1)i+1-q)) \end{aligned}$$

If p = 0, $1-\delta(p)pos((p+q-1)i+1-q) = 1-pos((q-1)i+1-q)$, which yields,

$$CE(Y(j\omega)) = \left[\,1\ c\cdot(1-pos(q-N))\,\right]$$

else, $1-\delta(p)pos((p+q-1)i+1-q) = 1$, which yields

$$\begin{aligned} CE(Y(j\omega)) &= \bigoplus_{i=0}^{\left\lfloor N-1/_{p+q-1}\right\rfloor} CE(H_{(p+q-1)i+1}(\cdot)) \\ &= \left[1\ c\ c^2 \cdots c^{\left\lfloor N-1/_{p+q-1}\right\rfloor}\right] \end{aligned}$$

This completes the proof. □

References

1. X.J. Jing, Z.Q. Lang, S.A. Billings and G. R. Tomlinson, The parametric characteristic of frequency response functions for nonlinear systems. *International Journal of Control*, 79(12), 1552–1564, 2006
2. Research Report No. 932, Department of Automatic Control and Systems Engineering, The University of Sheffield, August 2006: chrome-extension://efaidnbmnnnibpcajpcglclefindmkaj/https://eprints.whiterose.ac.uk/74592/1/932.pdf
3. R.K. Pearson, *Discrete time dynamic models*, Oxford University Press, 1994.
4. P.W.J.M., Nuij, O.H.Bosgra, Steinbuch M. Higher-order sinusoidal input describing functions for the analysis of non-linear systems with harmonic responses. *Mechanical Systems and Signal Processing*. 20, 1883–1904, 2006
5. P. Orlowski, Frequency domain analysis of uncertain time-varying discrete-time systems, *Circuits Systems Signal Processing*, 26(3), 293–310, 2007

6. J.C. Peyton Jones Automatic computation of polyharmonic balance equations for non-linear differential systems, Int. J. Control, 76(4), 355–365, 2003
7. M. A. Shah, M. A. Franchek, Frequency-based controller design for a class of nonlinear systems, *International Journal of Robust and Nonlinear Control*, 9(12) , 825 – 840, 1999
8. M. Solomou, C. Evans, D. Rees, N. Chiras, Frequency domain analysis of nonlinear systems driven by multiharmonic signals, *Proceedings of the 19th IEEE conference on Instrumentation and Measurement Technology*, 1, 799- 804, 2002
9. W. J. Rugh, *Nonlinear system theory --- the Volterra/Wiener approach.* Baltimore and London: the Johns Hopkins University Press. 1981
10. V. Volterra, Theory of Functionals of Integral and Integrodifferential Equations, Dover, New York, 1959.
11. S. Boyd and L. O. Chua, Fading memory and the problem of approximating nonlinear operators with Volterra series, *IEEE Trans. Circuits Syst.*, CAS-32, 1150—1161, 1985
12. I. W. Sandberg, Expansions for nonlinear systems, *Bell Syst. Tech. J.*, 61(2), 159-199, Feb. 1982
13. D.A. George, Continuous nonlinear systems, *Technical Report 355*, MIT Research Laboratory of Electronics, Cambridge, Mass. Jul. 24, 1959
14. E. Bedrosian and S. O. Rice, The output properties of Volterra systems driven by harmonic and Gaussian inputs. *Proceedings of the Institute of Electrical and Electronics Engineering*, 59, 1688-1707, 1971
15. J.S. Bendat, Nonlinear System Analysis and Identification from Random Data, New York: Wiley, 1990.
16. K.I. Kim, and E.J. Powers, A digital method of modelling quadratic nonlinear systems with a general random input. *IEEE Trans. Acoustics, Speech and Signal Processing*, 36, 1758–1769, 1988
17. S.W. Nam and E.J. Powers, Application of higher-order spectral analysis to cubically nonlinear-system identification, *IEEE Transactions on Signal Processing*, 42(7), 1746–1765, Jul. 1994.
18. S. Sastry, Nonlinear system: analysis, stability, and control. New York: Springer, 1999
19. S.A. Billings and J.C. Peyton-Jones, Mapping nonlinear integro-differential equation into the frequency domain, *International Journal of Control*, 54, 863-879, 1990
20. X. J. Jing, Z.Q. Lang, and S.A. Billings, New Bound Characteristics of NARX Model in the Frequency Domain, *International Journal of Control*, 80(1), 140-149, 2007
21. Z. Q. Lang, and S. A. Billings, Output frequency characteristics of nonlinear systems. *International Journal of Control*, 64, 1049-1067, 1996.
22. Z. Q. Lang, and S. A. Billings, Output frequencies of nonlinear systems. *International Journal of Control*, 67, 713-730, 1997
23. Z.Q. Lang, and S.A.Billings, Energy transfer properties of nonlinear systems in the frequency domain, *International Journal of Control*, 78, 345-362, 2005
24. R.Yue, S. A. Billings and Z.-Q. Lang, An investigation into the characteristics of non-linear frequency response functions. Part 1: Understanding the higher dimensional frequency spaces. *International Journal of Control*, 78(13), 1031–1044, 2005
25. Z.Q. Lang, S.A. Billings, R. Yue and J. Li, Output frequency response functions of nonlinear Volterra systems, *Automatica*, 43, 805-816, 2007
26. M. Solomou, D. Rees, and N. Chiras, Frequency Domain Analysis of Nonlinear Systems Driven by Multiharmonic Signals. *IEEE Transactions on Instrumentation and Measurement*, 53 (2), 243-250, 2004
27. SA Billings, and ZQ Lang. Nonlinear systems in the frequency domain: Energy transfer filters. *International Journal of Control*. 75(14): 1066-1081, 2002
28. W. Van Moer, Y. Rolain, and A. Geens, Measurement-Based Nonlinear Modeling of Spectral Regrowth, *IEEE Transactions on Instrumentation and Measurement*, 50(6), p1711-1716, 2001

29. X.J. Jing, Z.Q. Lang and S.A. Billings, Frequency domain analysis for suppression of output vibration from periodic disturbance using nonlinearities, *Journal of Sound and Vibration*, 314, 536–557, 2008
30. XJ Jing, ZQ Lang, SA Billings, Parametric characteristic analysis for generalized frequency response functions of nonlinear systems, Circuits, Systems & Signal Processing 28, 699-733, 2009

Chapter 4
The Truncation Error Theory

Abstract The Volterra series-based frequency-domain methodology, along with the associated frequency response functions (FRFs)of nonlinear systems, fundamentally depends on the accurate computation of the Generalized Frequency Response Functions (GFRFs) and the corresponding output spectra. While analytical derivation of these FRFs is feasible through recursive algorithms, as previously discussed, they can also be obtained via numerical techniques—provided that the system model is known or sufficient experimental data are available. A critical challenge in the numerical implementation (refer to Section 3 of Chapter 2) lies in the accurate estimation of each order of the output spectrum, particularly in the presence of truncation errors.This raises the essential question of whether it is possible to reliably compute low-order output spectra (typically up to second or third order in practical applications) with limited truncation while maintaining high fidelity. To address this issue, a truncation error theory is developed and introduced in this chapter. This theoretical framework establishes the foundational principles for efficiently and accurately computing each order of the output spectrum using a minimal truncation order. Unlike conventional curve-fitting approaches, this theory enables a principled and systematic treatment of truncation effects, thereby opening new avenues for advanced applications of Volterra series-based frequency-domain analysis in nonlinear system modeling and design.

Keywords Volterra series · Frequency domain analysis · Nonlinear systems · Truncation error

4.1 Introduction

4.1.1 About This Chapter

As discussed in Chaps. 1–3, the Volterra series based frequency domain method and associated frequency response functions of nonlinear systems are highly relying on the calculation of the GFRF and output spectrum of a given nonlinear system. The

X. Jing, *The Bio-Inspired X-Structure/Mechanism Approach for Exploring Nonlinear Benefits in Engineering*, https://doi.org/10.1007/978-981-96-5769-8_4

frequency response functions can be obtained analytically based on several recursive algorithms as mentioned, but can also be done with a numerical method if the system model is known, or some experimental data are obtained. To apply the numerical method (see Session 3 of Chap. 2), the other big issue is related to how to accurately obtain each order output spectrum subject to the truncation error. That is, whether it is possible to obtain accurately each order output spectrum of a given nonlinear system with a limited truncation order, say, 2 or 3 in most application cases. To solve this problem, a truncation error theory is established and introduced in this chapter. This theory provides the most fundamental for fast and accurately obtaining each order output spectrum with a very limited truncation order instead of just a curve fitting and leads to the possibility of a series of new applications related to the Volterra series frequency domain theory.

Therefore, key findings have been established regarding the truncation order and its impact in the Volterra class of nonlinear systems. These findings offer significant insights into the analytical relationships among truncation order, truncation error, and estimation error of the output spectrum of nonlinear systems. This also leads to the development of innovative methods for selecting the truncation order and excitation magnitudes to achieve accurate estimation of nonlinear frequency response components. These results should be highly relevant in kernel estimation and frequency domain analysis of nonlinear systems based on Volterra series, as well as in the structure and order selection of block-oriented nonlinear models with polynomial function. The findings should also be significant to related topics and areas that involve a multi-level excitation method for system analysis and identification. Theoretical analysis, case studies, and examples are provided to illustrate these results.

4.1.2 Research Background

For a certain class of nonlinear systems, the input–output relationship can be expressed as a Volterra series expansion [1–3]. In practice, the estimation of Volterra kernels often requires the determination of the truncation order [4–6]. The nonlinear output spectrum of the Volterra class of nonlinear systems can be broken down into different orders of frequency response components, which correspond to different orders of nonlinearity present in the system [3, 7, 8]. These components can offer valuable insights into the nonlinear dynamics caused by different nonlinearities, such as output frequency characteristics, nonlinear energy transfer properties, and fault detection, among others [8–17]. Clearly, the accurate estimation of these frequency response components is crucial for system analysis and identification [16, 18, 19]. However, the estimation of nonlinear frequency response components typically involves the precise determination of the truncation order of the Volterra series. An incorrect selection of the truncation order could potentially lead to completely erroneous estimation.

Choosing the right model order is a crucial aspect of system identification, as highlighted by Leonowicz et al. [20] and Bindlish [21]. A variety of nonlinear system models, such as nonlinear block-oriented models (for instance, Wiener models and Hammerstein models), can be directly converted into a Volterra-type series expansion if the nonlinear component of the block-oriented models is a polynomial function. This concept is well discussed in [15, 16, 18, 22].

The maximum degree of nonlinearity in the polynomial function is linked to the truncation order in the Volterra series expansion. Moreover, the structure of the polynomial function is essentially determined by the structure of the output frequency response components. Therefore, the issue of model order/structure selection in the identification of these nonlinear systems can be alternatively examined through the truncation order and the output spectrum components that should be included in the Volterra series expansion.

In this study, we focus on the importance of assessing the truncation order/error of the Volterra series and its impact on kernel or output spectrum estimations, which is crucial for nonlinear system identification and modeling. We look into the effects of the Volterra series' truncation order in the frequency domain. We establish some key findings on how truncation orders affect the estimates of different order nonlinear output spectrum components. We discover that the accuracy of output spectrum estimation can be significantly enhanced by choosing appropriate excitation magnitudes, regardless of how the truncation order is applied. This leads to precise estimation of the truncation order and frequency response components. We provide a theoretical analysis on the selection of excitation magnitudes and the determination of truncation influence, which is further illustrated with case studies and examples.

4.2 Background and Problem Formulation

4.2.1 Nonlinear Systems in the Frequency Domain

The input–output relationship for a class of nonlinear systems can be approximated by a Volterra series of order N as [1, 3]

$$y(t) = \sum_{n=1}^{N} y_n(t) \tag{4.1a}$$

$$y_n(t) = \int_{-\infty}^{\infty} \cdots \int_{-\infty}^{\infty} h_n(\tau_1, \ldots, \tau_n) \prod_{i=1}^{n} u(t - \tau_i) d\tau_i \tag{4.1b}$$

where $h_n(\tau_1, \ldots, \tau_n)$ is a real valued function of $\tau_1, \ldots, \tau_n$ known as the nth order Volterra kernel, and N is the maximum truncation order. It should be noted that the Volterra series expansion in (4.1) actually represents a considerably large class of nonlinear systems. Based on the Volterra series expansion, the nth order generalized frequency response function (GFRF) is defined as [2]

$$H_n(j\omega_1, \ldots, j\omega_n) = \int_{-\infty}^{\infty} \cdots \int_{-\infty}^{\infty} h_n(\tau_1, \cdots, \tau_n) \exp(-j(\omega_1\tau_1 + \cdots + \omega_n\tau_n))\mathrm{d}\tau_1 \cdots \mathrm{d}\tau_n \tag{4.2}$$

which is the multidimensional Fourier transform of $h_n(\tau_1, \cdots, \tau_n)$. With the inverse Fourier transform of the nth order GFRF, (4.1b) can be written as

$$y_n(t) = \frac{1}{(2\pi)^n} \int_{-\infty}^{\infty} \cdots \int_{-\infty}^{\infty} H_n(j\omega_1, \cdots, j\omega_n) \prod_{i=1}^{n} U(j\omega_i) e^{j(\omega_1+\cdots+\omega_n)t} \mathrm{d}\omega_1 \cdots \mathrm{d}\omega_n \tag{4.3}$$

denoting $\omega_n = \omega - \omega_1 - \cdots - \omega_{n-1}$, which can be further written as

$$y_n(t) = \frac{1}{2\pi} \int_{-\infty}^{\infty} \left[\frac{1}{(2\pi)^{n-1}} \underbrace{\int_{-\infty}^{\infty} \cdots \int_{-\infty}^{\infty}}_{n-1} H_n(j\omega_1, \ldots, j\omega_{n-1}, j(\omega - \omega_1 - \cdots - \omega_{n-1})) U_n(j\omega_1, \ldots j\omega_{n-1})\mathrm{d}\omega_1 \cdots \mathrm{d}\omega_{n-1} \right] e^{j\omega t} \mathrm{d}\omega$$

where $U_n(j\omega_1, \ldots, j\omega_{n-1}) = U(j\omega_1) \cdots U(j\omega_{n-1}) U(j(\omega - \omega_1 - \cdots - \omega_{n-1}))$. Therefore the Fourier transform of $y_n(t)$ is obtained as

$$Y_n(j\omega) = \frac{1}{(2\pi)^{n-1}} \underbrace{\int_{-\infty}^{\infty} \cdots \int_{-\infty}^{\infty}}_{n-1} H_n(j\omega_1, \cdots, j\omega_{n-1}, j(\omega - \omega_1 - \cdots - \omega_{n-1})) U_n(j\omega_1, \cdots, j\omega_{n-1})\mathrm{d}\omega_1 \cdots \mathrm{d}\omega_{n-1} \tag{4.4}$$

which is referred to as the nth-order output spectrum (nth-order output frequency response component). The output spectrum of the nonlinear system in (4.1) can then be given by

$$Y(j\omega) = \sum_{n=1}^{N} Y_n(j\omega) \tag{4.5}$$

Considering the truncation error, Eq. (4.4) can be written as

$$Y(j\omega) = \sum_{n=1}^{N} Y_n(j\omega) + \sigma_{[N]}(j\omega) \tag{4.6}$$

where $\sigma_{[N]}(j\omega)$ represents the truncation error, including all the remaining higher order output spectrum components in Volterra series expansion.

From Eq. (4.6), it's evident that the truncation order N is crucial for the precision of the Volterra series expansion in relation to the original nonlinear system. This truncation order is equivalent to the highest degree of the polynomial in the static nonlinear function within block-oriented nonlinear models (such as Wiener models, Hammerstein models) when the nonlinear segment of these models is approximated using a polynomial [15, 16, 18, 22]. An inappropriate truncation order could result in a larger truncation error and therefore lead to biased estimates for the model. The appropriateness of a specific truncation order can be assessed by the truncation influence on the estimate of each order output spectrum component $Y_n(j\omega)$ ($n = 1, 2, 3, \ldots$). The best truncation order should be chosen so that the slightest or satisfactory estimation error of $Y_n(j\omega)$ can be obtained.

4.2.2 Estimation of Output Spectrum Components

Considering the general nonlinear output spectrum in (4.4–4.5) with an input signal whose Fourier transform is $\rho R(j\omega)$, where ρ is a constant which is used to represent different magnitude of the input, the output spectrum in (4.6) can then be written as

$$\begin{aligned} Y(j\omega)_\rho &= \sum_{n=1}^{\infty} \rho^n Y_n(j\omega) \\ &= \rho Y_1(j\omega) + \rho^2 Y_2(j\omega) + \rho^3 Y_3(j\omega) + \cdots \\ &= \sum_{n=1}^{N} \rho^n Y_n(j\omega) + \sigma_{[N,\rho]}(j\omega) \end{aligned} \tag{4.7}$$

where $Y_n(j\omega)$ is the nth-order output spectrum component given in (4.3) with $U(j\omega_i) = R(j\omega_i)$ ($i = 1, 2, 3\ldots$), and the truncation error $\sigma_{[N,\rho]}(j\omega)$ is now given by

$$\begin{aligned} \sigma_{[N,\rho]}(j\omega) &= Y(j\omega) - \sum_{i=1}^{N} \rho^i Y_i(j\omega) \\ &= \rho^{N+1} Y_{N+1}(j\omega) + \rho^{N+2} Y_{N+2}(j\omega) \\ &\quad + \rho^{N+3} Y_{N+3}(j\omega) + \cdots \end{aligned}$$

$$= \rho^{N+1}\overline{\sigma}_{[N,\rho]}(j\omega) \tag{4.8}$$

where $\overline{\sigma}_{[N,\rho]}(j\omega) = Y_{N+1}(j\omega) + \rho Y_{N+2}(j\omega) + \rho^2 Y_{N+3}(j\omega) + \cdots$.

To estimate $Y_n(j\omega)$ for $n \in \aleph = \{1, 2, 3, \ldots N\}$, a multi-level excitation method is adopted here [4–6, 8, 14]. The system can be excited by the same input $R(j\omega)$ of different magnitudes $\rho_0, \rho_1, \rho_2, \ldots, \rho_{N-1}$, and there will be a series of output that can be obtained accordingly, which are denoted by $Y(j\omega)_{\rho 0}$, $Y(j\omega)_{\rho 1}$, $Y(j\omega)_{\rho 2}$, ..., $Y(j\omega)_{\rho N-1}$. It can be obtained that

$$\begin{bmatrix} 1 & 1 & \cdots & 1 \\ \rho_1 & \rho_1^2 & \cdots & \rho_1^N \\ \vdots & \vdots & \ddots & \vdots \\ \rho_{N-1} & \rho_{N-1}^2 & \cdots & \rho_{N-1}^N \end{bmatrix} \begin{bmatrix} \hat{Y}_1(j\omega) \\ \hat{Y}_2(j\omega) \\ \vdots \\ \hat{Y}_N(j\omega) \end{bmatrix} = \begin{bmatrix} Y(j\omega)_{\rho_0} \\ Y(j\omega)_{\rho_1} \\ \vdots \\ Y(j\omega)_{\rho_{N-1}} \end{bmatrix}$$

Note that $\rho_0, \rho_1, \ldots, \rho_{N-1}$ can be chosen such that the square matrix in the left side of the equation above is nonsingular (this can be easily guaranteed if $\rho_0 \neq \rho_1 \neq \ldots \neq \rho_{N-1}$). Therefore, from the equation above, different order output spectrum can then be estimated by a LS method, i.e.,

$$\begin{bmatrix} \hat{Y}_1(j\omega) \\ \hat{Y}_2(j\omega) \\ \vdots \\ \hat{Y}_N(j\omega) \end{bmatrix} = \begin{bmatrix} 1 & 1 & \cdots & 1 \\ \rho_1 & \rho_1^2 & \cdots & \rho_1^N \\ \vdots & \vdots & \ddots & \vdots \\ \rho_{N-1} & \rho_{N-1}^2 & \cdots & \rho_{N-1}^N \end{bmatrix}^{-1} \begin{bmatrix} Y(j\omega)_{\rho_0} \\ Y(j\omega)_{\rho_1} \\ \vdots \\ Y(j\omega)_{\rho_{N-1}} \end{bmatrix} \tag{4.9}$$

Given a specific input signal $R(j\omega)$, the output spectrum of each order $Y_n(j\omega)$ for $n \in \aleph = \{1, 2, 3, \ldots N\}$ can be estimated accurately if the truncation error $\sigma_{[N,\rho]}(j\omega)$ is trivial (e.g., $\left|\sigma_{[N,\rho]}(j\omega)\right| \leq \varepsilon$ a predefined small real number). Obviously, an inappropriate truncation order N would possibly result in significantly biased estimates of $Y_n(j\omega)$ for $n \in \aleph = \{1, 2, 3, \ldots N\}$ and thus poor performance in model prediction.

The aim of this research is to uncover how the truncation order influences the estimation in Eq. (4.9) and to determine the optimal truncation order N and excitation magnitudes $\rho 0, \rho 1, \ldots, \rho N - 1$ to achieve accurate estimation through Eq. (4.9). As previously mentioned, this will be beneficial for kernel estimation based on Volterra series and frequency domain analysis of nonlinear systems, as well as for the structure and order selection of block-oriented nonlinear models with a polynomial function. It's worth noting that the multi-level excitation method has been widely applied in previous studies [4–6, 8, 14]. Hence, the findings of this study should be significant to the related topics and fields.

4.3 Influence of Truncation Error, Truncation Order and Excitation Magnitudes

4.3.1 Influence of the Truncation Error

Define a parameter set

$$\mathrm{P} = \{\rho_0, \rho_1, \rho_2, \ldots, \rho_{N-1}\} \tag{4.10}$$

which includes all the parameters that could be freely tuned in practice (ρ_0 could be 1).

From (4.9), it can be obtained that for any $\rho \in P$

$$Y(j\omega)_\rho = \rho\hat{Y}_1(j\omega) + \rho^2\hat{Y}_2(j\omega) + \rho^3\hat{Y}_3(j\omega) + \cdots + \rho^N\hat{Y}_N(j\omega) \tag{4.11}$$

Define

$$\hat{Y}_n(j\omega) = Y_n(j\omega) + \Delta Y_n(j\omega) \tag{4.12}$$

where $Y_n(j\omega)$ denote the real nth-order output spectrum and $\Delta Y_n(j\omega)$ denotes the estimation error incurred by the truncation error $\sigma_{[N,\rho]}(j\omega)$. Substituting (4.12) into (4.11) gives

$$\begin{aligned} Y(j\omega)_\rho = &\rho Y_1(j\omega) + \rho^2 Y_2(j\omega) + \rho^3 Y_3(j\omega) \\ &+ \cdots + \rho^N Y_N(j\omega) \\ &+ \rho\Delta Y_1(j\omega) + \rho^2\Delta Y_2(j\omega) \\ &+ \rho^3\Delta Y_3(j\omega) + \cdots + \rho^N\Delta Y_N(j\omega) \end{aligned} \tag{4.13}$$

Noting (4.7), Eq. (4.13) yields

$$\begin{aligned} &\rho\Delta Y_1(j\omega) + \rho^2\Delta Y_2(j\omega) + \rho^3\Delta Y_3(j\omega) \\ &\quad + \cdots + \rho^N\Delta Y_N(j\omega) = \sigma_{[N,\rho]}(j\omega) \end{aligned}$$

using (4.8), which further gives

$$\begin{aligned} &\Delta Y_1(j\omega) + \rho^1\Delta Y_2(j\omega) + \rho^2\Delta Y_3(j\omega) \\ &\quad + \cdots + \rho^{N-1}\Delta Y_N(j\omega) = \rho^N\overline{\sigma}_{[N,\rho]}(j\omega) \end{aligned} \tag{4.14}$$

Therefore, for $\rho \in \mathrm{P}$, it can be obtained from (4.14) that

$$\begin{bmatrix} 1 & 1 & \cdots & 1 \\ 1 & \rho_1^1 & \cdots & \rho_1^{N-1} \\ \vdots & \vdots & \ddots & \vdots \\ 1 & \rho_{N-1}^1 & \cdots & \rho_{N-1}^{N-1} \end{bmatrix} \begin{bmatrix} \Delta Y_1(j\omega) \\ \Delta Y_2(j\omega) \\ \vdots \\ \Delta Y_N(j\omega) \end{bmatrix} = \begin{bmatrix} \overline{\sigma}_{[N,1]}(j\omega) \\ \rho_1^N \overline{\sigma}_{[N,\rho_1]}(j\omega) \\ \vdots \\ \rho_{N-1}^N \overline{\sigma}_{[N,\rho_{N-1}]}(j\omega) \end{bmatrix} \tag{4.15a}$$

which gives

$$\begin{bmatrix} \Delta Y_1(j\omega) \\ \Delta Y_2(j\omega) \\ \vdots \\ \Delta Y_N(j\omega) \end{bmatrix} = A \cdot \begin{bmatrix} \overline{\sigma}_{[N,1]}(j\omega) \\ \rho_1^N \overline{\sigma}_{[N,\rho_1]}(j\omega) \\ \vdots \\ \rho_{N-1}^N \overline{\sigma}_{[N,\rho_{N-1}]}(j\omega) \end{bmatrix} \tag{4.15b}$$

where

$$A = \begin{bmatrix} 1 & 1 & \cdots & 1 \\ 1 & \rho_1^1 & \cdots & \rho_1^{N-1} \\ \vdots & \vdots & \ddots & \vdots \\ 1 & \rho_{N-1}^1 & \cdots & \rho_{N-1}^{N-1} \end{bmatrix}^{-1} \tag{4.15c}$$

Theoretically, the information matrix A^{-1} in (4.15a, b) is the well-known Vandermonde matrix, which is nonsingular if $\rho_0 \neq \rho_1 \neq \ldots \neq \rho_{N-1} \neq 1$. However, (4.15b) cannot reveal explicitly the analytical relationships between $\Delta Y_n(j\omega)$ and $\sigma_{[N,\rho]}(j\omega)$ for $n \in \aleph$ and $\rho \in P$. Therefore, a more straightforward evaluation of the estimation error is preferable. To this aim, an estimate of the relationship between $\Delta Y_n(j\omega)$ and $\sigma_{[N,\rho]}(j\omega)$ for $\rho \in P$ is developed in the following section.

4.3.1.1 The Relationship Between $\Delta Y_n(j\omega)$ and $\sigma_{[N,\rho]}(j\omega)$ (for $\rho \in P$).

The following lemma is given, which provides an important basis for the computation and understanding of the truncation error and its effects.

Lemma 1. (Mutual-modulation function) *For* $\forall x_1, x_2, \ldots, x_N \in R\backslash 0$ *satisfying* $x_1 \neq x_2 \neq \ldots \neq x_N$ *and* $N > 1$, *it holds that*

$$\sum_{i=1}^{N} \frac{x_i^m}{\prod_{k=1,k\neq i}^{N}(x_i - x_k)} = \begin{cases} (x_1 + x_2 + \cdots + x_N)^{[m-N+1]} & m \geq N-1 \\ 0 & 0 \leq m < N-1 \end{cases} \tag{4.16a}$$

where

$$(x_1 + x_2 + \cdots + x_N)^{[j]} = \sum_{\substack{l_1 + l_2 + \cdots + l_N = j \\ l_1, l_2, \ldots, l_N \in \{0, 1, 2, \ldots, j\}}} x_1^{l_1} x_2^{l_2} \cdots x_N^{l_N} \quad (4.16b)$$

Proof See Appendix 1. □

To understand (4.16b), it can be computed that

$$(x_1 + x_2 + x_3)^{[j]} = \begin{cases} \sum_{\substack{l_1 + l_2 + l_3 = j \\ l_1, l_2, l_3 \in \{0, 1, 2, \ldots, j\}}} x_1^{l_1} x_2^{l_2} x_3^{l_3} \\ x_1^2 + x_1x_2 + x_1x_3 + x_2^2 + x_2x_3 + x_3^2 \\ x_1 + x_2 + x_3 \\ 1 \end{cases}$$

The function in (4.16a) demonstrates a nonlinear mutual modulation mechanism among different signals and the output can be tuned to different values by parameter m. With this lemma, it would be convenient to evaluate the truncation effect $\Delta Y_n(j\omega)$.

Lemma 2. $A_{i,j} = \frac{(-1)^{N-i} \sum_{\substack{m = N - i; \\ j_1 \neq \cdots \neq j_m \in \aleph_0 \backslash \{j-1\}}} \rho_{j_1} \rho_{j_2} \cdots \rho_{jm}}{\prod_{k \in \aleph_0 \backslash \{j-1\}} (\rho_{j-1} - \rho_k)}$, *where* $\aleph_0 = \{0, 1, 2, \cdots, N-1\}$ *and* $\rho_{j_0} = 1$.

Proof See Appendix 2. □

Lemma 2 provides an analytical solution to the inverse of matrix A. Different approaches to computation of the inverse Vandermonde matrix can also be referred to [23], where recursive algorithms are provided. This study focuses on the application of the analytical solution to evaluation of the truncation influence.

From Lemma 2, the estimation error, i.e. $\Delta Y_n(j\omega)$, incurred by the truncation, can be computed directly, i.e., for any $n \in \aleph$,

$$\begin{aligned} \Delta Y_n(j\omega) &= \sum_{j=1}^{N} A_{n,j} \cdot \rho_{j-1}^N \overline{\sigma}_{[N,\rho_{j-1}]}(j\omega) \\ &= \sum_{j=1}^{N} \left(\frac{(-1)^{N-n} \sum_{\substack{m = N - n; \\ j_1 \neq \cdots \neq j_m \in \aleph_0 \backslash \{j-1\}}} \rho_{j_1} \rho_{j_2} \cdots \rho_{jm}}{\prod_{k \in \aleph_0 \backslash \{j-1\}} (\rho_{j-1} - \rho_k)} \cdot \rho_{j-1}^N \overline{\sigma}_{[N,\rho_{j-1}]}(\cdot) \right) \end{aligned} \quad (4.17)$$

Equation (4.17) gives the truncation-incurred estimation error (i.e., truncation influence) as a function of the truncation residue $\overline{\sigma}_{[N,\rho_{j-1}]}(j\omega)$ and the actuation amplitude parameters $\rho_0, \rho_1, \rho_2, \ldots, \rho_{N-1}$. Obviously, subject to different parameter values for $\rho_0, \rho_1, \rho_2, \ldots, \rho_{N-1}$, the influence of $\overline{\sigma}_{[N,\rho_{j-1}]}(j\omega)$ on $\Delta Y_n(j\omega)$ could be significantly different given the same truncation order N. Note that all the parameters $\rho_0, \rho_1, \rho_2, \ldots, \rho_{N-1}$ in P can be freely tuned in practice in order to have an accurate estimation of $Y_n(j\omega)$ ($\forall n \in \aleph$), i.e., to make $\Delta Y_n(j\omega)$ as small as possible. With the result in (4.17), the task now is to apply Lemma 1 to find a set of appropriate parameter values for $\rho_0, \rho_1, \rho_2, \ldots, \rho_{N-1}$ in P so that $\Delta Y_n(j\omega)$ could be as small as possible for $\forall n \in \aleph$. This would be of significance to dealing with the truncation issues since $\overline{\sigma}_{[N,\rho_{j-1}]}(j\omega)$ is actually always difficult to know in practice.

For example, from (4.17) $\Delta Y_1(j\omega)$ can be computed as

$$\begin{aligned}
\Delta Y_1(j\omega) &= \sum_{j=1}^{N}\left(\frac{(-1)^{N-1}\sum\limits_{m=N-1;j_1\neq\cdots\neq j_m\in\aleph_0\backslash\{j-1\}}\rho_{j_1}\rho_{j_2}\cdots\rho_{j_m}}{\prod\limits_{k\in\aleph_0\backslash\{j-1\}}\left(\rho_{j-1}-\rho_k\right)}\cdot\rho_{j-1}^{N}\overline{\sigma}_{[N,\rho_{j-1}]}(\cdot)\right)\\
&= \sum_{j=1}^{N}\left(\frac{(-1)^{N-1}\rho_0\rho_1\cdots\rho_{N-1}}{\rho_{j-1}\prod\limits_{k\in\aleph_0\backslash\{j-1\}}\left(\rho_{j-1}-\rho_k\right)}\cdot\rho_{j-1}^{N}\overline{\sigma}_{[N,\rho_{j-1}]}(\cdot)\right)\\
&= \sum_{j=1}^{N}\left(\frac{(-1)^{N-1}\rho_0\rho_1\cdots\rho_{N-1}}{\prod\limits_{k\in\aleph_0\backslash\{j-1\}}\left(\rho_{j-1}-\rho_k\right)}\cdot\rho_{j-1}^{N-1}\overline{\sigma}_{[N,\rho_{j-1}]}(\cdot)\right)\\
&= (-1)^{N-1}\rho_0\rho_1\cdots\rho_{N-1}\sum_{j=1}^{N}\left(\frac{\rho_{j-1}^{N-1}\overline{\sigma}_{[N,\rho_{j-1}]}(\cdot)}{\prod\limits_{k\in\aleph_0\backslash\{j-1\}}\left(\rho_{j-1}-\rho_k\right)}\right)
\end{aligned} \tag{4.18}$$

From (4.18) and Lemma 1, it can be concluded that $\Delta Y_1(j\omega) = 0$ if $\rho_{j-1}^{N-1}\overline{\sigma}_{[N,\rho_{j-1}]}(\cdot)$ could be (approximately) a constant for some $\rho_0, \rho_1, \rho_2, \ldots, \rho_{N-1}$ and N.

Considering the nonlinear output spectrum in (4.7) and the estimation error in (4.15b), the following result can be obtained.

Proposition 1. *Consider the estimation error $\Delta Y_i(j\omega)$ in (4.15b). If there exist positive integers M and N satisfying $M \le N$, and some real numbers $\rho_0, \rho_1, \ldots, \rho_{N-1} \in R$ so that*

$$J(1) = J(2) = \ldots = J(N) \tag{4.19}$$

then $\Delta Y_i(j\omega) = 0$, where $M \ge 1$ if $i = 1$ else $M \ge 2$; $N \ge 2$ if $i = 1$ else $N \ge i$; for $j = 1, 2, \ldots, N$, $J(j) = (-1)^{N-i}\rho_{j-1}^{M}\overline{\sigma}_{[N,\rho_{j-1}]}(\cdot)\sum\limits_{\substack{m=N-i;\\ j_1\neq\cdots\neq j_m\in\aleph_0\backslash\{j-1\}}}\rho_{j_1}\rho_{j_2}\cdots\rho_{j_m}$.

Proof See Appendix 3. □

With the result in Proposition 1, the problem mentioned before (to find some ρ's so that the estimation error is zero) is now to find some real numbers for ρ so that

Eq. (4.9) could hold. To understand Proposition 1, for $i = 1$, it can be obtained that

$$J(j) = (-1)^{N-1} \rho_{j-1}^{M} \overline{\sigma}_{[N,\rho_{j-1}]}(\cdot) \prod_{k \in \aleph_0} (\rho_k)$$

Note that given specifically $\rho_0, \rho_1, \ldots, \rho_{N-1}$, $(-1)^{N-1} \prod_{k \in \aleph_0} (\rho_k)$ is constant for any j and $\rho_{j-1}^{M} \overline{\sigma}_{[N,\rho_{j-1}]}(\cdot)$ is only a function of ρ_{j-1}. Therefore, (4.19) in Proposition 1 can be written as

$$\rho_0^M \overline{\sigma}_{[N,\rho_0]}(\cdot) = \rho_1^M \overline{\sigma}_{[N,\rho_1]}(\cdot) = \cdots = \rho_{N-1}^M \overline{\sigma}_{[N,\rho_{N-1}]}(\cdot)$$

It can be verified that similar result also hold for i = N. The following result can be summarized.

Corollary 1. *Consider the estimation error $\Delta Y_i(j\omega)$ for $i = 1$ and N. If there exists $\varepsilon \in C$ so that $\rho^M \overline{\sigma}_{[N,\rho]}(\cdot) = \varepsilon$ has 2 or N solutions, then $\Delta Y_1(j\omega) = 0$ or $\Delta Y_N(j\omega) = 0$ correspondingly, where $M \geq 1$ if $i = 1$ else $M \geq 2$.*

Analytical computation of $\overline{\sigma}_{[N,\rho]}(\cdot)$ would be complicated. However, it can be seen from (4.8) that $\rho^M \overline{\sigma}_{[N,\rho]}(\cdot)$ would be around zero with sufficiently small ρ. Together with Corollary 1, this implies that ρ could be chosen properly in (0, 1) so that $\rho^M \overline{\sigma}_{[N,\rho]}(\cdot)$ is around a small value ε and thus both the estimation errors $\Delta Y_1(j\omega) = 0$ and $\Delta Y_N(j\omega)$ would be approximate to zero. Moreover, ρ could also be chosen around a value (e.g., 1) so that $\rho^M \overline{\sigma}_{[N,\rho]}(\cdot)$ would be around a value ε. In this way, the estimation error $\Delta Y_i(j\omega)$ could be approximate to zero for all i based on Proposition 1. The significance of the results above is to provide a manner to tune $\rho_0, \rho_1, \ldots, \rho_{N-1}$ so that the estimation error incurred by series truncation could be minimized by applying Lemma 1. Obviously, different methods could be tried to this end by using Lemma 1. This would be explored further in future studies.

Alternatively, to facilitate the analysis of estimation error incurred by the truncation, the analytical relationship between the estimation error and excitation parameters $\rho_0, \rho_1, \ldots, \rho_{N-1}$ as demonstrated in (4.17) can be revealed more clearly. Considering $\Delta Y_1(j\omega)$ and $\Delta Y_N(j\omega)$, the following results can be obtained. Similar result could be extended to the other orders.

Proposition 2.

$$\Delta Y_1(j\omega) = (-1)^{N-1} \rho_0 \rho_1 \cdots \rho_{N-1} \sum_{k=0}^{\infty} (\rho_0 + \rho_1 + \cdots + \rho_{N-1})^{[k]} Y_{N+k+1}(j\omega) \quad (4.20)$$

$$\Delta Y_N(j\omega) = \sum_{k=1}^{\infty} (\rho_0 + \rho_1 + \cdots + \rho_{N-1})^{[k]} Y_{N+k}(j\omega) \quad (4.21)$$

Proof See Appendix 4. □

Proposition 2 provides a more straightforward insight into the estimation error incurred by excitation magnitudes and truncation order. It can be seen that not only the truncation order affects the estimation errors but also the excitation magnitudes have significant effects. For the same truncation order N, different values of the excitation magnitudes could bring very different estimation errors. The significance of this result is to provide a fundamental basis for designing $\rho_0\rho_1\cdots\rho_{N-1}$ and N so that the estimation errors (i.e., $\Delta Y_1(j\omega)$, $\Delta Y_N(j\omega)$) can be minimized in practice. This will be demonstrated in the next section.

4.3.2 Selection of Truncation Order and Excitation Magnitudes: Case Studies

To demonstrate the theoretical results, the truncation orders $N = 2, 3, 4$ are discussed in this section. Similar results can be extended to the other orders. For convenience in discussion, the following definition is needed.

Definition 1. Give an input U($j\omega$) (the magnitude index $\rho = 1$) and a threshold ε, the best truncation order under the given input is denoted by $N^*(\mathrm{U(j\omega)},\varepsilon) = N*$ satisfying $\left|\sigma_{[N^*]}(j\omega)\right| \le |Y_{N^*+1}(j\omega)| + |Y_{N^*+2}(j\omega)| + |Y_{N^*+3}(j\omega)| + \cdots \le \varepsilon$.

Case 1: N = 2

Considering N = 2 firstly, the output spectrum is truncated at N = 2, which can be written as

$$Y(j\omega) = Y_1(j\omega) + Y_2(j\omega) + \sigma_{[2]}(j\omega) \tag{4.22}$$

where $\sigma_{[2]}(j\omega) = Y_3(j\omega) + Y_4(j\omega) + Y_5(j\omega) + \cdots$ is the truncation error. Using (4.9) to estimate the different orders of output spectra, the estimation errors are given by Proposition 2 as

$$\Delta Y_1(j\omega) = -\rho_0\rho_1 \sum_{k=0}^{\infty} (\rho_0 + \rho_1)^{[k]} Y_{k+3}(j\omega).$$

$$\Delta Y_2(j\omega) = \sum_{k=1}^{\infty} (\rho_0 + \rho_1)^{[k]} Y_{k+2}(j\omega).$$

It can be computed that

$$\begin{aligned}
(\rho_0 + \rho_1)^{[0]} &= 1 \\
(\rho_0 + \rho_1)^{[1]} &= \rho_0 + \rho_1 \\
(\rho_0 + \rho_1)^{[2]} &= \rho_0^2 + \rho_0\rho_1 + \rho_1^2 \\
(\rho_0 + \rho_1)^{[3]} &= \rho_0^3 + \rho_0\rho_1^2 + \rho_0^2\rho_1 + \rho_1^3
\end{aligned}$$

......

Obviously, ρ_0 and ρ_1 can be properly designed so that the influence incurred by truncation error can be suppressed as small as possible. For example, if $\rho_0 = -\rho_1$, it can be obtained that

$$(\rho_0 + \rho_1)^{[k]} = \begin{cases} \rho_1^k & k = 2i \\ 0 & k = 2i + 1 \end{cases}$$

where i = 0, 1, 2, ... Therefore, it can be obtained that

$$\Delta Y_1(j\omega) = \rho_1^2 Y_3(j\omega) + \rho_1^4 Y_5(j\omega) + \rho_1^6 Y_7(j\omega) + \cdots$$
$$\Delta Y_2(j\omega) = \rho_1^2 Y_4(j\omega) + \rho_1^4 Y_6(j\omega) + \rho_1^6 Y_8(j\omega) + \cdots$$

From the equations above, it can be concluded that

(a1) the even order nonlinearity in the system has no effect on the estimation error $\Delta Y_1(j\omega)$, while the odd order nonlinearity has no effect on the estimation error $\Delta Y_2(j\omega)$;
(a2) $|\Delta Y_1(j\omega)| < \rho_1^2 \varepsilon$ if $N^* = 2$ and $|\rho_1| \leq 1$, while $|\Delta Y_2(j\omega)| < \rho_1^2 \varepsilon$ if $N^* = 3$ and $|\rho_1| \leq 1$;
(a3) there exists $|\rho_1| \leq 1$ so that $|\Delta Y_1(j\omega)| < \varepsilon$ and $|\Delta Y_2(j\omega)| < \varepsilon$ irrespectively of the value of N^*.

Case 2: N = 3

Similar results can be derived for this case. Using (4.9) to estimate the different orders of output spectra, the estimation errors are given by Proposition 2 as

$$\Delta Y_1(j\omega) = -\rho_0\rho_1\rho_2 \sum_{k=0}^{\infty} (\rho_0 + \rho_1 + \rho_2)^{[k]} Y_{k+4}(j\omega)$$

$$\Delta Y_3(j\omega) = \sum_{k=1}^{\infty} (\rho_0 + \rho_1 + \rho_2)^{[k]} Y_{k+3}(j\omega)$$

The parameters ρ_0, ρ_1, ρ_2 can be designed to reduce the estimation error. For example, if choosing $\rho_0 + \rho_1 + \rho_2 = 0$, it can be obtained that

$$\begin{aligned} \Delta Y_1(j\omega) &= -\rho_0\rho_1\rho_2\{Y_4(j\omega) + (\rho_1^2 + \rho_1\rho_2 + \rho_2^2)Y_6(j\omega) \\ &\quad + \rho_0\rho_1\rho_2 Y_7(j\omega) + \cdots\} \\ \Delta Y_3(j\omega) &= (\rho_1^2 + \rho_1\rho_2 + \rho_2^2)Y_5(j\omega) + \rho_0\rho_1\rho_2 Y_6(j\omega) \\ &\quad + (\rho_1^2\rho_0^2 + \rho_1^2\rho_2^2 + \rho_0^2\rho_2^2)Y_7(j\omega) + \cdots \end{aligned}$$

If additionally choosing $\rho_0 = 1$, it can be obtained that

$$\Delta Y_1(j\omega) = -\rho_1\rho_2\{Y_4(j\omega) + (\rho_1^2 + \rho_1\rho_2 + \rho_2^2)Y_6(j\omega) + \rho_1\rho_2 Y_7(j\omega) + \cdots\}$$
$$\Delta Y_3(j\omega) = (\rho_1^2 + \rho_1\rho_2 + \rho_2^2)Y_5(j\omega) + \rho_1\rho_2 Y_6(j\omega) + (\rho_1^2 + \rho_1^2\rho_2^2 + \rho_2^2)Y_7(j\omega) + \cdots$$

If additionally letting $\rho_1\rho_2 = \alpha$ and $|\alpha|<1/4$, there will be real solutions for $\rho_1+\rho_2 = -\rho_0 = -1$. In this case, it can be obtained that

$$\Delta Y_1(j\omega) = -\alpha Y_4(j\omega) - \alpha(1-\alpha)Y_6(j\omega) - \alpha^2 Y_7(j\omega) + \cdots$$
$$\Delta Y_3(j\omega) = (1-\alpha)Y_5(j\omega) + \alpha Y_6(j\omega) + (1-\alpha)^2 Y_7(j\omega) + \cdots$$

From the equations above with the mentioned parameter setting, it can be concluded that

(b1) $|\Delta Y_1(j\omega)| < \alpha\varepsilon$ if $N^* \leq 3$, while $|\Delta Y_3(j\omega)| < \varepsilon$ if $N^* \leq 4$;
(b2) there exists $\alpha < 1/4$ so that $|\Delta Y_1(j\omega)| < \varepsilon$ irrespectively of the value of N^*.

From Case 1, $Y_1(j\omega)$ and $Y_2(j\omega)$ could be estimated accurately and from Case 2, $Y_1(j\omega)$ and $Y_3(j\omega)$ could be estimated. Following this method, more high order output spectrum could be estimated accurately until an appropriate truncation order N is found. These will be demonstrated in the following section.

4.4 An Example

To demonstrate the results above, consider a Wiener system given by

$$u(t) = g \circ r(t) \text{ and } y(t) = f(u(t)) \tag{4.23ab}$$

where "∘" represents the convolution operator, $g(t)$ is the impulse response of the linear part, and $f(u(t))$ is the static nonlinear polynomial function. The linear part and nonlinear part are given by

$$\mathrm{G(s)} = \frac{26}{s^2 + 2s + 26} \text{ and } f(u) = u + 2u^2 - u^4 \tag{4.24ab}$$

where G(s) is the Laplace transform of $g(t)$. It can be checked that the nonlinear output spectrum can be written as [10, 18]

$$Y(\mathrm{j}\omega) = Y_1(\mathrm{j}\omega) + Y_2(\mathrm{j}\omega) + Y_4(\mathrm{j}\omega) \tag{4.25}$$

That is, the 3rd order and the other high order output spectra are all zero. Obviously the best truncation order for this system is $N^* = 4$, which corresponds the maximum degree in polynomial function (4.24b).

Given only some input–output data, the truncation order N^* and system output frequency response components ($Y_i(j\omega)$ for $i = 1, 2, \ldots, N^*$) can be estimated with the methods discussed in this study. Note that for a Wiener system structured by (4.23ab), the truncation order N^* corresponds the maximum degree in polynomial function (4.23b); $Y_i(j\omega) = 0$ implies that the coefficient of the polynomial term of degree i is zero, and the accurate estimate of $Y_i(j\omega)$ implies the accurate estimate of the corresponding coefficient [5, 18].

Following the method discussed in the last section, the estimation is conducted from $N = 2$ and then on using the algorithm in (4.9) but with different selections of excitation magnitude ρ. When choosing $N = 2$, the excitation magnitude is chosen as $\rho_0 = 0.005$ and $\rho_1 = -0.005$ (referred to as Case 1) according to the discussion in Sect. 3.3 Case 1. When choosing $N = 3$, the excitation magnitude is chosen according to Sect. 3.3 Case 2 as $\rho_0 = 1$, $\rho_1 = -0.99995$ and $\rho_2 = -0.00005$ satisfying $\rho_0 + \rho_1 + \rho_2 = 0$ (referred to as Case 2). For comparison, ρ_0, ρ_1 and ρ_2 can also be chosen randomly in a range (0, 10) as for example 1.0000, 1.1112 and 2.5806, respectively. The latter method for the selection of excitation magnitudes is referred to as Case 3. For $N = 4$ and $N = 5$, the excitation magnitudes are all chosen with this method. The estimated results are given in nonparametric forms and shown in Figs. 4.1 and 4.2.

As shown in Fig. 4.1, all cases, except for when $N = 3$ in case3, can accurately estimate the transfer function of the system's linear part. It's important to note that the estimate using $N = 3$ with case 3 is significantly biased. This clearly demonstrates that the algorithm in Eq. (4.9), which is frequently used in the literature without

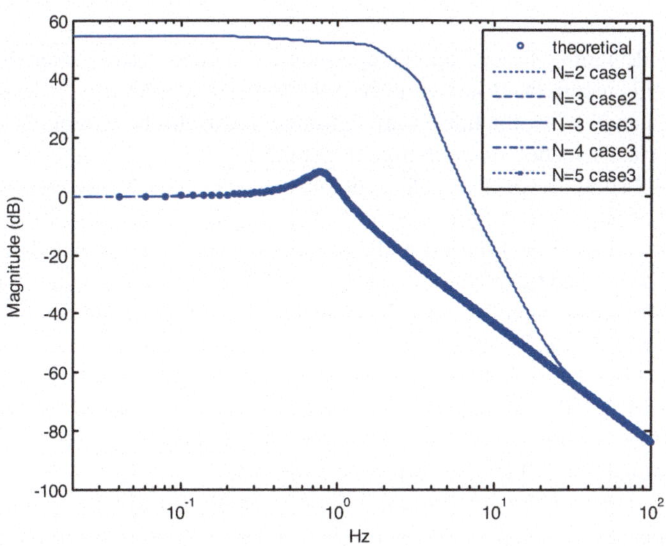

Fig. 4.1 The estimates of the linear part in different cases

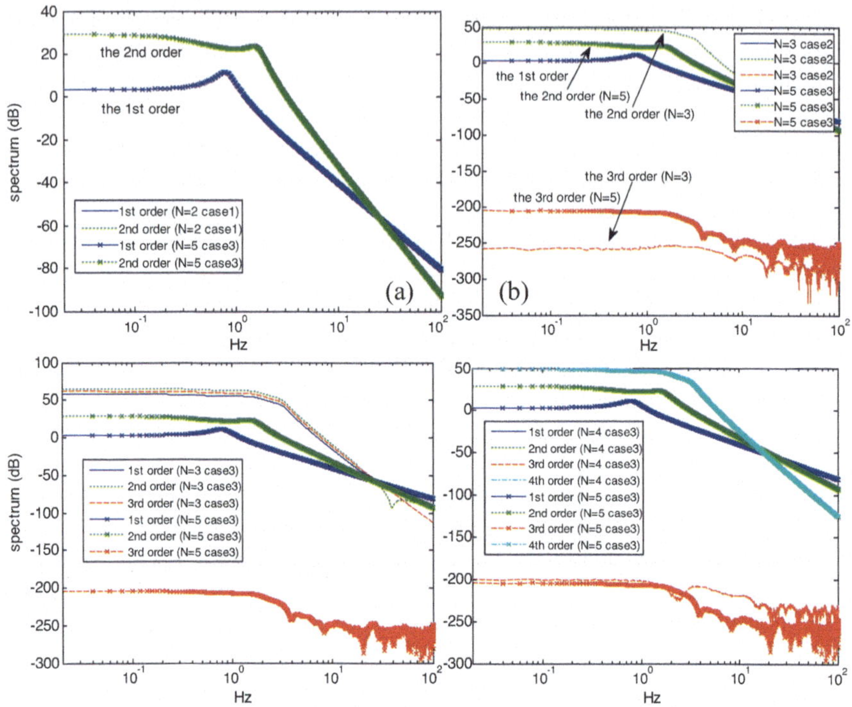

Fig. 4.2 The estimates of different order output spectrum in different cases

specifically designing the excitation magnitudes, is not dependable when the truncation order N is incorrect. However, even if the truncation order used is less than the actual one, i.e., $N < N^*$, the linear part's estimation can still be accurately produced with the correct selection of excitation magnitudes.

Figure 4.2 shows respectively the estimation results in each case compared with the estimates using $N = 5$ (with case3). In Fig. 2a, $N = 2$ and Case 1 discussed in Sect. 3.3 are adopted, and the 1st and 2nd output spectra are estimated accurately. In Fig. 2b, $N = 3$ and Case 2 discussed in Sect. 3.3 are adopted, and the 1st and 3rd output spectra are estimated accurately. However, the estimated 2nd output spectrum is biased since it is not considered in the selection of the excitation magnitudes. When $N = 3$ but the excitation magnitudes are not properly chosen as discussed in Case 2 of Sect. 3.3, the estimated output spectra are obviously biased and wrong (see Fig. 2c). When N takes the correct truncation order or a larger number, i.e., $N \geq N^* = 4$, the estimates are of course correct (see Fig. 2d).

It can be seen that, the algorithm in Eq. (4.9), combined with the methods introduced in this study, is capable of precisely estimating the truncation order (or the highest degree of nonlinearity) and the output spectrum of each order in nonlinear systems.

To this end, the following general procedure, as a summary, can be used.

(1) Given a truncation order N, use Eq. (4.17) or Proposition 2 to find the best excitation magnitudes $\rho_0, \rho_1 \cdots \rho_{N-1}$ so that $Y_1(j\omega)$, $Y_N(j\omega)$ and others can be accurately estimated.
(2) The estimation can start from N = 2 and then go on until N = N* when the estimates with any randomly-chosen parameters $\rho_0, \rho_1 \cdots \rho_{N-1}$ are completely consistent (e.g., the error is less than a predefined threshold) with the previous estimates using properly-designed ρ_i by Proposition 2.

4.5 Conclusions

This study examines the truncation order and its impact in a category of nonlinear systems, establishing some key findings. These provide a substantial understanding of the analytical connections between truncation order, truncation error, and the estimation error of the output spectrum of nonlinear systems. These findings are notably crucial in kernel estimation based on Volterra series and frequency domain analysis of nonlinear systems, as well as in the structure and order selection of block-oriented nonlinear models with a polynomial function. The results obtained should also hold importance for related topics and areas that involve the use of the multi-level excitation method for system analysis and identification.

It should be emphasized that the work of this chapter places an important basis for the following nCOS method and its related applications, especially those in fault detection, which can be referred to the following review paper and the references inside—Li et al. [24].

Acknowledgements The main results of this chapter can also be referred to the published paper [25].

Appendix 1: Proof of Lemma 1

Proof The mutual-modulation function can be proved by firstly showing $\sum_{i=1}^{N} \frac{1}{\prod_{k=1,k\neq i}^{N}(x_i-x_k)} = 0$, which corresponds to the power degree $m = 0$, and then proving that the mutual-modulation function holds for $m = 1, 2, \ldots$ N, ... □

To prove $\sum_{i=1}^{N} \frac{1}{\prod_{k=1,k\neq i}^{N}(x_i-x_k)} = 0$, the mathematical induction can be used. It can be checked that it holds for N = 2, 3, ... Now supposing that $\sum_{i=1}^{N} \frac{1}{\prod_{k=1,k\neq i}^{N}(x_i-x_k)} = 0$ for N = n, i.e.,

$$\sum_{i=1}^{n} \frac{1}{\prod_{k=1,k\neq i}^{n}(x_i - x_k)} = \frac{1}{(x_1 - x_2)(x_1 - x_3)\cdots(x_1 - x_n)}$$

$$+\frac{1}{(x_2-x_1)(x_2-x_3)\cdots(x_2-x_n)}+\ldots$$
$$+\frac{1}{(x_n-x_1)(x_n-x_2)\cdots(x_n-x_{n-1})}=0 \tag{4.26}$$

one needs to prove the equation holds for $N=n+1$.

$$\sum_{i=1}^{n+1}\frac{1}{\prod_{k=1,k\neq i}^{n+1}(x_i-x_k)}=\frac{1}{(x_1-x_2)(x_1-x_3)\cdots(x_1-x_n)(x_1-x_{n+1})}$$
$$+\frac{1}{(x_2-x_1)(x_2-x_3)\cdots(x_2-x_{n+1})}$$
$$+\ldots+\frac{1}{(x_n-x_1)(x_n-x_2)\cdots(x_n-x_{n-1})(x_n-x_{n+1})}$$
$$+\frac{1}{(x_{n+1}-x_1)(x_{n+1}-x_2)\cdots(x_{n+1}-x_n)} \tag{4.27}$$

Substituting (4.26) into the first term of the right side of Eq. (4.27) and after some manipulations, it can be readily obtained that $\sum_{i=1}^{N}\frac{1}{\prod_{k=1,k\neq i}^{N}(x_i-x_k)}=0$ for $N=n+1$.

With $\sum_{i=1}^{N}\frac{1}{\prod_{k=1,k\neq i}^{N}(x_i-x_k)}=0$ holds for $N>1$, the next step is to show $\sum_{i=1}^{N}\frac{x_i}{\prod_{k=1,k\neq i}^{N}(x_i-x_k)}=0$, that is the mutual-modulation function holds for $m=1$. From (4.2), it can be obtained that

$$\frac{x_n}{(x_1-x_2)(x_1-x_3)\cdots(x_1-x_n)}+\frac{x_n}{(x_2-x_1)(x_2-x_3)\cdots(x_2-x_n)}$$
$$+\ldots+\frac{x_n}{(x_n-x_1)(x_n-x_2)\cdots(x_n-x_{n-1})}=0 \tag{4.28}$$

Note that $\sum\limits_{i=1}^{n-1}\frac{1}{\prod_{k=1,k\neq i}^{n-1}(x_i-x_k)}=0$, i.e.,

$$\frac{1}{(x_1-x_2)(x_1-x_3)\cdots(x_1-x_{n-1})}+\frac{1}{(x_2-x_1)(x_2-x_3)\cdots(x_2-x_{n-1})}$$
$$+\ldots+\frac{1}{(x_{n-1}-x_1)(x_{n-1}-x_2)\cdots(x_{n-1}-x_{n-2})}=0 \tag{4.29}$$

Adding (4.29) to (4.28), the equation still holds. The first term in the left side of (4.28) is

$$\frac{x_n}{(x_1-x_2)(x_1-x_3)\cdots(x_1-x_{n-1})(x_1-x_n)}$$

$$+\frac{1}{(x_1-x_2)(x_1-x_3)\cdots(x_1-x_{n-1})}$$
$$=\frac{x_1}{(x_1-x_2)(x_1-x_3)\cdots(x_1-x_{n-1})(x_1-x_n)}$$

Similar equations hold for the next n – 2 terms in the new equation by adding (4.29) to (4.28). Therefore, it can be obtained that $\sum_{i=1}^{N}\frac{x_i}{\prod_{k=1,k\neq i}^{N}(x_i-x_k)}=0$, that is, the mutual-modulation function holds for $m=1$.

With the mutual-modulation function (4.1) holds for $m=1$, one can then prove that it holds for $m=2$ by following the same method above. Following this process, it can be proved that the mutual-modulation function (4.1) holds for m = 2, 3, …, N – 2.

The next task is to show the mutual-modulation function holds for m = N – 1, i.e., $\sum\limits_{i=1}^{N}\frac{x_i^{N-1}}{\prod_{k=1,k\neq i}^{N}(x_i-x_k)}=1$. By mathematical induction, one can check that it holds for N = 2, 3, …. Supposing it holds for N = n – 1, i.e.,

$$\frac{x_1^{n-2}}{(x_1-x_2)(x_1-x_3)\cdots(x_1-x_{n-1})}+\frac{x_2^{n-2}}{(x_2-x_1)(x_2-x_3)\cdots(x_2-x_{n-1})}+\ldots+\frac{x_n^{n-2}}{(x_{n-1}-x_1)(x_{n-1}-x_2)\cdots(x_{n-1}-x_{n-2})}=1 \tag{4.30}$$

one needs to show that it still holds for N = n. Note that the mutual-modulation function holds for m = N – 2. That is, one has

$$\frac{x_1^{n-2}}{(x_1-x_2)(x_1-x_3)\cdots(x_1-x_n)}+\frac{x_2^{n-2}}{(x_2-x_1)(x_2-x_3)\cdots(x_2-x_n)}+\ldots+\frac{x_n^{n-2}}{(x_n-x_1)(x_n-x_2)\cdots(x_n-x_{n-1})}=0$$

which yields

$$\frac{x_1^{n-2}x_n}{(x_1-x_2)(x_1-x_3)\cdots(x_1-x_n)}+\frac{x_2^{n-2}x_n}{(x_2-x_1)(x_2-x_3)\cdots(x_2-x_n)}+\ldots+\frac{x_n^{n-1}}{(x_n-x_1)(x_n-x_2)\cdots(x_n-x_{n-1})}=0 \tag{4.31}$$

Adding (4.30) to (4.31) gives a new equation. The first term in the left side of the new equation is

$$\frac{x_1^{n-2}x_n}{(x_1-x_2)(x_1-x_3)\cdots(x_1-x_n)}+\frac{x_1^{n-2}}{(x_1-x_2)(x_1-x_3)\cdots(x_1-x_n)}$$
$$=\frac{x_1^{n-1}}{(x_1-x_2)(x_1-x_3)\cdots(x_1-x_n)}$$

Similar equations hold for the other n − 2 terms. Therefore, it can be obtained that $\sum_{i=1}^{N}\frac{x_i^{N-1}}{\prod_{k=1,k\neq i}^{N}(x_i-x_k)}=1$, that is, the mutual-modulation function holds for $m =$ N − 1. With the same method, one can show the mutual-modulation function holds for $m =$ N and $m >$ N. This completes the proof.

Appendix 2: Proof of Lemma 2

Proof From [26], it can be obtained that

$$\mathrm{A}=\mathrm{U}\cdot\mathrm{L} \tag{4.32}$$

where $\rho_0 = 1$,

$$U=\begin{bmatrix} 1 & -\rho_0 & \rho_0\rho_1 & -\rho_0\rho_1\rho_2 & \cdots \\ 0 & 1 & -(\rho_0+\rho_1) & \rho_0\rho_1+\rho_0\rho_2+\rho_1\rho_2 & \cdots \\ 0 & 0 & 1 & -(\rho_0+\rho_1+\rho_2) & \cdots \\ 0 & 0 & 0 & 1 & \cdots \\ \vdots & \vdots & \vdots & \vdots & \ddots \end{bmatrix}_{N\times N} \tag{4.33a}$$

i.e.,

$$\mathrm{U_{i,i}}=1;\ \mathrm{U_{i,1}}=0;\ \mathrm{U_{i,j}}=\mathrm{U_{i-1,j-1}}-\mathrm{U_{i,j-1}}\rho_{j-2} \tag{4.33b}$$

$$L=\begin{bmatrix} 1 & 0 & 0 & 0 & \cdots \\ \frac{1}{\rho_0-\rho_1} & \frac{1}{\rho_1-\rho_0} & 0 & 0 & \cdots \\ \frac{1}{(\rho_0-\rho_1)(\rho_0-\rho_2)} & \frac{1}{(\rho_1-\rho_0)(\rho_1-\rho_2)} & \frac{1}{(\rho_2-\rho_0)(\rho_2-\rho_1)} & 0 & \cdots \\ \frac{1}{(\rho_0-\rho_1)\cdots(\rho_0-\rho_3)} & \frac{1}{(\rho_1-\rho_0)\cdots(\rho_1-\rho_3)} & \frac{1}{(\rho_2-\rho_0)\cdots(\rho_2-\rho_3)} & \frac{1}{(\rho_3-\rho_0)\cdots(\rho_3-\rho_2)} & \cdots \\ \vdots & \vdots & \vdots & \vdots & \ddots \end{bmatrix}_{N\times N} \tag{4.34a}$$

i.e.,

$$Li,\mathrm{j}=0(i<j);\ L1,1=1;\ L_{i,j}=\prod_{k=0;k\neq j-1}^{i-1}\frac{1}{\rho_{j-1}-\rho_k} \tag{4.34b}$$

Therefore, it can be obtained from (4.33–4.34) that

$$A_{i,j} = \sum_{k=1}^{N} U_{i,k} L_{k,j} = \sum_{k=i}^{N} U_{i,k} L_{k,j} = L_{i,j} + \sum_{k=i+1}^{N} U_{i,k} L_{k,j} \tag{4.35}$$

Noting that (4.34b), (4.35) can be further written as

$$A_{i,j} = L_{i,j} + \sum_{k=\max(i+1,j)}^{\min(N,j)} U_{i,k} L_{k,j} \tag{4.36}$$

Based on (4.33b), (4.34b) and (4.35), a mathematical induction leads to

$$A_{i,j} = \frac{(-1)^{N-i} \sum_{n=N-j; j_1 \neq \cdots \neq j_n \neq j-1; j_n \in \aleph_0 \backslash j-1} \rho_{j_1} \rho_{j_2} \cdots \rho_{j_n}}{\prod_{k \in \aleph_0 \backslash j-1} (\rho_{j-1} - \rho_k)}$$

This completes the proof.

Appendix 3: Proof of Proposition 1

Proof From (4.17) it can be obtained that

$$\Delta Y_i(j\omega) = \sum_{j=1}^{N} \left(\frac{(-1)^{N-i} \sum\limits_{\substack{m = N-i; \\ j_1 \neq \cdots \neq j_m \in \aleph_0 \backslash \{j-1\}}} \rho_{j_1} \rho_{j_2} \cdots \rho_{j_m}}{\prod\limits_{k \in \aleph_0 \backslash \{j-1\}} (\rho_{j-1} - \rho_k)} \cdot \rho_{j-1}^{N} \bar{\sigma}_{[N,\rho_{j-1}]}(\cdot) \right)$$

$$= \sum_{j=1}^{N} \left(\frac{\rho_{j-1}^{N-M}}{\prod\limits_{k \in \aleph_0 \backslash \{j-1\}} (\rho_{j-1} - \rho_k)} \cdot (-1)^{N-i} \rho_{j-1}^{M} \bar{\sigma}_{[N,\rho_{j-1}]}(\cdot) \right.$$

$$\left. \sum_{\substack{m = N-i; \\ j_1 \neq \cdots \neq j_m \in \aleph_0 \backslash \{j-1\}}} \rho_{j_1} \rho_{j_2} \cdots \rho_{j_m} \right) \tag{4.37}$$

From (4.19), the term J(i) in (4.37) can be regarded as a constant J_i for all j. Thus (4.37) gives

$$\Delta Y_i(j\omega) = \sum_{j=1}^{N}\left(\frac{\rho_{j-1}^{N-M}}{\prod_{k\in\aleph_0\backslash\{j-1\}}\left(\rho_{j-1}-\rho_k\right)}\cdot J_i\right)$$
$$= J_i\sum_{j=1}^{N}\left(\frac{\rho_{j-1}^{N-M}}{\prod_{k\in\aleph_0\backslash\{j-1\}}\left(\rho_{j-1}-\rho_k\right)}\right) \tag{4.38}$$

By Lemma 1, $\Delta Y_i(j\omega) = 0$ if M > 1. Specifically, for i = 1, that is

$$\Delta Y_1(j\omega) = \sum_{j=1}^{N}\left(\frac{(-1)^{N-1}\sum\limits_{\substack{m=N-1;\\ j_1\neq\cdots\neq j_m\in\aleph_0\backslash\{j-1\}}}\rho_{j_1}\rho_{j_2}\cdots\rho_{j_m}}{\prod_{k\in\aleph_0\backslash\{j-1\}}\left(\rho_{j-1}-\rho_k\right)}\cdot\rho_{j-1}^{N}\overline{\sigma}_{[N,\rho_{j-1}]}(\cdot)\right)$$
$$= \sum_{j=1}^{N}\left(\frac{(-1)^{N-1}\prod_{k\in\aleph_0\backslash\{j-1\}}(\rho_k)}{\prod_{k\in\aleph_0\backslash\{j-1\}}\left(\rho_{j-1}-\rho_k\right)}\cdot\rho_{j-1}^{N}\overline{\sigma}_{[N,\rho_{j-1}]}(\cdot)\right)$$
$$= \sum_{j=1}^{N}\left(\frac{(-1)^{N-1}\prod_{k\in\aleph_0}(\rho_k)}{\prod_{k\in\aleph_0\backslash\{j-1\}}\left(\rho_{j-1}-\rho_k\right)}\cdot\rho_{j-1}^{N-1}\overline{\sigma}_{[N,\rho_{j-1}]}(\cdot)\right)$$
$$= \sum_{j=1}^{N}\left(\frac{\rho_{j-1}^{N-1-M}}{\prod_{k\in\aleph_0\backslash\{j-1\}}\left(\rho_{j-1}-\rho_k\right)}\cdot(-1)^{N-1}\rho_{j-1}^{M}\overline{\sigma}_{[N,\rho_{j-1}]}(\cdot)\prod_{k\in\aleph_0}(\rho_k)\right) \tag{4.39}$$

Note that $\prod_{k\in\aleph_0}(\rho_k)$ must be a constant for all j. Therefore, in order to apply Lemma 1, the condition in (4.19) for M is now that M = 1 for i = 1. This completes the proof.

Appendix 4: Proof of Proposition 2

Proof From (4.8),

$$\rho_{j-1}^{N-1}\overline{\sigma}_{[N,\rho_{j-1}]}(j\omega) = \rho_{j-1}^{N-1}Y_{N+1}(j\omega) + \rho_{j-1}^{N}Y_{N+2}(j\omega) + \rho_{j-1}^{N+1}Y_{N+3}(j\omega) + \cdots \tag{4.40}$$

From (4.18) and using (4.40),

$$
\begin{aligned}
\Delta Y_1(j\omega) &= (-1)^{N-1}\rho_0\rho_1\cdots\rho_{N-1}\sum_{j=1}^{N}\frac{\rho_{j-1}^{N-1}\overline{\sigma}_{[N,\rho_{j-1}]}(j\omega)}{\prod_{k\in\aleph_0\{j-1\}}(\rho_{j-1}-\rho_k)} \\
&= (-1)^{N-1}\rho_0\rho_1\cdots\rho_{N-1}\left(\sum_{j=1}^{N}\frac{\rho_{j-1}^{N-1}Y_{N+1}(j\omega)}{\prod_{k\in\aleph_0\{j-1\}}(\rho_{j-1}-\rho_k)}\right. \\
&\quad \left.+\sum_{j=1}^{N}\frac{\rho_{j-1}^{N}Y_{N+2}(j\omega)}{\prod_{k\in\aleph_0\{j-1\}}(\rho_{j-1}-\rho_k)}+\cdots\right) \\
&= (-1)^{N-1}\rho_0\rho_1\cdots\rho_{N-1}\left(\begin{array}{c}(\rho_0+\rho_1+\cdots+\rho_{N-1})^{[0]}Y_{N+1}(j\omega)\\ +(\rho_0+\rho_1+\cdots+\rho_{N-1})^{[1]}Y_{N+2}(j\omega)\\ +(\rho_0+\rho_1+\cdots+\rho_{N-1})^{[2]}Y_{N+3}(j\omega)\cdots\end{array}\right)
\end{aligned}
\tag{4.41}
$$

In (4.41), Lemma 1 is used. Obviously, (4.41) gives Eq. (4.20). Similarly, from (4.17),

$$
\begin{aligned}
\Delta Y_N(j\omega) &= \sum_{j=1}^{N}\left(\frac{\rho_{j-1}^{N}\overline{\sigma}_{[N,\rho_{j-1}]}(\cdot)}{\prod_{k\in\aleph_0\backslash\{j-1\}}(\rho_{j-1}-\rho_k)}\right) \\
&= \sum_{j=1}^{N}\left(\frac{\rho_{j-1}^{N}Y_{N+1}(j\omega)+\rho_{j-1}^{N+1}Y_{N+2}(j\omega)+\cdots}{\prod_{k\in\aleph_0\backslash\{j-1\}}(\rho_{j-1}-\rho_k)}\right) \\
&= \left(\sum_{j=1}^{N}\frac{\rho_{j-1}^{N}Y_{N+1}(j\omega)}{\prod_{k\in\aleph_0\{j-1\}}(\rho_{j-1}-\rho_k)}\right. \\
&\quad \left.+\sum_{j=1}^{N}\frac{\rho_{j-1}^{N+1}Y_{N+2}(j\omega)}{\prod_{k\in\aleph_0\{j-1\}}(\rho_{j-1}-\rho_k)}+\cdots\right) \\
&= \left(\begin{array}{c}(\rho_0+\rho_1+\cdots+\rho_{N-1})^{[1]}Y_{N+1}(j\omega)\\ +(\rho_0+\rho_1+\cdots+\rho_{N-1})^{[2]}Y_{N+2}(j\omega)\\ +(\rho_0+\rho_1+\cdots+\rho_{N-1})^{[3]}Y_{N+3}(j\omega)\cdots\end{array}\right)
\end{aligned}
\tag{4.42}
$$

Note (4.42) is obtained by applying Lemma 1, and it gives Eq. (4.21). This completes the proof.

References

1. Boyd S. and Chua L.O., Fading memory and the problem of approximating nonlinear operators with Volterra series. IEEE Trans. Circuits & Systems,CAS32(11),1150–1161, 1985

2. George D.A., Continuous nonlinear systems, Technical Report 355, MIT Research Laboratory of Electronics, Cambridge, Mass. Jul. 24, 1959
3. Rugh W.J., Nonlinear System Theory: the Volterra/Wiener Approach, Baltimore, Maryland, U.S.A.: Johns Hopkins University Press, 1981
4. Doyle III F.J., Pearson R.K., Ogunnaike B.A., Identification of control using Volterra models, Springer-Verlag, London, 2002
5. Lee G.M., Estimation of nonlinear system parameters using higher-order frequency response functions, Mechanical Systems and Signal Processing, 11(2), 219-228, 1997
6. Sam Shanmugam K, Jong M.T., Identification of nonlinear systems in frequency domain, IEEE Trans. on Aerospace and Electronic Systems, 11(6), 1218–1225, 1975
7. Lang Z.Q., and Billings S. A. Output frequency characteristics of nonlinear systems. International Journal of Control, Vol. 64, 1049-1067, 1996
8. Jing X. J., Lang Z.Q. Billings S.A., Output Frequency Response Function based Analysis for Nonlinear Volterra Systems. Mechanical Systems & Signal Processing, 22,102-120, 2008a
9. Friston K.J., Mechelli A., Turner R., Price C.J., Nonlinear responses in fMRI: The Balloon model, Volterra kernels, and other hemodynamics, NeuroImage, 12, 466–477, 2000
10. Jing X. J., Lang Z. Q., Billings S. A. and Tomlinson G. R., The parametric characteristic of frequency response functions for nonlinear systems. International Journal of Control, Vol. 79, No. 12, December 2006, 1552–1564
11. Jing X.J., Lang Z.Q. and Billings S.A., Frequency Domain Analysis for Suppression of Output Vibration from Periodic Disturbance using Nonlinearities. Journal of sound and vibration 314, 536–557, 2008b
12. Jing X.J., Lang Z.Q., Billings S.A., Output Frequency Properties of Nonlinear Systems, International Journal of Non-Linear Mechanics, 45(7), September 2010, pp 681-690
13. Jing X.J., Lang Z.Q. and Billings S.A., Nonlinear influence in the frequency domain: Alternating series, Systems & Control Letters, 60,295–309, 2011a
14. Lang Z.Q., Billings S.A., Yue R., Li J., Output frequency response functions of nonlinear Volterra systems, Automatica, 43, 805–816, 2007
15. Vazquez Feijoo J.A., Worden K., Stanway R. Associated Linear Equations for Volterra operators, Mechanical Systems and Signal Processing, 19, 57–69, 2005
16. Worden K. and Tomlinson G.R.. Nonlinearity in Structural Dynamics, Detection, Identification and Modeling, Institute of Physics Publishing, Bristol, Philadelphia, 2001
17. Yang J, Tan S. X.-D., Nonlinear transient and distortion analysis via frequency domain Volterra series, Circuits Systems Signal Process. 25(3), 295-314, 2006
18. Jing X.J., Frequency Domain Analysis and Identification of Block-Oriented Nonlinear Systems, to appear in Journal of Sound and Vibration, 2011b
19. Silva S.D., Cogan S, Foltete E, Nonlinear identification in structural dynamics based on Wiener series and Kautz filters, Mechanical Systems and Signal Processing, 24:52-58, 2010
20. Leonowicz Z, Karvanen J, Tanaka T, Rezmer J, Model order selection criteria: comparative study and applications, International workshop on Computational Problems of Electrical Engineering, Zakopane, 2004
21. Bindlish R, Model order selection for process identification applied to an industrial ethylene furnace, Journal of Process Control, 13, 569-577, 2003
22. Billings S. A. and Tsang K. M., Spectral analysis of block structured non-linear systems, Mechanical Systems and Signal Processing, 4(2), Pages 117-130, March 1990
23. Eisinberg A., Fedele G., On the inversion of the Vandermonde matrix, Applied Mathematics and Computation. 174:1384–1397, 2006
24. Li Q, Liao M, Jing X, Transmissibility function-based fault diagnosis methods for beam-like engineering structures: a review of theory and properties, Nonlinear Dynamics, Volume 106, pages 2131–2163, 2021
25. Jing, XJ, Truncation order and its effect in a class of nonlinear systems, Automatica 48 (11), 2978-2985, 2012
26. Turner, L. R. Inverse of the Vandermonde matrix with applications. NASA *technical note* (TN D-3547), 1966

Chapter 5
The nCOS Function

Abstract The nonlinear Characteristic Output Spectrum (nCOS) function provides an explicit and analytically tractable representation of the relationship between a nonlinear system's output spectrum and its key parametric characteristics—namely, excitation frequency, nonlinear coefficients, and input amplitude. This formulation significantly advances the theoretical understanding and practical implementation of nonlinear system analysis and design in the frequency domain. Given simulation or experimental output data, the nCOS function can be accurately constructed to capture contributions from arbitrarily high-order nonlinearities. Compared to conventional simulation-based or purely theoretical approaches, this method requires substantially fewer simulation trials and incurs lower computational overhead, thereby enhancing efficiency without compromising accuracy. Moreover, the nCOS framework facilitates the precise decomposition of the nonlinear output spectrum into its linear and nonlinear components, enabling detailed insight into the system's dynamic behavior. These capabilities are particularly valuable for a wide range of engineering applications, including nonlinear system identification, signal processing, fault diagnosis, and performance optimization.

Keywords Frequency domain analysis · Nonlinear systems · Nonlinear characteristic output spectrum (nCOS)

5.1 Introduction

5.1.1 About This Chapter

This chapter is to introduce the important concept **n**onlinear **C**haracteristic **O**utput **S**pectrum (nCOS) function and its associated identification method from experimental excitation data based on the truncation error theory in Chap. 4. The key is to accurately define the structure of the n-th order nCOS function and to identify it from experimental data with specially designed excitation amplitudes such that the obtained limited order, e.g., the 2nd or 3rd order output spectrum of a given

X. Jing, *The Bio-Inspired X-Structure/Mechanism Approach for Exploring Nonlinear Benefits in Engineering*, https://doi.org/10.1007/978-981-96-5769-8_5

nonlinear system can be sufficiently accurate instead of a curve-fitting result as discussed before. These results, together with Chap. 4 and the next few chapters lay down a significant basis for another important topic—structural health monitoring or fault diagnosis based on vibration signals.

The nCOS function offers an analytical and clear representation of the relationship between the nonlinear output spectrum and the system's characteristic parameters of interest, which include frequency, nonlinear parameters, and excitation magnitude. This function can significantly enhance our understanding of nonlinear analysis and design in the frequency domain. When provided with simulation or experimental output data from a nonlinear system, the nCOS function can be accurately determined up to any high nonlinear orders. This method requires fewer simulation trials and less computational cost compared to a purely simulation-based study or traditional theoretical computations. Furthermore, this approach can be used to accurately identify the linear and nonlinear components in the nonlinear output frequency response (or output spectrum) of a nonlinear system. These findings are undoubtedly important for practical applications such as nonlinear analysis and design, nonlinear signal processing, system identification, and fault detection. To demonstrate these results, examples and case studies, including the analysis of a nonlinear vehicle suspension system, are provided.

5.1.2 Research Background

Nonlinear analysis plays a pivotal role in practical system analysis and design, ranging from engineering challenges to biological systems. A variety of methods are available for this purpose, including the perturbation method, averaging method, and harmonic balance method, among others [1–3]. Nonlinear analysis can also be performed in the frequency domain using the Volterra series theory [4, 5]. A significant class of nonlinear systems permits a Volterra series expansion [6, 7]. The generalized frequency response function (GFRF), which is a multi-variate Fourier transform of the Volterra kernels, is defined based on the Volterra series expansion [8]. This concept is useful for nonlinear analysis in the frequency domain, extending the transfer function concept from linear to nonlinear systems. For a nonlinear system characterized by nonlinear differential equations (NDE), nonlinear auto-regressive with exogenous input (NARX) models, or nonlinear block-oriented (NBO) models, the nth-order GFRF can be computed recursively [9–12]. This allows for the evaluation of the output spectrum, output frequency characteristics, and the boundedness of the GFRFs and output spectrum, among other things [13–16]. These findings form a fundamental basis for nonlinear frequency domain analysis, with applications in various fields such as signal processing, control, analysis and design of damping systems, crack detection, and understanding of biological data [4, 5, 11, 17–20].

Nonlinear analysis based on the GFRFs typically involves complex computational costs, particularly for orders >3. While traditional recursive algorithms for the GFRFs are simple to implement, they actually complicate the relationship between the GFRFs and model parameters. Recent studies have shown that both the GFRF and the output spectrum can be formulated as a polynomial in relation to the system model's nonlinear parameters [21, 22]. Notably, through parametric characteristic analysis and a mapping function, the GFRFs and output spectrum can be expressed as a straightforward polynomial function with respect to any nonlinear parameters of interest [21, 23]. Therefore, for a nonlinear system described by an NDE or NARX model, these results could offer significant insights or a powerful approach into the nonlinear influence on the system output frequency response [24, 25]. However, the quantitative analysis of nonlinear dynamics and its impact on system dynamic response still faces challenges due to computational complexity. Although a numerical method could be used for estimating the output frequency response function (OFRF) [18, 21, 22], it's often difficult to avoid biased or even incorrect estimates since the truncation order of the underlying Volterra series expansion for the nonlinear system under study is hard to predict in advance (it may also vary with different input magnitudes). This can reduce the effectiveness and reliability of the OFRF-based analysis.

This study introduces a systematic frequency domain method for the analysis, design, and estimation of nonlinear systems, building upon previous findings [18, 21, 22]. Firstly, this method enables the precise identification of linear and nonlinear components in the system output spectrum of a given nonlinear system, described by NDE, NARX, or NBO models, using simulation or experimental output data. These output spectrum components can subsequently be utilized for system identification or nonlinear analysis for various purposes, such as fault detection. Secondly, the OFRF in [18, 21, 22] is transformed into an enhanced polynomial function, referred to here as the nonlinear characteristic output spectrum (nCOS) function. This function provides a clear expression for the relationship between the nonlinear output spectrum and system characteristic parameters of interest, including nonlinear parameters, frequency variable, and input excitation magnitude, with a more universal parametric structure. With the accurate identification of system output spectrum components in the initial step, the nCOS function can be accurately determined up to any high orders, requiring fewer simulation trials and less computational cost compared to a purely simulation-based study or traditional theoretical computation [15, 23]. Therefore, these findings could offer a significant methodology for the qualitative and quantitative analysis and design of nonlinear dynamics in the frequency domain. Examples, including a nonlinear vehicle suspension system, are provided to illustrate these new findings.

5.2 Nonlinear Characteristic Output Spectrum (nCOS)

5.2.1 Nonlinear Output Spectrum (nOS)

The input–output relationship for a considerably large class of nonlinear systems can be approximated by a Volterra series of order N as [4, 6–8]

$$y(t) = \sum_{n=1}^{N} y_n(t) \tag{5.1a}$$

$$y_n(t) = \int_{-\infty}^{\infty} \cdots \int_{-\infty}^{\infty} h_n(\tau_1, \ldots, \tau_n) \prod_{i=1}^{n} u(t - \tau_i) \mathrm{d}\tau_i \tag{5.1b}$$

where $h_n(\tau_1, \cdots, \tau_n)$ is the nth order Volterra kernel. The nth order GFRF is defined as

$$H_n(j\omega_1, \ldots, j\omega_n) = \int_{-\infty}^{\infty} \cdots \int_{-\infty}^{\infty} h_n(\tau_1, \ldots, \tau_n) \exp(-j(\omega_1\tau_1 + \cdots + \omega_n\tau_n))\mathrm{d}\tau_1 \cdots \mathrm{d}\tau_n \tag{5.2}$$

With the inverse Fourier transform of the nth order GFRF, (5.1b) can be written as

$$y_n(t) = \frac{1}{(2\pi)^n} \int_{-\infty}^{\infty} \cdots \int_{-\infty}^{\infty} H_n(j\omega_1, \ldots, j\omega_n) \prod_{i=1}^{n} U(j\omega_i) e^{j(\omega_1 + \cdots + \omega_n)t} \mathrm{d}\omega_1 \cdots \mathrm{d}\omega_n$$

where $\omega_n = \omega - \omega_1 - \cdots - \omega_{n-1}$, which can be further written as

$$y_n(t) = \frac{1}{2\pi} \int_{-\infty}^{\infty} \left[\frac{1}{(2\pi)^{n-1}} \underbrace{\int_{-\infty}^{\infty} \cdots \int_{-\infty}^{\infty} \Xi \mathrm{d}\omega_1 \cdots \mathrm{d}\omega_{n-1}}_{n-1} \right] e^{j\omega t} \mathrm{d}\omega$$

where $\Xi = H_n(j\omega_1, \ldots, j\omega_{n-1}, j(\omega - \omega_1 - \cdots - \omega_{n-1}))U_n(j\omega, j\omega_1, \ldots j\omega_{n-1})U_n(j\omega, j\omega_1, \ldots, j\omega_{n-1}) = U(j\omega_1) \cdots U(j\omega_{n-1})U(j(\omega - \omega_1 - \cdots - \omega_{n-1}))$. Therefore the Fourier transform of $y_n(t)$ is obtained as

$$Y_n(j\omega) = \frac{1}{(2\pi)^{n-1}} \underbrace{\int_{-\infty}^{\infty} \cdots \int_{-\infty}^{\infty}}_{n-1} \Xi \mathrm{d}\omega_1 \cdots \mathrm{d}\omega_{n-1} \tag{5.3a}$$

which is referred to as the nth-order output spectrum. Equation (5.3a) is defined by a $(n-1)$-time integration of the multiplication of the nth-order GFRF and an input function $U_n(j\omega, j\omega_1, \ldots, j\omega_{n-1})$.. The output spectrum of system (5.1) can then be computed by

$$Y(j\omega) = \sum_{n=1}^{N} Y_n(j\omega) \tag{5.3b}$$

Nonlinear output spectrum involves the computation of GFRFs. Given a nonlinear system described by NDE, NARX and NBO models, the GFRFs can be derived with the probing method [4] and recursive algorithms can be referred to [9–12].

5.2.2 Nonlinear Characteristic Output Spectrum (nCOS)

In practice, nonlinear systems can typically be identified or modeled into a parametric model, such as NDE, NARX, or Wiener models [5, 11, 26–28]. The nonlinear output spectrum of these systems is a complex-valued function of frequency variables, but it also depends on model parameters and the magnitude of the input of interest (collectively referred to as characteristic parameters in this study). Establishing a clear relationship between the system output spectrum and these characteristic parameters is crucial for system analysis and design. Consider NDE systems as,

$$\sum_{m=1}^{M}\sum_{p=0}^{m}\sum_{k_1,k_{p+q}=0}^{K} c_{p,q}(k_1,\ldots,k_{p+q})\prod_{i=1}^{p}\frac{d^{k_i}y(t)}{dt^{k_i}}\prod_{i=p+1}^{p+q}\frac{d^{k_i}u(t)}{dt^{k_i}}=0 \tag{5.4}$$

where $\left.\frac{d^k x(t)}{dt^k}\right|_{k=0} = x(t)$, $p+q=m$, $\sum\limits_{k_1,k_{p+q}=0}^{K}(\cdot)=\sum\limits_{k_1=0}^{K}(\cdot)\cdots\sum\limits_{k_{p+q}=0}^{K}(\cdot)$, M is the maximum degree of nonlinearity in terms of $y(t)$ and $u(t)$, and K is the maximum order of the derivative. In this model, the parameters such as $c_{0,1}(.)$ and $c_{1,0}(.)$ are linear parameters corresponding to the linear terms in the model, i.e., $\frac{d^k y(t)}{dt^k}$ and $\frac{d^k u(t)}{dt^k}$ for k = 0, 1, …, L; and $c_{p,q}(\cdot)$ for $p+q>1$ are referred to as nonlinear parameters corresponding to nonlinear terms in the model of the form $\prod\limits_{i=1}^{p}\frac{d^{k_i}y(t)}{dt^{k_i}}\prod\limits_{i=p+1}^{p+q}\frac{d^{k_i}u(t)}{dt^{k_i}}$, e.g., $y(t)^p u(t)^q$. $p+q$ is referred to as the nonlinear degree of parameter $c_{p,q}(\cdot)$. It is shown in [23] that nonlinear output spectrum of NDE models can be written into an explicit polynomial function of system characteristic parameters as

$$Y(j\omega)=\sum_{n=1}^{N}\chi_n\cdot\phi_n(j\omega)^T \tag{5.5}$$

where χ_n denotes the nth-order characteristic parameter vector composed of nonlinear parameters and $\phi_n(j\omega)$ its associated complex-valued function of the nth-order output spectrum, both of which can be (analytically) determined with the method in [21, 23]. Any nonlinear parameters of interest in analysis and design will be included in χ_n, given by

$$\chi_n = C_{n,0} \oplus \bigoplus_{\substack{k=0 \\ \sum_{i=0}^{k}(p_i+q_i)=n+k \\ 1\le p_0 \le n-k,\ 2\le p_i+q_i\le n-k}}^{n-2} \left(\bigotimes_{i=0}^{k} C_{p_i,q_i} \right) \tag{5.6}$$

where $C_{p,q} = [c_{p,q}(0,\ldots,0), c_{p,q}(0,\ldots,1),\ldots,c_{p,q}(\underbrace{K,\ldots,K}_{p+q=m})]$, $\oplus$ and $\otimes$ are two operators defined for symbolic manipulation (see the details in [21, 23]). For example, for two vectors C_1 and C_2 consisting of some symbolic variables, $C_1 \oplus C_2$ is a vector including all the elements in C_1 and C_2 without repetition, and $C_1 \otimes C_2$ is a vector including all the elements produced by the Kronecker product without repetition. If some model parameters will not be considered in the analysis and design, they can be set to 1 by default in (5.6). If there is an element 1 in χ_n, it means that there will be a pure frequency-dependent term in the polynomial (5.5). Similar results hold for the NARX model [23].

Roughly speaking, the GFRFs are polynomial functions of all those model parameters of nonlinear order larger than 1 [even though in recursive relationship in (5.5)]. The separable property is still kept after manipulation by inserting the GFRFs in (5.3b) (see Chap. 3]). Therefore, the nCOS function of a nonlinear system can always be expressed as (5.5), where $\boldsymbol{\chi}_\mathbf{n}$ denotes the nth-order characteristic parameter vector, whose component is composed by a monomial function of model parameters (e.g., $c_1^3c_2c_3^2$) and $\phi_\mathbf{n}(j\omega)$ is the associated complex-valued function. For example, assuming that there are 3 parameters denoted as c_1, c_2, c_3 of the same nonlinearity degree, the output spectrum can then be given as

$$\begin{aligned} Y(j\omega) &= \varphi_1^1(j\omega)c_1 + \varphi_2^1(j\omega)c_2 + \varphi_3^1(j\omega)c_3 + \varphi_{11}^2(j\omega)(c_1)^2 \\ &\quad + \varphi_{12}^2(j\omega)c_1c_2 + \varphi_{13}^2(j\omega)c_1c_3 + \varphi_{22}^2(j\omega)(c_2)^2 + \cdots \end{aligned}$$

where $\varphi_{(\cdot)}^{(\cdot)}(j\omega)$ is the coefficient of the polynomial function, which is independent of the parameters of interest, i.e., c_1, c_2, c_3.

Obviously, Eq. (5.5) is an analytical and straightforward function of system characteristic parameters, which could considerably facilitate the analysis and design of nonlinear systems in the frequency domain. To emphasize the parametric relationship between nOS and the characteristic parameters (including nonlinear parameters, excitation magnitude and frequency), Eq. (5.5) is referred to here as nonlinear characteristic output spectrum (nCOS) function and the nth-order component as the nth-order nCOS. In order to conduct a nonlinear analysis based on the nCOS function in (5.5), both χ_n and $\phi_n(j\omega)$ must be determined up to a sufficiently high order. Since nonlinear systems can always be identified into a NARX, NDE or NBO model with experiment data in practice [5, 11, 26, 27], this study suppose that a nonlinear model of the system of interest is already known. Therefore χ_n is known from (5.6) with the nonlinear model, and thus only $\phi_n(j\omega)$ is yet to be determined. Although $\phi_n(j\omega)$ can be computed analytically with the method in [23], the computation cost is usually high and it is even worse for high orders (>5). Therefore, the objective is to develop an effective method so that $\phi_n(j\omega)$ can be determined accurately and directly with only some simulation data.

From (5.5), a simple least square algorithm could be applied in order to compute $\phi_n(j\omega)$ under different parameter excitations as discussed in [18, 21, 22]. However, the difficulty is that the maximum truncation order N is not known and it is also varying with different input magnitudes. A larger input magnitude or a larger range of a specific model parameter of interest would result in a larger truncation order N (for an accurate Volterra series expansion). An inappropriate guessed truncation order N will lead to large error in the computation of $\phi_n(j\omega)$. A sufficiently larger N could be attempted to avoid this problem (if realistic), but will easily result in singularity of the matrix inverse in the LS algorithm. If $\phi_n(j\omega)$ is not computed correctly, the simple least square algorithm can only guarantee accurate fitting at the training points, and the generalization of (5.5) could be very worse.

To solve the problems mentioned above, new algorithms will be developed so that $\phi_n(j\omega)$ can be determined accurately with only some simulation data and without knowing the best truncation order N^*.

5.3 Accurate Determination of the nCOS Function

For determination of the nCOS function (5.5), the first step of the proposed method is to compute the nth-order output spectrum (for any n) based on numerical or experimental data, and then to determine the nth-order nCOS function.

A. **Computation of the Nth-Order Output Spectrum**

The nOS in (5.3b) can be rewritten by considering the truncation error $\sigma_{[N]}(j\omega)$ and input function $\rho R(j\omega)$ as

$$\begin{aligned} Y(j\omega)_\rho &= \sum_{n=1}^{\infty} \rho^n Y_n(j\omega) \\ &= \rho Y_1(j\omega) + \rho^2 Y_2(j\omega) + \rho^3 Y_3(j\omega) + \cdots \\ &= \sum_{n=1}^{N} \rho^n Y_n(j\omega) + \sigma_{[N,\rho]}(j\omega) \end{aligned} \tag{5.7}$$

where $\sigma_{[N,\rho]}(j\omega)$ represents the truncation error, including all the remaining higher order output spectrum components in Volterra series expansion; ρ is a constant which is used to represent different magnitude of the input.

To determine $Y_n(j\omega)$ for $n \in \aleph = \{1, 2, 3, \ldots N\}$, a multi-level excitation method can be adopted [18, 22, 28, 29]. The system can be excited by the same input $R(j\omega)$ of different magnitudes $\rho_0, \rho_1, \rho_2, \ldots, \rho_{N-1}$, and there will be a series of output obtained accordingly, which are denoted by $Y(j\omega)_{\rho 0}$, $Y(j\omega)_{\rho 1}$, $Y(j\omega)_{\rho 2}$, ..., $Y(j\omega)_{\rho N-1}$. By a LS method, it gives

$$\begin{bmatrix} \hat{Y}_1(j\omega) \\ \hat{Y}_2(j\omega) \\ \vdots \\ \hat{Y}_N(j\omega) \end{bmatrix} = \begin{bmatrix} \rho_0 & \rho_0^2 & \cdots & \rho_0^N \\ \rho_1 & \rho_1^2 & \cdots & \rho_1^N \\ \vdots & \vdots & \ddots & \vdots \\ \rho_{N-1} & \rho_{N-1}^2 & \cdots & \rho_{N-1}^N \end{bmatrix}^{-1} \begin{bmatrix} Y(j\omega)_{\rho_0} \\ Y(j\omega)_{\rho_1} \\ \vdots \\ Y(j\omega)_{\rho_{N-1}} \end{bmatrix} \tag{5.8}$$

The square matrix above is nonsingular if $\rho_0 \neq \rho_1 \neq \ldots \neq \rho_{N-1}$. Note that the computation of $Y_n(j\omega)$ for $n \in \aleph = \{1, 2, 3, \ldots N\}$ is basically accurate if the truncation error $\sigma_{[N,\rho]}(j\omega)$ is trivial. Otherwise, the computation for $Y_n(j\omega)$ could be significantly biased, and the LS in (5.8) only results in a good fitting at the training points. To overcome this problem, an alternative method can be employed by choosing the excitation magnitudes $\rho_0, \rho_1, \ldots, \rho_{N-1}$ so that the accurate computation through (5.8) can be achieved for any given N. The following results are derived for this purpose.

Proposition 1.

$$\Delta Y_1(j\omega) = (-1)^{N-1}\rho_0\rho_1\cdots\rho_{N-1} \cdot \sum_{k=0}^{\infty}(\rho_0 + \rho_1 + \cdots + \rho_{N-1})^{[k]} Y_{N+k+1}(j\omega) \tag{5.9a}$$

$$\Delta Y_N(j\omega) = \sum_{k=1}^{\infty}(\rho_0 + \rho_1 + \cdots + \rho_{N-1})^{[k]} Y_{N+k}(j\omega) \tag{5.9b}$$

where for $N(>1)$ *nonzero distinct real numbers and for any non-negative integer* r,

$$(\mathrm{x}_1 + \mathrm{x}_2 + \cdots + \mathrm{x}_\mathrm{N})^{[\mathrm{r}]} = \sum_{\substack{l_1 + l_2 + \cdots + l_N = r \\ l_1, l_2, \ldots, l_N \in \{0, 1, 2, \ldots, \mathrm{r}\}}} \mathrm{x}_1^{l_1}\mathrm{x}_2^{l_2}\cdots \mathrm{x}_\mathrm{N}^{l_\mathrm{N}} \tag{5.9c}$$

Proof. See Appendix 1. □

Proposition 1 provides a straightforward insight into the computation error incurred by excitation magnitudes and truncation order. Given a truncation order N, different values of the excitation magnitudes could bring very different computation error. Proposition 1 demonstrates an effective method for designing $\rho_0\rho_1\cdots\rho_{N-1}$ so that the computation errors (i.e., $\Delta Y_1(j\omega)$, $\Delta Y_N(j\omega)$) can be mitigated for any N.

In (5.9ab), the computation of $(\rho_0 + \rho_1 + \cdots + \rho_{N-1})^{[k]}$ is involved. Lemma 1 gives a recursive method for this.

Lemma 1. .

$$(\rho_0 + \rho_1 + \cdots + \rho_{N-1})^{[k]} = \sum_{r=0}^{N-1}\rho_r(\rho_r + \cdots + \rho_{N-1})^{[k-1]}$$

Proof. By a mathematical induction, it is easy to have this conclusion. □

Based on Proposition 1, different excitation methods can be used to minimize the computation error. The following results can be obtained by applying Proposition 1.

Corollary 1. *Case 1:* $N = 3$, $\rho_1 = -\rho$, $\rho_2 = -\rho/\alpha$, $\rho_0 = -\rho_1 - \rho_2$, $0 < \rho < 1$, $\alpha > 1$.

$$\Delta Y_1(j\omega) = \frac{\alpha+1}{\alpha^2}\rho^3 \cdot \left\{ Y_4(j\omega) + \frac{\alpha^2+\alpha+1}{\alpha^2}\rho^2 Y_6(j\omega) + \frac{\alpha+1}{\alpha^2}\rho^3 Y_7(j\omega) + \cdots \right\} \quad (5.10a)$$

$$\Delta Y_3(j\omega) = \frac{\alpha^2+\alpha+1}{\alpha^2}\rho^2 Y_5(j\omega) + \frac{\alpha+1}{\alpha^2}\rho^3 Y_6(j\omega) + \frac{\alpha^4+2\alpha^3+3\alpha^2+2\alpha+1}{\alpha^4}\rho^4 Y_7(j\omega)\cdots \quad (5.10b)$$

Case 2: $N = 5$, $\rho_0 = \rho$, $\rho_1 = -\rho$, $\rho_2 = -\rho/\beta$, $\rho_3 = -\rho_2/\alpha$, $\rho_4 = -\rho_2 - \rho_3$, $0 < \rho < 1$, $\beta > 1$, $\alpha > 1$.

$$\Delta Y_1(j\omega) = -\frac{1}{\alpha\beta}\left(\frac{1}{\alpha}+\frac{1}{\beta}\right)\rho^5\left\{Y_6(j\omega) + \left(1+\frac{1}{\alpha^2}+\frac{1}{\beta^2}+\frac{1}{\alpha\beta}\right)\rho^2 Y_8(j\omega) + \frac{1}{\alpha\beta}\left(\frac{1}{\alpha}+\frac{1}{\beta}\right)\rho^3 Y_9(j\omega) + \cdots\right\} \quad (5.10c)$$

$$\Delta Y_5(j\omega) = \left(1+\frac{1}{\alpha^2}+\frac{1}{\beta^2}+\frac{1}{\alpha\beta}\right)\rho^2 Y_7(j\omega) + \frac{1}{\alpha\beta}\left(\frac{1}{\alpha}+\frac{1}{\beta}\right)\rho^3 Y_8(j\omega) + \cdots \quad (5.10d)$$

Proof. See Appendix 2. □

Corollary 1 indicates that properly choosing the excitation magnitudes as specified would produce accurate computation of $Y_1(j\omega)$ and $Y_3(j\omega)$ or $Y_5(j\omega)$, respectively, although the truncation order is chosen as N = 3 or 5. The result in Proposition 1 actually provides many choices for this purpose. For example, in Case 2 of Corollary 1, if $\rho = 0.001$, $\beta = 2$, and $\alpha = 10$, then (5.10cd) can be written as

$$\Delta Y_1(j\omega) = -0.03 \cdot 0.001^5\{Y_6(j\omega) + 1.31 \cdot 0.001^2 Y_8(j\omega) + 0.03 \cdot 0.001^3 Y_9(j\omega) + \cdots\}$$

$$\Delta Y_5(j\omega) = 1.31 \cdot 0.001^2 Y_7(j\omega) + 0.03 \cdot 0.001^3 Y_8(j\omega) + \cdots$$

From the equations above it can be seen that higher order output spectra (>5) would have very limited effect on the determination of lower order output spectra. The computation error in $Y_5(j\omega)$ incurred by $Y_7(j\omega)$ would be $1.31 \cdot 0.001^2 Y_7(j\omega)$. Only when the magnitude of $Y_7(j\omega)$ is larger than 10^6, the influence could be significant. Similar conclusions hold for the other higher orders. Moreover, it can be checked by a mathematical induction that the other orders of output spectra between 1 and 5 can also be computed accurately if $Y_5(j\omega)$ can be computed sufficiently accurately. For different order N, similar results can be obtained.

Corollary 1 leads to the following estimation algorithm for the nth-order output spectrum (referred to as nth-order Output Spectrum Estimation (nth-OSE) algorithm).

Step 1. Choose N = 3 or 5 etc., and properly set ρ, α and β, satisfying $0 < \rho < 1$, $\beta > 1$, $\alpha > 1$, e.g., $\rho = 0.001$, $\alpha = 10$ and $\beta = 2$.

Step 2. Use Eq. (5.8) to find the estimates for $Y_n(j\omega)$, i.e., $\hat{Y}_n(j\omega)$, for n = 1, 2, …, N. Note that the computation error for n = 1 or N is given by (5.10ab) or (5.10cd).

Step 3. Note that Eq. (5.7) can be written as $Y(j\omega)_\rho / \rho - Y_1(j\omega) = \rho Y_2(j\omega) + \rho^2 Y_3(j\omega) + \cdots$ Therefore, replacing $Y(j\omega)_{\rho_i}$ by $Y(j\omega)_{\rho_i} / \rho_i - \hat{Y}_1(j\omega)$ in Eq. (5.8) and re-applying Eq. (5.8) with a larger ρ (usually 5–10 times than the previous one) lead to the estimation of $Y_n(j\omega)$, i.e., $\hat{Y}_n(j\omega)$ for n = 2, …, N + 1. The estimation error is still given by (5.10a–d) accordingly by replacing $Y_i(j\omega)$ with $Y_{i+1}(j\omega)$.

Step 4. Similarly, Eq. (5.7) can be written as $\left(Y(j\omega)_\rho - \rho Y_1(j\omega) - \rho^2 Y_2(j\omega)\right) / \rho^2 = \rho Y_3(j\omega) + \rho^2 Y_4(j\omega) + \cdots$. Therefore, replacing $Y(j\omega)_{\rho_i}$ by $\left(Y(j\omega)_\rho - \rho Y_1(j\omega) - \rho^2 Y_2(j\omega)\right) / \rho^2$ in Eq. (5.8) and re-applying Eq. (5.8) with a larger ρ (usually 5–10 times than the previous one) leads to the estimation of $Y_n(j\omega)$, i.e., $\hat{Y}_n(j\omega)$ for $n = 3, \ldots, N + 2$.

Step 5. Follow a similar process as Step 3 and Step 4 until a sufficiently high order N^*. The truncation order N* is not known but can be determined by evaluating the magnitude of $\hat{Y}_{N^*}(j\omega)$ according to a predefined threshold (e.g., $\left|\hat{Y}_{N^*}(j\omega)\right| < \varepsilon$).

With the nth-OSE algorithm, the nth-order output spectrum can be determined accurately (by properly choosing ρ (discussed later)) into a nonparametric form and an appropriate truncation order N^* can also be obtained. This estimation process does not necessarily need a system model but input–output data only from simulations or experiments.

B. **Determination of the Nth-Order nCOS Function**

From (5.5), the nth-order nCOS function can be written as

$$Y_n(j\omega) = \chi_n \cdot \phi_n(j\omega)^T \tag{5.11}$$

where χ_n is the nth-order characteristic parameter vector given by (5.6). If the system model under study is not known, several methods [5, 11, 26, 27] can be employed to identify a NDE, NARX or NBO model from experimental data. Here supposes that the NDE model of the system is known and similar results hold for other models. With the NDE model, χ_n is known explicitly for any characteristic parameters [21].

At any frequency ω, an identification method can be used to determine $\phi_n(j\omega)$. Let τ denote the dimension of χ_n, and suppose there are m characteristic parameters in χ_n, i.e., $c_1, c_2, \ldots, c_m$. Denote $\overline{c} = (c_1, c_2, \ldots, c_m)$. Then,

Step 1: Compute the nth-order characteristic vector χ_n using (5.6) with respect to $\overline{c}$. If $\chi_n = 1$ implying that this order output spectrum has no relationship with $\overline{c}$, then let $n \to n + 1$ and repeat Step 1; otherwise, go to Step 2–5;

Step 2: Randomly generate τ distinct points $\overline{c}$ =(c_1, c_2, ..., c_m) so that $\left[[1\chi_{n|\overline{c}_1}]; [1\chi_{n|\overline{c}_2}]; \ldots; [1\chi_{n|\overline{c}_m}]\right]$ is nonsingular.

Step 3: At each point $\overline{c}$ =(c_1, c_2, ..., c_m), the system can be simulated subject to a specific input and applying the nth-OSE algorithm leads to determination of the nth-order output spectrum denoted by $\hat{Y}_n(j\omega)_{|\overline{c}}$, i.e., $\hat{Y}_n(j\omega)_{|\overline{c}} = \hat{Y}_n(j\omega)_{|\overline{c}=0} + \chi_{n\overline{c}} \cdot \phi_n(j\omega)^T$.

Step 4: It can be obtained that

$$\left[\hat{Y}_n(j\omega)_{|\overline{c}=0}, \hat{\varphi}_n(j\omega)\right]^T = \left[[1\chi_{n|\overline{c}_1}]; [1\chi_{n|\overline{c}_2}]; \ldots; [1\chi_{n|\overline{c}_m}]\right]^{-1} \cdot \left[\hat{Y}_n(j\omega)_{|\overline{c}_1}, \hat{Y}_n(j\omega)_{|\overline{c}_2}, \ldots, \hat{Y}_n(j\omega)_{|\overline{c}_m}\right]^T \tag{5.12}$$

Step 5: Thus, the nth-order nCOS function (5.11) is achieved, and finally the nCOS function can be obtained as

$$\hat{Y}(j\omega) = \rho\hat{Y}_1(j\omega) + \sum_{n=2,3,\ldots}^{\chi_{n|\overline{c}}=1} \rho^n \hat{Y}_n(j\omega) + \sum_{n=2,3,\ldots}^{\chi_{n|\overline{c}}\neq 1} \rho^n \left(\hat{Y}_n(j\omega)_{|\overline{c}=0} + \chi_{n|\overline{c}}\hat{\varphi}_n(j\omega)^T\right)$$

The method above is referred to as nth-order nCOS estimation (nth-COSE) algorithm. Because the elements in χ_n is symbolically independent, there must exist τ distinct points $\overline{c}$ =($c_1, c_2, \ldots, c_m$) so that the matrix $\left[\chi_{n|\overline{c}_1}; \chi_{n|\overline{c}_2}; \ldots; \chi_{n|\overline{c}_m}\right]$ is nonsingular [30]. If the estimation of $\hat{Y}_n(j\omega)_{|\overline{c}_i}$ is unbiased, the determination of $\hat{\varphi}_n(j\omega)$ in (5.12) will be unbiased. Compared with the result in [18, 22], (a) the nCOS function here is determined as not only a polynomial function of model parameters but also an explicit function of the input magnitude ρ; (b) The best truncation order N* is not necessarily known in advance (this must be given in the previous result); (c) With the help of the nth-OSE algorithm, the complex-valued polynomial coefficients can be determined accurately, while the previous algorithm may result in biased (if not wrong) estimation; (d) The parameter-independent terms such as $\hat{Y}_n(j\omega)_{|\overline{c}=0}$ are explicitly considered in the nCOS function, which are incurred by those which have no relationship with model parameter $\overline{c}$; Thus the characteristic parametric structure of the nCOS function is more generic.

5.4 Example and Case Studies

The findings of this study have various applications. The nth-OSE algorithm can be employed for the estimation of nonlinear polynomials similar to Eq. (5.7). The precise identification of each order of nonlinear output spectrum components can be utilized for nonlinear detection in non-destructive evaluation [11, 17, 20]. The methodology can also be applied to accurately estimate the linear portion of a nonlinear system, even when the nonlinearity is complex. Crucially, the precise determination of the nCOS function can be leveraged for nonlinear analysis and design. These applications will be exemplified using a nonlinear vehicle suspension system.

Example 1 (Identification of a Polynomial Function)

Given a polynomial function

$$\begin{aligned} f(x) = {} & 10x - 20x^2 + 300x^3 \\ & - 40x^4 + 500x^5 - 600x^6 \\ & + 700x^7 - 800x^8 + 900x^9 \end{aligned}$$

For convenience in understanding, the equation above can be regarded as a polynomial in (5.7) with $\rho = x$, $Y(j\omega)_\rho = f(x)$ and $Y_i(j\omega) = 10, -20, \ldots$ for i = 1, 2, ... The coefficients can be estimated accurately with any guessed truncation order N (the real one is N = 9) using the nth-OSE algorithm. For example, by Corollary 1, taking N = 5, $\rho = 0.001$, $\alpha = 10$ and $\beta = 2$, the coefficients of the first 5 orders are given in Fig. 1a by applying Steps 1–2 of the nth-OSE algorithm. By applying Step 3 with $\rho = 0.01$, the estimation for the coefficients from the 2nd to the 6th orders is shown in Fig. 1b. Both clearly show that the estimated coefficients accurately match the real values.

One issue with the nth-OSE algorithm is how to choose the excitation magnitude ρ for both α and β are given heuristically satisfying $\alpha > 1$ and $\beta > 1$. From Table 5.1, it can be seen that given $\alpha = 10$ and $\beta = 2$, different values for ρ result in a little difference in the estimation of lower order coefficients (especially for order 1), but could lead to very different estimates for higher order coefficients such as the coefficient of order 5. Table 5.1 shows that there is a large range for ρ to choose (e.g., $0 < \rho < 0.1$) to have an accurate estimation for each coefficient especially for lower order coefficients. It also clearly indicates that the estimation could be greatly biased without a proper value for ρ (e.g., $\rho = 0.5$).

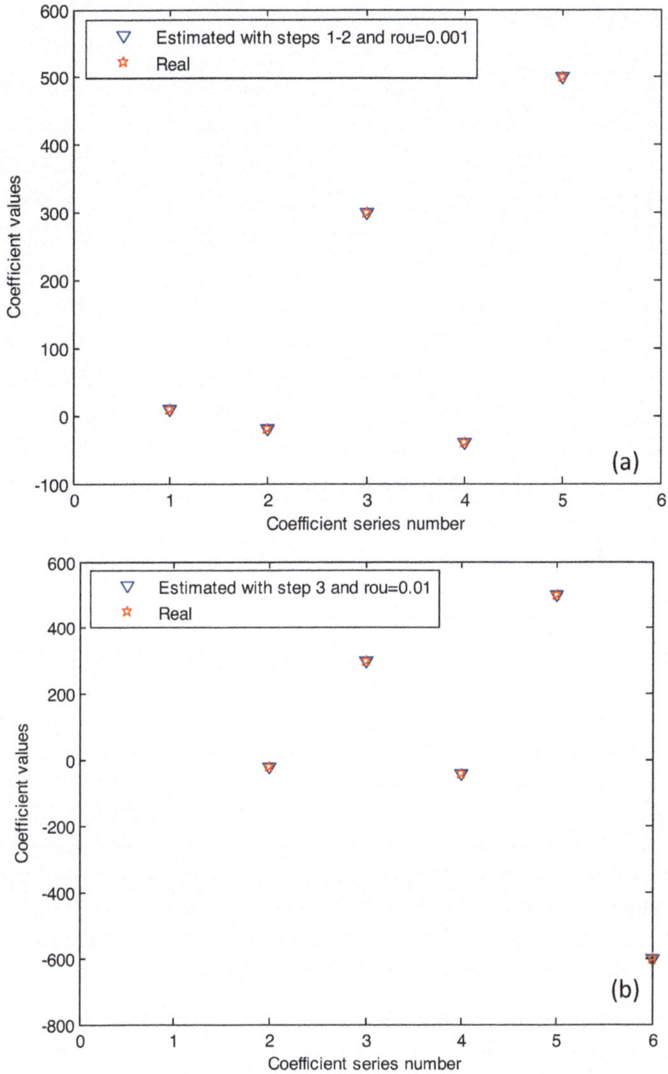

Fig. 5.1 Estimation with the nth-OSE algorithm with **a** steps 1–2, **b** step 3

To find the best excitation magnitude ρ, define the following excitation sensitivity function

$$S(\rho) = \left|\frac{\partial Y_\rho}{\partial \rho}\right| = \left|\frac{Y_\rho - Y_{\rho+\Delta}}{\Delta}\right|$$

where Y_ρ denotes the variable to be estimated under the excitation magnitude ρ, and Δ is a small positive number. By the definition, for each order of the coefficients there

Table 5.1 Estimation with different excitation magnitude

ρ	Real				
	10	−20	30	−40	500
0.00001	*10.000*	*−20.000*	*300.0000*	*−36.0000*	−1,048,576.0000
0.00005	*10.000*	*−20.000*	*300.0000*	*−40.0156*	512.0000
0.0001	*10.000*	*−20.000*	*300.0000*	*−40.0156*	512.0000
0.0005	*10.000*	*−20.000*	*300.0000*	*−40.0002*	*500.0625*
0.001	*10.000*	*−20.000*	*300.0000*	*−40.0008*	*500.0000*
0.005	*10.000*	*−20.000*	*300.0000*	*−40.0192*	*500.0224*
0.01	*10.000*	*−20.000*	*300.0000*	*−40.0767*	*500.0894*
0.05	*10.000*	*−19.999*	*299.9977*	−41.9218	502.2419
0.1	*10.000*	*−19.983*	*299.9720*	−47.7635	509.0534
0.5	10.3669	−5.6135	281.7501	−297.5458	798.3785
The effective range for ρ	$0 < \rho < 0.1$	$0 < \rho < 0.1$	$0 < \rho < 0.1$	$0.00005 < \rho < 0.01$	$0.0005 < \rho < 0.01$

will be a V-shape or U-shape curve given by the sensitivity function (Fig. 5.2), and the best excitation magnitude should locate around the bottom of the curve which corresponds to the slightest estimation error for the corresponding order coefficient. In Fig. 5.2a, the effective excitation range for the orders 4 and 5 in Table 5.1 are highlighted, which are clearly around the bottom or turning corner of the V-shape sensitivity function; for the first 3 orders the effective excitation ranges cover the whole testing excitation range, having very small sensitivity values (below 1). Similar cases hold for different other values of α and β, for example $\alpha = 5, \beta = 5$ in Fig. 2b, indicating that ρ is actually the sensitive parameter to tune in the nth-OSE algorithm.

From the discussions above, considering Table 5.1 and Fig. 5.2 and based on some other testing results, it can be concluded for the nth-OSE algorithm that

(a) The parameters α and β can be chosen heuristically satisfying $\alpha > 1$ and $\beta > 1$. The excitation magnitude ρ is the key factor to be tuned properly.
(b) Testing can be done to generate plots of the excitation sensitivity function, in which the best excitation magnitude locates around the bottom or turning corner of the V-shape curve, corresponding to the estimation with smallest error (e.g., the orders 4–5 in Fig. 5.2).
(c) If the V-shape curve is very flat or very low in values at all testing range, this indicates that all the testing magnitudes are effective for the estimation of the corresponding order (e.g., the orders 1–3 in Fig. 5.2). Moreover, a flat curve around the bottom of the V-shape curve implies the estimation with the corresponding excitation magnitudes is consistently accurate.
(d) The best ρ^* obtained in this method implies that the obtained lower order polynomial estimated by the nth-OSE algorithm will be applicable to the excitation

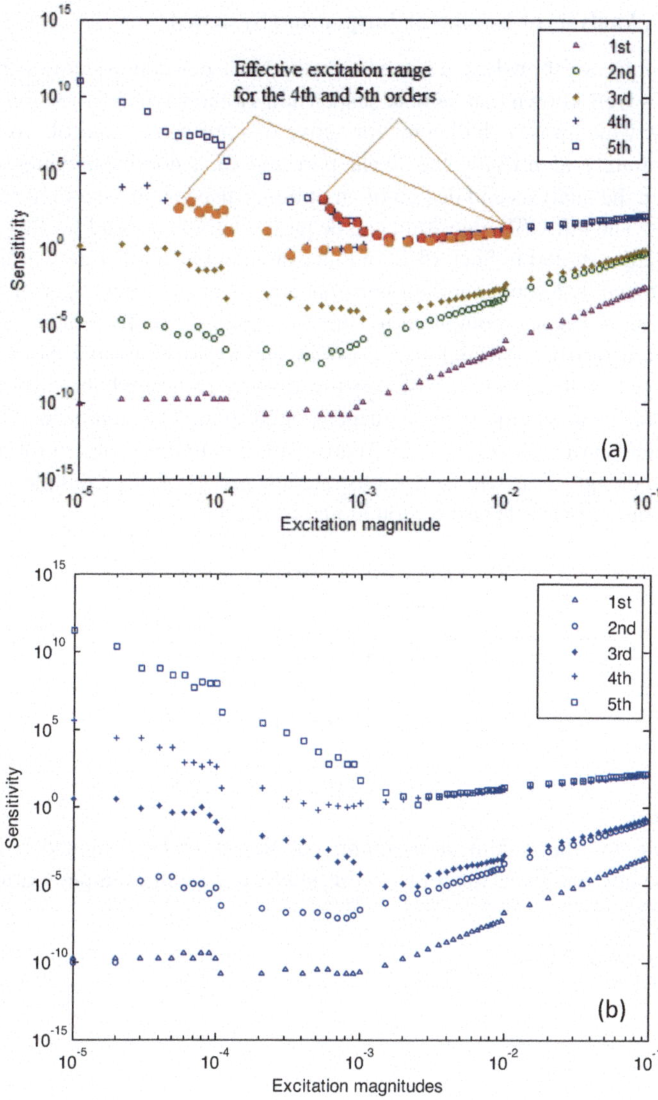

Fig. 5.2 Sensitivity to excitation in estimation of different order coefficients, **a** $\alpha = 10$, $\beta = 2$, **b** $\alpha = 5$, $\beta = 5$

input $\rho R(j\omega)$ of magnitude $\rho \in [0, \overline{\rho}]$, where $\overline{\rho} = \rho^*$ when N = 5 and $\overline{\rho} = \frac{1+\alpha}{\alpha}\rho^*$ when N = 3 by Corollary 1.

This could be used as a useful guidance in selection of the excitation magnitude ρ in the nth-OSE algorithm (referred to as the ρ-selection method). The value of ρ in step 1 and 3 of the nth-OSE algorithm can both be obtained with this method.

Example 2 (Analysis of Nonlinear Suspension Systems)

In this example, we'll analyze a nonlinear vehicle suspension system (as shown in Fig. 5.3). It's well-known that vehicle suspension systems typically exhibit inherent nonlinearity, which poses challenges for active or semi-active control. To a certain degree, accurately identifying the linear part and each nonlinear component can greatly aid in the analysis and design of an optimal suspension system in real-world applications. The nth-OSE algorithm can be particularly beneficial for this purpose, as will be demonstrated in Sect. (5.1) of this example. On another note, it's increasingly recognized that nonlinear damping characteristics can offer superior vibration suppression performance compared to linear damping [25]. To achieve adjustable damping characteristics, MR dampers are often employed as an ideal shock absorber [31, 32]. However, the question arises on how to design a desired nonlinear damping characteristic for a specific vehicle suspension system. The nonlinear COS function mentioned earlier can offer an alternative solution to this problem for nonlinear analysis and design. This will be demonstrated in case II of this example.

The one quarter vehicle suspension model is given as

$$m_s\ddot{x}_s = -f_s - b(t)(\dot{x}_s - \dot{x}_u) \tag{5.13a}$$

$$m_u\ddot{x}_u = f_s + b(t)(\dot{x}_s - \dot{x}_u) - k_t(x_u - x_0) \tag{5.13b}$$

where the spring force is a nonlinear function given by

$$f_s = k_{s1}(x_s - x_u) + k_{s2}(x_s - x_u)^2 + k_{s3}(x_s - x_u)^3 \tag{5.13c}$$

and b(t) denotes the nonlinear damping coefficient to be designed, which can generally be any nonlinear function. In this study, it is as an example written as

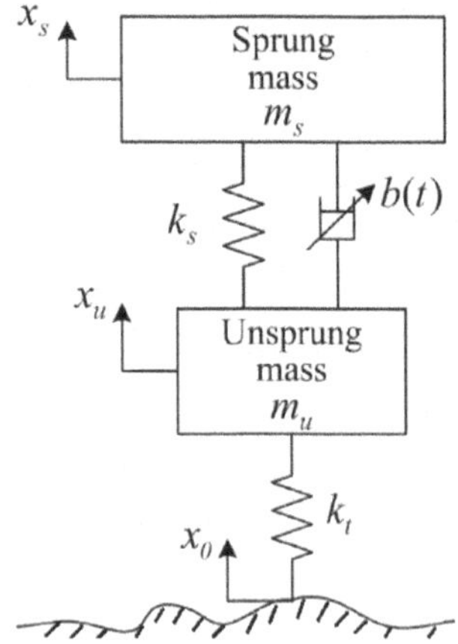

Fig. 5.3 A nonlinear vehicle suspension system

$$b(t) = b_0 + b_1 z^2 \tag{5.13d}$$

where $z = x_s - x_u$. Previous results have studied the cubic nonlinear damping $b_1 \dot{z}^2$[18, 19, 25]. For the model parameters above, a model in [33] is adopted here as: m_s = 240 kg, k_{s1} = 12,394 N/m, k_{s2} = − 73,696 N/m^2, k_{s3} = 3,170,400 N/m^3, m_u = = 25 kg, k_t = 160,000 N/m, and b_0 = 1385.4 Ns/m.

Case I: The estimation of the linear part and the nonlinear output spectrum without knowing the truncation order N and system model. In this part, it is supposed that the system model is not known and $b_1 = 10^6$ in (5.13d). It can be seen that the suspension system has strong nonlinearity both in stiffness and damping elements. The nth-OSE algorithm can be used to estimate the linear part of the system and therefore to estimate how the nonlinearity takes a role in the dynamic response using the method discussed in [11]. To this aim, the input excitation is considered as a multi-tone function given by

$$u(t) = \rho \sum_{i=0}^{11} \sin((0.5 + 0.1i)t) \tag{5.14}$$

Note that the multi-tone signal above can actually include more frequency points and thus provide a sufficient excitation to the system at different frequencies of interest. The parameter ρ is to be determined in the estimation.

Usually the nonlinear output spectrum of system (5.13) would have an infinite order of output spectrum components but could be truncated at an appropriate order N to approximate the original nonlinear dynamic response in practice. With the nth-OSE algorithm, the nth-order output spectrum of any given nonlinear systems could be estimated in a non-parametric form up to any high order without knowing the truncation order and system model. If only the estimation of the linear part of the system is interested, it would be much easier to choose a proper excitation magnitude ρ. To estimate accurately all the output spectrum components simultaneously, ρ must be chosen using the ρ-selection method above. For example, if taking N = 3, the corresponding parameters can be chosen as $\rho = 0.0005$ and $\alpha = 10$ based on the ρ-selection method, and the estimated results for the nth-order output spectrum up to order 3 are given in Figs. 5.4 and 5.5.

In Fig. 5.4, it can be seen that the output frequencies in the first order are exactly the same as the input frequencies, those in the second order are doubled and in the third order tripled. This is completely consistent with the known results in [13, 14] that is, the output frequencies in the nth-order output spectrum can only appear in the range [0, nb] given that the input frequencies are in [a, b]. To further validate the estimation results in Fig. 5.4, the real output spectrum of the linear part is given in Fig. 5.5. It is clearly shown that the estimated values of the first order output spectrum exactly match the real ones with overall root-mean-square error 8.5068e−004.

Similarly, if taking N = 5 and the other parameters by Corollary 1 and the ρ–selection method as $\rho = 0.001$, $\alpha = 10$ and $\beta = 2$, the estimation results of the output spectrum up to order 5 are shown in Fig. 5.6. Comparisons indicate that the

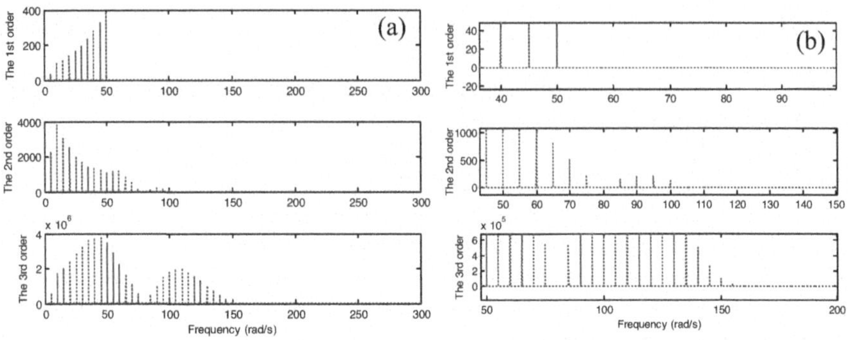

Fig. 5.4 The estimated output spectra up to order 3: **a** the estimated results, **b** the zoomed results of (**a**)

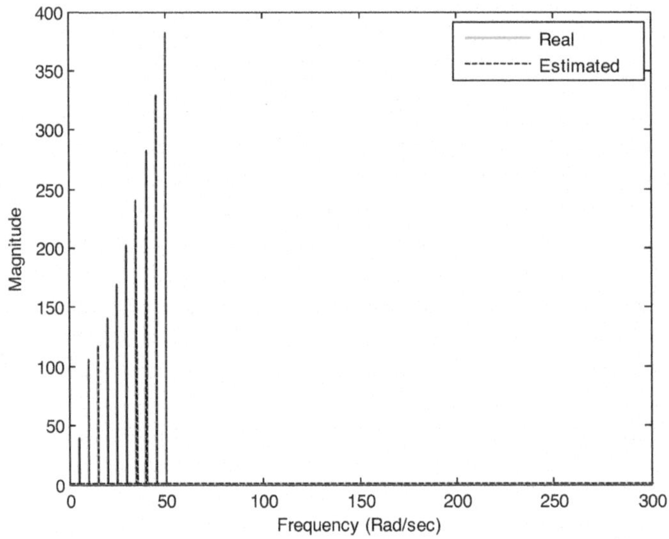

Fig. 5.5 The estimated and real first order output spectrum

estimated first three orders of output spectrum are exactly the same in Figs. 5.4 and 5.6, implying the consistence and validation of the nth-OSE algorithm in estimation of the nth-order output spectrum for the system. Figures 5.4 and 5.6 both show that the magnitude of the higher order output spectrum is becoming larger with the increase of order n. Recalling Eq. (5.7), this implies that when the input magnitude is becoming larger, the nonlinearity will take more dominant roles in system dynamic response and the vibration performance could become worse (see Fig. 5.7). Figure 5.7 clearly confirms that when the input magnitude is increased to $\rho = 0.1$, the nonlinear response is dominant and vibration transmissibility becomes obviously worse. Therefore, a proper design of the linear or nonlinear damping coefficient would be very crucial in vibration control. This consequently incurs a need for a systematic method for the nonlinear analysis and design, which would be focused in the following section.

It should be noted that if the excitation magnitude ρ_1, ρ_2, ρ_3 are not properly chosen according to Corollary 1, the estimation results would be greatly biased (for

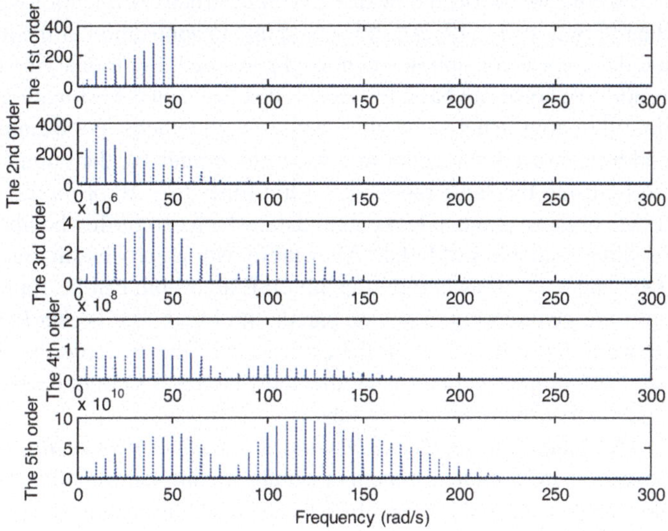

Fig. 5.6 The estimated output spectrum up to order 5

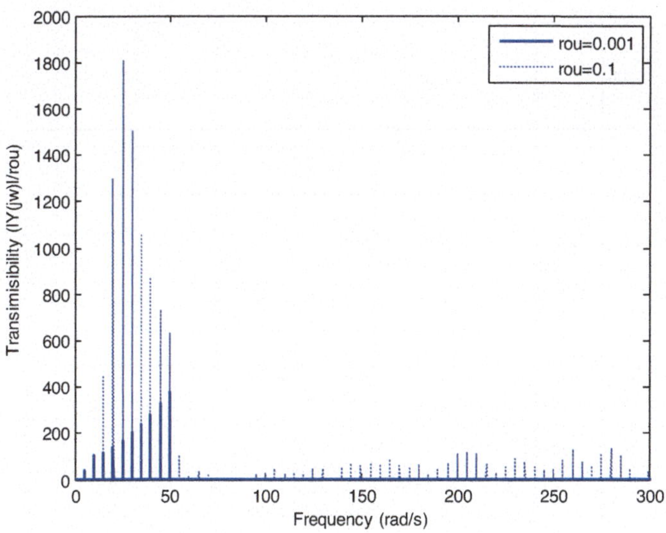

Fig. 5.7 Transmissibility $\left(\frac{|Y(j\omega)|}{\rho}\right)$ for input magnitude $\rho = 0.001$ and 0.1

simplicity of discussion, this is not illustrated here). With the accurately estimated first order output spectrum, many methods are available to estimate a linear parametric model [34–36]. This is not demonstrated here either. Moreover, it is interesting to mention that the nth-OSE algorithm developed here provides an effective tool to accurately estimate the linear and nonlinear components in the dynamical response of a system. This would also be of significance to non-destructive evaluation and similar topics can be referred to [11, 17, 20]. It is noted that in application of the method to experimental data, noise corruption can never be avoided. Preliminary results indicate that the additive noise can basically have little effects on the estimation with the nth-OSE algorithm after using some noise processing methods. For example, if the output has additive noise with signal to noise ratio around 10, the output data can be generated by using the same input for several times (say 10 or more), and then take the average of these independent output data, which can further be filtered with a smoothing filter available in Matlab. After these data preprocessing, taking $N = 3$, $\rho = 0.0005$ and $\alpha = 10$ as before, the nth-OSE algorithm can be applied again and the results are given in Fig. 5.8. Compared with Fig. 5.4, it is clearly seen that the noise basically has only effect on the estimation of the 3rd order but in a very limited level. With proper preprocessing methods, the results could be even better. This topic will be focused in a further study.

Case II: The estimation of the nCOS function with respect to the nonlinear damping. To explore the nonlinear benefit in vibration control, the nonlinear damping characteristic can be analyzed by regarding $\ddot{x}_s$ as the system output, i.e.,

$$y = \ddot{x}_s \tag{5.15}$$

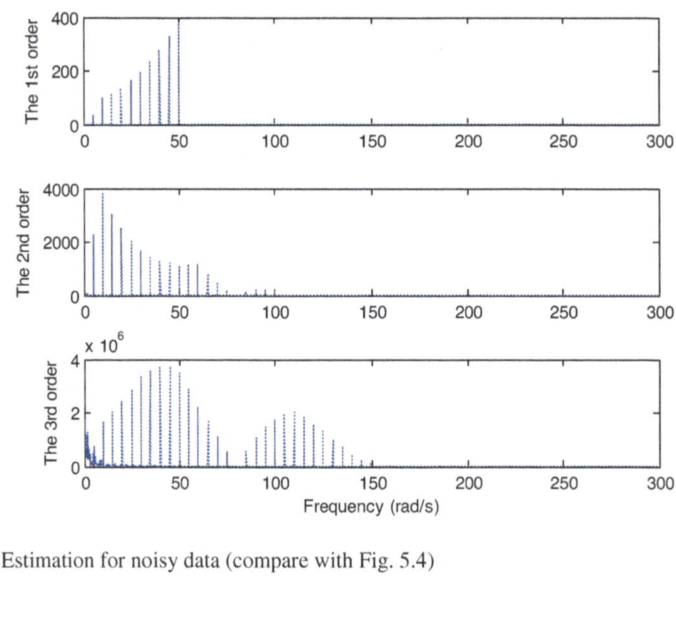

Fig. 5.8 Estimation for noisy data (compare with Fig. 5.4)

and deriving the nonlinear COS function in (5.5) in terms of the nonlinear parameter b_1. To this aim, consider the input as

$$u(t) = \rho\sin(\Omega t) \tag{5.16}$$

where Ω could be any frequency of interest. Since only a cubic order nonlinear term $b_1 z^2\dot{z}$ would be considered, it can be obtained using the parametric characteristic analysis [21] that

$$\begin{aligned}&\chi_2 = 1;\ \chi_3 = b_1;\ \chi_4 = b_1;\ \chi_5 = \left[b_1, b_1^2\right];\\&\chi_6 = \left[b_1, b_1^2\right];\ \chi_7 = \left[b_1, b_1^2, b_1^3\right];\ \chi_8 = [b_1, b_1^2, b_1^3];\ \cdots\end{aligned}$$

Then in this case the nonlinear COS function can be written as

$$\begin{aligned}Y(j\omega) = {}& Y_1(j\omega) + Y_2(j\omega) + Y_3(j\omega)_{|b_1=0}\\&+ b_1\cdot\phi_3(b_1;j\omega) + Y_4(j\omega)_{|b_1=0} + b_1\cdot\phi_4(b_1;j\omega)\\&+ Y_5(j\omega)_{|b_1=0} + b_1\cdot\phi_5(b_1;j\omega) + b_1^2\cdot\phi_5(b_1^2;j\omega) + \cdots\end{aligned} \tag{5.17}$$

Note that $Y_1(j\omega)$ and $Y_2(j\omega)$ are the 1st and 2nd order output spectrum, $Y_3(j\omega)_{|b_1=0}$, $Y_4(j\omega)_{|b_1=0}$, and$Y_5(j\omega)_{|b_1=0}$ are all incurred by the 2nd and 3rd order nonlinearity in the stiffness, which have no relationship with the nonlinear damping term. By the parametric characteristic analysis [21], there will be a term in χ_5 given by the multiplication between b_1 and the coefficient of the 3rd order nonlinearity in the stiffness. While the latter is a constant, it thus yields the term b_1 in χ_5. This results in the term $b_1\varphi_5(b_1;j\omega)$ in (5.17). By applying the nth-COSE algorithm, the estimation of (5.17) can be obtained as

$$\begin{aligned}\hat{Y}(j\omega) = {}& \rho\hat{Y}_1(j\omega) + \rho^2\hat{Y}_2(j\omega)\\&+ \rho^3\underbrace{\left[\hat{Y}_3(j\omega)_{|b_1=0} + b_1\cdot\hat{\phi}_3(b_1;j\omega)\right]}_{\hat{Y}_3(j\omega,b_1)}\\&+ \rho^4\underbrace{\left[\hat{Y}_4(j\omega)_{|b_1=0} + b_1\cdot\hat{\phi}_4(b_1;j\omega)\right]}_{\hat{Y}_4(j\omega,b_1)}\\&+ \rho^5\underbrace{\left[\hat{Y}_5(j\omega)_{|b_1=0} + b_1\cdot\hat{\phi}_5(b_1;j\omega) + b_1^2\cdot\hat{\phi}_5(b_1^2;j\omega)\right]}_{\hat{Y}_5(j\omega,b_1)} + \cdots\end{aligned} \tag{5.18}$$

Note that the nCOS function in (5.18) is an explicit function of the frequency ω, nonlinear parameter b_1 and excitation magnitude ρ. In (5.16), consider $\Omega = 18$ rad/s as the input frequency, which is around the resonant frequency of the system and also a sensitive frequency to human body. If in (5.18), take $\omega = \Omega$, the even order output spectrum at this frequency must be zero [13, 21]. Thus, (5.18) can be written as

$$\hat{Y}(j\Omega) = \rho \hat{Y}_1(j\Omega) + \rho^3 \underbrace{\left[\hat{Y}_3(j\Omega)_{|b_1=0} + b_1 \cdot \hat{\phi}_3(b_1; j\Omega)\right]}_{\hat{Y}_3(j\Omega, b_1)}$$
$$+ \rho^5 \underbrace{\left[\hat{Y}_5(j\Omega)_{|b_1=0} + b_1 \cdot \hat{\phi}_5(b_1; j\Omega) + b_1^2 \cdot \hat{\phi}_5(b_1^2; j\Omega)\right]}_{\hat{Y}_5(j\Omega, b_1)} + \cdots \quad (5.19)$$

By the nth-COSE algorithm, to estimate (5.19), the following steps can be used (referred to as the nCOSE-procedure):

(a) Setting $b_1 = 10^6$ (or any other value), apply the nth-OSE algorithm to estimate $Y_n(j\omega)$ for n = 1, 2, 3, …, 5…;
(b) Take the estimated values of the output spectrum $Y_n(j\omega)$ at the frequency Ω, denoted by $\hat{Y}_1(j\Omega)$, $\hat{Y}_3(j\Omega)$ and $\hat{Y}_5(j\Omega)$ ($\hat{Y}_2(j\Omega)$ and $\hat{Y}_4(j\Omega)$ must be zero);
(c) Take $b_1 = 10^5$ (or any other value) and $b_1 = 0$ (preferable), and repeat (a-b);
(d) Equation (5.19) can therefore be estimated using a least square method for each component.

Based on Steps (a–c), Table 5.2 can be obtained with the parameter setting in the nth-OSE algorithm as $\rho = 0.001$, $\alpha = 10$ and $\beta = 2$ as suggested in the previous case study.

In Table 5.2, the estimates for $Y_1(j\Omega)$ under each parameter value are consistent since $Y_1(j\Omega)$ is the dynamic response of the linear part of the system, independent of system nonlinearity. This confirms again the effectiveness and reliability of the nth-OSE algorithm.

For the estimation of $\hat{Y}_3(j\Omega, b_1)$ in (5.19), Table 5.2 yields

$$\hat{Y}_3(j\Omega)_{|b_1=0} = 5.3988e + 002 + 1.2252e + 004i$$

Therefore,

$$\hat{\phi}_3(b_1; j\Omega) = \left(\hat{Y}_3(j\Omega, b_1) - \hat{Y}_3(j\Omega)_{|b_1=0}\right)/b_1$$
$$= \begin{cases} (2.8583e + 003 - 1.2300e + 002i)10^{-5} \\ (2.8583e + 004 - 1.2310e + 003i)10^{-6} \end{cases}$$

Table 5.2 The estimated output spectrum at Ω rad/s for different values of b_1

b_1	$\hat{Y}_1(j\Omega)$	$\hat{Y}_3(j\Omega)$	$\hat{Y}_5(j\Omega)$
0	1.0876e+002 + 7.2310e+001i	5.3988e+002 + 1.2252e+004i	1.8958e+005 + 2.5757e+006i
10^5	1.0876e+002 + 7.2310e+001i	3.3982e+003 + 1.2129e+004i	4.3776e+005 + 2.4906e+006i
10^6	1.0876e+002 + 7.2310e+001i	2.9123e+004 + 1.1021e+004i	1.5670e+006 + 1.7068e+006i

for $b_1 = 10^5$ and 10^6, respectively, which are almost the same. The averaged value can be used

$$\hat{\phi}_3(b_1; j\Omega) = (2.8583e + 003 - 1.2305e + 002i)10^{-5}$$

Similarly, it can be obtained that

$$\begin{bmatrix} \hat{\phi}_5(b_1; j\Omega) \\ \hat{\phi}_5(b_1^2; j\Omega) \end{bmatrix} = \begin{bmatrix} 10^5 & 10^{10} \\ 10^6 & 10^{12} \end{bmatrix}^{-1} \begin{bmatrix} \hat{Y}_5(j\Omega, 10^5) - \hat{Y}_5(j\Omega)_{|b_1=0} \\ \hat{Y}_5(j\Omega, 10^6) - \hat{Y}_5(j\Omega)_{|b_1=0} \end{bmatrix}$$
$$= \begin{bmatrix} 2.6045 - 0.8490i \\ -1.2271e - 007 - 1.9889e - 009i \end{bmatrix}$$

Therefore, the nCOS function (up to the 5th order) of the system with respect to input (5.16) and nonlinear damping parameter b_1 is obtained. To verify the nCOS function in output prediction, different excitation magnitudes and different nonlinear damping coefficient b_1 can be used for a validation, which is shown in Fig. 5.9 and Table 5.3, indicating clearly the exact prediction by using the estimated nCOS function (5.19) in the excitation range $\rho \in [0,0.001]$.

Importantly, it can also be checked that the prediction of the estimated nCOS (up to the 5th order) is still reliable even when the excitation magnitude is larger for example up to 0.05 (see Fig. 5.10 and Table 5.3), which is much larger than the excitation range [0, 0.001] used in the estimation. Similar conclusion can be drawn for a larger parameter value b_1. This demonstrates clearly the reliability and advantage using the

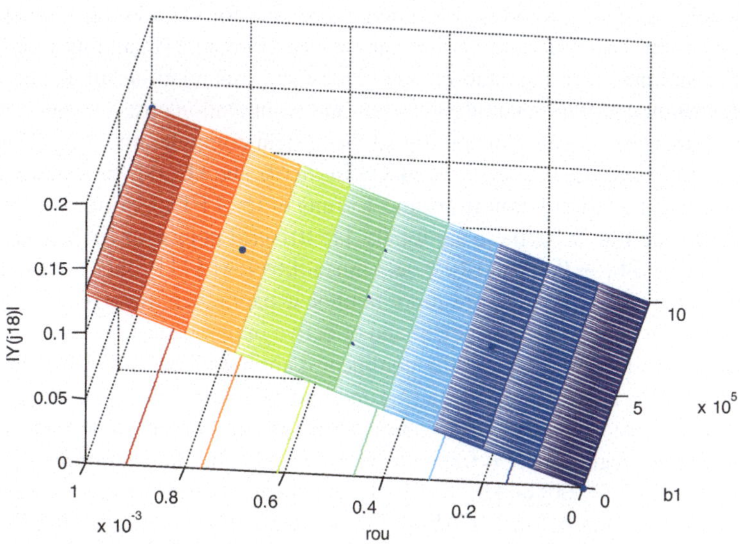

Fig. 5.9 The estimated nCOS function and its validation for $\rho \in [0, 0.001]$

Table 5.3 Prediction of the estimated nCOS function with Eq. (5.19)

ρ	b_1	Real magnitude	Prediction by (5.19)
0	0	0	0
$0.5*10^{-3}$	$5*10^5$	0.0653	0.0653
$0.5*10^{-3}$	$2.5*10^5$	0.0653	0.0653
$0.5*10^{-3}$	$7.5*10^5$	0.0653	0.0653
$0.25*10^{-3}$	$5*10^5$	0.0327	0.0327
$0.75*10^{-3}$	$5*10^5$	0.098	0.098
10^{-3}	10^6	0.1306	0.1306
0.01	$2*10^5$	1.3181	1.3181
0.01	$6*10^5$	1.3274	1.3274
0.03	$6*10^5$	4.5275	4.5504
0.05	$6*10^5$	9.5741	9.9726
0.05	10^6	10.655	11.3523

nCOS function in system analysis. Moreover, comparing Figs. 5.10 with 9, it can also be seen that the nonlinear dynamics become more obviously when the input magnitude becomes larger. Using the nth-COSE algorithm, it is easy to estimate the nCOS function up to any high order. This therefore provides a useful tool for the analysis and design of nonlinear damping in vehicle suspension systems, and the parameter optimization would also be possible in terms of any nonlinear parameters of interest to achieve a desired output spectrum in vibration suppression. The advantages of the nCOS function based method proposed in this study for nonlinear analysis and design could be that it provides an explicit expression for the relationship between nonlinear output spectrum and system parameters of interest (including frequency, nonlinear parameters and excitation magnitude), and this relationship can be accurately determined with less simulation trials and computation cost compared with a pure simulation based study or traditional theoretical computations. Regarding the last point, for example, to obtain the nCOS in (5.19), the nCOSE-procedure above involves at most 15 simulation trials and 15 runs of the FFT algorithms. However, only to generate Fig. 5.10, there are totally 11*101 grid points for (ρ, b_1) (i.e., $\rho = 0{:}0.005{:}0.05$ and $b_1 = 0{:}10{,}000{:}10^6$), and thus it involves 1111 times of simulation trials and 1111 FFTs. If there are more nonlinear parameters involved, the nCOS based method would be more efficient and effective in practice.

Moreover, to further illustrate the nonlinear damping given in (5.13d), comparisons between the nonlinear damping (b_1) and linear damping (b_0) are conducted for the peak suppression of the vehicle suspension system. For the same input (5.16) with excitation magnitude $\rho = 0.05$ and $\Omega = 18$ rad/s, the linear damping is used with $b_0 \in [0, 1385.4*100]$ and $b_1 = 0$ firstly, and the nonlinear damping is then used with $b_0 = 1385.4$ and $b_1 \in [0, 1385.4*100]$ secondly. The output magnitudes at Ω for both cases are shown in Fig. 5.11.

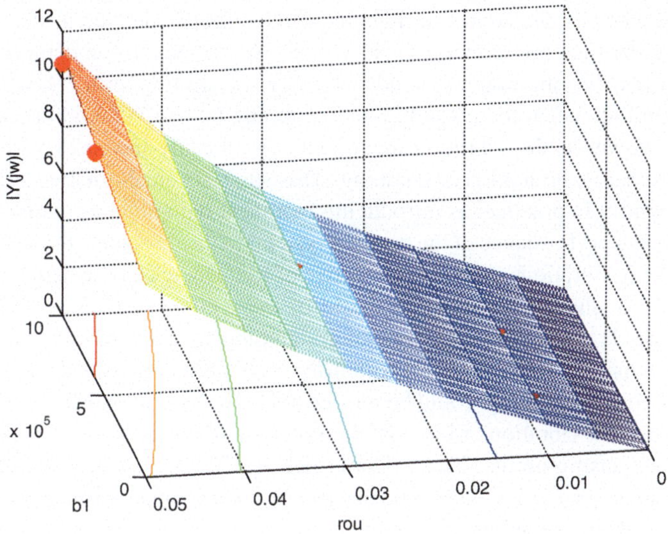

Fig. 5.10 The estimated nCOS function and its validation for $\rho \in [0, 0.05]$

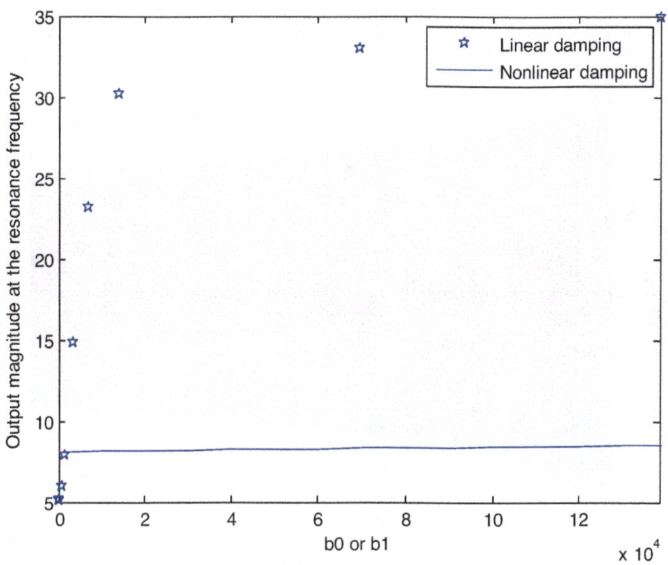

Fig. 5.11 Linear and nonlinear damping in suppression of resonance peak

The outcomes for the nonlinear damping can be directly derived using the nCOS function (5.19) that was previously developed. As the nonlinear damping coefficient increases, the vibration magnitude slightly rises, starting from 7.9 (which is the output magnitude when there's only linear damping b0 = 1385.4 and no nonlinear damping). However, the vibration magnitude significantly increases with the rise of b0 when there's no nonlinear damping. This suggests an alternative strategy for vibration control in practice, as introducing nonlinear damping has minimal impact on the damping performance of the original systems at the resonance frequency, but it could suppress vibrations across a wide range of frequencies. For instance, Fig. 5.11 suggests that a small linear damping could be more effective in controlling vibrations at a steady state. However, too small damping would certainly cause strong vibrations considering the transient response. A suitable nonlinear damping, along with a small linear damping, could be introduced to the system, which could quickly suppress transient vibrations while maintaining a small linear damping performance in vibration transmissibility. This is confirmed by Fig. 5.12, where a small linear damping with b0 = 1385.4*0.0005 results in a very slow suppression speed in vibration control, while a nonlinear damping b1 = 10,000 could lead to a much faster convergence speed and thus better vibration suppression. Further discussions about nonlinear damping can be found in references [25, 32, 37, 38].

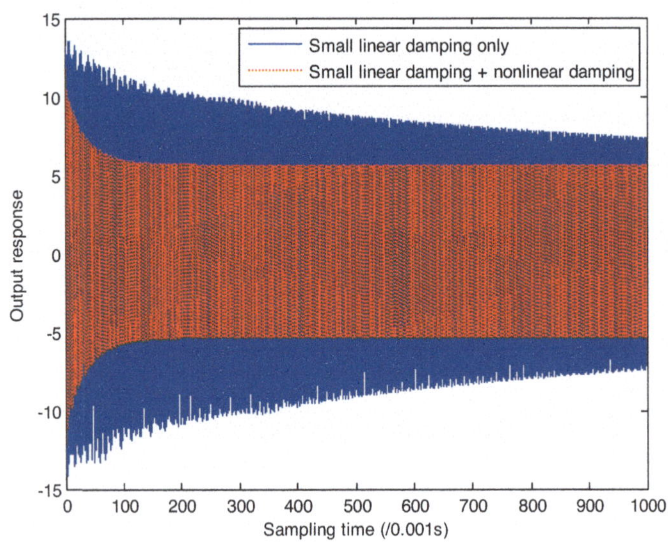

Fig. 5.12 Vibration suppression with linear and nonlinear damping

5.5 Conclusions

A comprehensive frequency domain method for the analysis, design, and estimation of nonlinear systems has been developed using the nonlinear characteristic output spectrum (nCOS) function. This method enables the precise identification of the linear and nonlinear components in the nonlinear output spectrum of a given nonlinear system, which can be described by NDE, NARX, or NBO models, given some simulation or experimental output data. These output spectrum components can subsequently be utilized for system identification or nonlinear analysis for various applications, such as signal processing and fault detection. Notably, the nCOS function can be developed, providing a clear expression for the relationship between the nonlinear output spectrum and system characteristic parameters of interest, including nonlinear parameters, frequency, and excitation magnitude. This relationship can be accurately determined with fewer simulation trials and less computational cost compared to a purely simulation-based study or traditional theoretical computation, offering a significant and straightforward approach for nonlinear analysis and design. The nCOS function can be seen as a significantly enhanced version of the OFRF function established in [18, 21, 22]. The case study on a vehicle suspension system clearly illustrates the effectiveness and potential value of the proposed nCOS-based method and its corresponding algorithms. Future studies will concentrate more on application issues, taking into account noise and disturbance effects.

Acknowledgements The main results of this chapter can also be referred to the published work [39].

Appendices

1. **Proof of Proposition 1**

The proof needs three more lemmas. An outline of the proof is given here and the detailed proof for Lemmas 2–4 can be referred to [40]. The inverse matrix in (5.8) is denoted by A whose element is denoted by $A_{\mathrm{i,r}}$. An explicit expression for $A_{\mathrm{i,r}}$ is given in Lemma 2. Lemma 3 provide an analytical computation of the estimation error in terms of the truncation order N and the excitation magnitudes $\rho_1, \rho_2, \rho_3 \ldots$ based on Lemma 2 and using (5.7), while Lemma 4 provides a method for the simplification of the estimation error in Lemma 3 for different excitation magnitudes $\rho_1, \rho_2, \rho_3 \ldots$. With Lemmas 3 and 4, the result of this proposition can be established.

Lemma 2. $A_{i,r} = \frac{(-1)^{N-i} \sum_{m=N-i; j_1 \neq \cdots \neq j_m \in \aleph_0 \backslash \{r-1\} \text{with no repetition}} \rho_{j_1} \cdot \rho_{j_2} \cdots \rho_{jm}}{\prod_{k \in \aleph_0 \backslash \{r-1\}} (\rho_{r-1}) - \rho_k}$, *where* $1 \leq i, r \leq N$, $\aleph_0 = \{0, 1, 2, \ldots, N-1\}$, $\rho_{j_0} = 1$.

Lemma 3. *Given the truncation order N, the estimation error with* (5.8) *for the nth-order output spectrum, i.e.,* $\Delta Y_n(j\omega)$, *is*

$$\Delta Y_n(j\omega) = \sum_{j=1}^{N} A_{n,j} \cdot \rho_{j-1}^{N} \overline{\sigma}_{\left[N,\rho_{j-1}\right]}(j\omega)$$
$$= \sum_{j=1}^{N} \left(\frac{(-1)^{N-n} \sum_{\substack{m=N-n; \\ j_1 \neq \cdots \neq j_m \in \aleph_0 \backslash \{j-1\}}} \rho_{j_1}\rho_{j_2}\cdots\rho_{j_m}}{\prod_{k\in\aleph_0\backslash\{j-1\}} (\rho_{j-1}-\rho_k)} \cdot \rho_{j-1}^{N} \overline{\sigma}_{\left[N,\rho_{j-1}\right]}(\cdot) \right)$$

where $A_{i,j} = \frac{(-1)^{N-i} \sum_{\substack{m=N-i; \\ j_1 \neq \cdots \neq j_m \in \aleph_0 \backslash \{j-1\}}} \rho_{j_1}\rho_{j_2}\cdots\rho_{j_m}}{\prod_{k\in\aleph_0\backslash\{j-1\}}(\rho_{j-1}-\rho_k)}$, $\aleph_0 = \{0, 1, 2, \cdots, N-1\}$ *and* $\rho_{j_0} = 1$; $\overline{\sigma}_{[N,\rho]}(j\omega) = Y_{N+1}(j\omega) + \rho Y_{N+2}(j\omega) + \rho^2 Y_{N+3}(j\omega) + \cdots$, *and* $\sigma_{[N,\rho]}(j\omega) = \rho^{N+1}\overline{\sigma}_{[N,\rho]}(j\omega)$.

Lemma 4. *For* $\forall x_1, x_2, \ldots, x_N \in R\backslash 0$ *satisfying* $x_1 \neq x_2 \neq \ldots \neq x_N$ *and* $N > 1$, *it holds that*

$$\sum_{i=1}^{N} \frac{x_i^m}{\prod_{k=1,k\neq i}^{N}(x_i - x_k)} = \begin{cases} (x_1 + x_2 + \cdots + x_N)^{[m-N+1]} & m \geq N-1 \\ 0 & 0 \leq m < N-1 \end{cases}$$

The following proof is given. From (5.7),

$$\begin{aligned} \rho_{r-1}^{N-1}\overline{\sigma}_{\left[N,\rho_{r-1}\right]}(j\omega) = \rho_{r-1}^{N-1} Y_{N+1}(j\omega) \\ + \rho_{r-1}^{N} Y_{N+2}(j\omega) \\ + \rho_{r-1}^{N+1} Y_{N+3}(j\omega) + \cdots \end{aligned} \tag{5.20}$$

Using Lemmas 3–4 and Eq. (5.20),

$$\begin{aligned} \Delta Y_1(j\omega) &= (-1)^{N-1}\rho_0\rho_1\cdots\rho_{N-1}\sum_{r=1}^{N}\frac{\rho_{r-1}^{N-1}\overline{\sigma}_{\left[N,\rho_{r-1}\right]}(j\omega)}{\prod_{k\in\aleph_0\backslash\{r-1\}}(\rho_{r-1}-\rho_k)} \\ &= (-1)^{N-1}\rho_0\rho_1\cdots\rho_{N-1}\left(\sum_{r=1}^{N}\frac{\rho_{r-1}^{N-1}Y_{N+1}(j\omega)}{\prod_{k\in\aleph_0\backslash\{r-1\}}(\rho_{r-1}-\rho_k)}\right. \\ &\quad \left.+\sum_{r=1}^{N}\frac{\rho_{r-1}^{N}Y_{N+2}(j\omega)}{\prod_{k\in\aleph_0\backslash\{r-1\}}(\rho_{r-1}-\rho_k)}+\cdots\right) \\ &= (-1)^{N-1}\rho_0\rho_1\cdots\rho_{N-1}\begin{pmatrix} (\rho_0+\rho_1+\cdots+\rho_{N-1})^{[0]}Y_{N+1}(j\omega) \\ +(\rho_0+\rho_1+\cdots+\rho_{N-1})^{[1]}Y_{N+2}(j\omega) \\ +(\rho_0+\rho_1+\cdots+\rho_{N-1})^{[2]}Y_{N+3}(j\omega)\cdots \end{pmatrix} \end{aligned} \tag{5.21}$$

This gives (5.9a). Similarly, Eq. (5.9b) can be obtained.

2. **Proof of Corollary 1**

Considering N = 3 firstly, the output spectrum is truncated at N = 3, which can be written as

$$Y(j\omega) = Y_1(j\omega) + Y_2(j\omega) + Y_3(j\omega) + \sigma_{[3]}(j\omega)$$

where $\sigma_{[3]}(j\omega) = Y_4(j\omega) + Y_5(j\omega) + Y_6(j\omega) + \cdots$ is the truncation error. Using (5.8) to estimate the different orders of output spectra, the estimation errors are given by Proposition 1

$$\Delta Y_1(j\omega) = -\rho_0\rho_1\rho_2 \sum_{k=0}^{\infty} (\rho_0 + \rho_1 + \rho_2)^{[k]} Y_{k+3}(j\omega)$$

$$\Delta Y_3(j\omega) = \sum_{k=1}^{\infty} (\rho_0 + \rho_1 + \rho_2)^{[k]} Y_{k+2}(j\omega)$$

If choosing $\rho_0 + \rho_1 + \rho_2 = 0$, $\rho_1 = -\rho$, $\rho_2 = -\rho/\alpha$, and using Lemma 1, computation of $(\rho_0 + \rho_1 + \rho_2)^{[k]}$ can be obtained

$$\begin{aligned}
(\rho_0 + \rho_1 + \rho_2)^{[1]} &= \rho_0 + \rho_1 + \rho_2 = 0 \\
(\rho_0 + \rho_1 + \rho_2)^{[2]} &= \rho_0(\rho_0 + \rho_1 + \rho_2)^{[1]} + \rho_1(\rho_1 + \rho_2)^{[1]} \\
+\rho_2^2 &= (\rho_1 + \rho_2)^2 - \rho_1\rho_2 \\
&= \frac{\alpha^2 + \alpha + 1}{\alpha^2}\rho^2 \\
(\rho_0 + \rho_1 + \rho_2)^{[3]} &= \rho_0\rho_1\rho_2 = \frac{\alpha + 1}{\alpha^2}\rho^3 \\
(\rho_0 + \rho_1 + \rho_2)^{[4]} &= \rho_0\rho_0\rho_1\rho_2 + \rho_1^4 + \rho_1^3\rho_2 \\
&\quad + \rho_1^2\rho_2^2 + \rho_1\rho_2^3 + \rho_2^4 \\
&= \frac{\alpha^4 + 2\alpha^3 + 3\alpha^2 + 2\alpha + 1}{\alpha^4}\rho^4 \\
&\cdots\cdots
\end{aligned}$$

Therefore, Eqs. (5.10a–b) can be obtained. Following a similar process, Eqs. (5.10c–d) can be derived.

References

1. Judd, K.L., Numerical methods in economics, Cambridge, Massachussets: MIT Press, 1998.
2. A.I. Mees, Dynamics of Feedback Systems, Wiley, 1981
3. Gilmore, R. J.; Steer, M. B. (1991). "Nonlinear circuit analysis using the method of harmonic balance—A review of the art. Part I. Introductory concepts". Int. J. Microw. Mill.-Wave Comput.-Aided Eng. 1: 22–27
4. Rugh W.J., Nonlinear System Theory: the Volterra/Wiener Approach, Baltimore, Maryland, U.S.A.: Johns Hopkins University Press, 1981
5. K. Worden and G. R. Tomlinson. Non-linearity in structural dynamics: detection, identification and modeling. Institute of Physics Publishing, Bristol and Philadelphia, Sheffield, 2001
6. Boyd S. and Chua L.O., Fading memory and the problem of approximating nonlinear operators with Volterra series. IEEE Trans. Circuits & Systems,CAS32(11),1150–1161, 1985
7. Thomas Hélie and Béatrice Laroche, Computation of Convergence Bounds for Volterra Series of Linear-Analytic Single-Input Systems, IEEE Trans. On Automatic Control, 56(9): 2062-2072, 2011
8. George D.A., Continuous nonlinear systems, Technical Report 355, MIT Research Laboratory of Electronics, Cambridge, Mass. Jul. 24, 1959
9. Peyton Jones J.C. and Billings S.A. Recursive algorithm for computing the frequency response of a class of nonlinear difference equation models. International Journal of Control, 50(5), 1925–1940. 1989
10. Billings S.A. and Peyton-Jones J.C., Mapping nonlinear integro-differential equation into the frequency domain, International Journal of Control, Vol 54, 863-879, 1990
11. Jing X.J., Frequency Domain Analysis and Identification of Block-Oriented Nonlinear Systems, Journal of Sound and Vibration, 330 (22), 5427-5442, 2011
12. Swain A.K. and Billings S.A.. Generalized frequency response function matrix for MIMO nonlinear systems. International Journal of Control. Vol. 74. No. 8, 829-844, 2001
13. Jing X.J., Lang Z.Q., Billings S.A., Output Frequency Properties of Nonlinear Systems, International Journal of Non-Linear Mechanics, 45(7), September 2010, pp 681-690
14. Lang Z.Q., Billings S. A. Output frequency characteristics of nonlinear systems. International Journal of Control, 64:1049-1067, 1996
15. R. Yue, S.A. Billings and Z.-Q. Lang, An investigation into the characteristics of nonlinear frequency response functions. Part 1: understanding the higher dimensional frequency spaces, Int. J. Contr., 78, pp. 1031–1044
16. Zhang H., Billings S.A., Gain bounds of higher order nonlinear transfer functions, International Journal of Control, 64(4):767-773, 1996
17. Chatterjee A, Structural damage assessment in a cantilever beam with a breathing crack using higher order frequency response functions, Journal of Sound and Vibration, 329:3325-3334, 2010
18. Jing X. J., Lang Z.Q. Billings S.A., Output Frequency Response Function based Analysis for Nonlinear Volterra Systems. Mechanical Systems & Signal Processing, 22,102-120, 2008
19. J.C. Peyton-Jones, Simplified computation of the Volterra frequency response functions of nonlinear systems, Mechanical Systems and Signal Processing 21 (3) (2007) 1452–1468.
20. Lang Z Q, Peng Z K, A novel approach for nonlinearity detection in vibrating systems. Journal of Sound and Vibration, 314:603-615, 2008
21. Jing X. J., Lang Z. Q., Billings S. A. and Tomlinson G. R., The parametric characteristic of frequency response functions for nonlinear systems. International Journal of Control, 79(12):1552–1564, 2006
22. Lang Z.Q., Billings S.A., Yue R., Li J. Output frequency response functions of nonlinear Volterra systems. Automatica 43, 805-816, 2007
23. Jing X. J., Lang X.Q., Billings S. A., Mapping from parametric characteristics to generalized frequency response functions of nonlinear systems. International Journal of Control, 81(7), 1071–1088, July 2008

24. A.K. Swain, E.M.A.M. Mendes, S.K. Nguang, Analysis of the effects of time delay in nonlinear systems using generalised frequency response functions, Journal of Sound and Vibration, 294(1–2): 341–354, 27 June 2006
25. Jing X. J., Lang Z.Q. and Billings S.A., Nonlinear influence in the frequency domain: Alternating series, Systems and Control Letters, 60 (5), 295-309, 2011
26. Ahn K. K., Anh H. P. H., Inverse Double NARX Fuzzy Modeling for System Identification, IEEE/ASME Transactions on Mechatronics, 15(1): 136 – 148, 2010
27. Wei, H. L. and Billings, S. A., "Model structure selection using an integrated forward orthogonal search algorithm assisted by squared correlation and mutual information", Int. J. Modelling, Identification and Control, 3(4), pp. 341–356, 2008
28. Doyle III F, Pearson R, Ogunnaike B. Identification of control using Volterra models, Springer-Verlag,London, 2002
29. Lee G.M., Estimation of nonlinear system parameters using higher-order frequency response functions, Mechan Syst and Signal Process, 11(2), 219-228, 1997
30. Jing X. J., Lang Z.Q. and Billings S.A., Determination of the analytical parametric relationship for output spectrum of Volterra systems based on its parametric characteristics. Journal of Mathematical Analysis and Applications, 351, pp 694-706, 2009
31. Case, D., Taheri, B., Richer, E., Design and Characterization of a Small-Scale Magnetorheological Damper for Tremor Suppression, IEEE/ASME Transactions on Mechatronics, https://doi.org/10.1109/TMECH.2011.2151204 , 2012
32. Zapateiro, M., Pozo, F., Karimi, H.R., Ningsu Luo, Semiactive Control Methodologies for Suspension Control With Magnetorheological Dampers, IEEE/ASME Transactions on Mechatronics, 17(2): 370 - 380, 2012
33. Dixit R.K., Buckner G.D., Sliding mode observation and control for semiactive vehicle suspensions, Vehicle System Dynamics, 43(2): 83-105, 2011
34. Levi E.C., Complex curve fitting, IRE Transaction on Automatic Control, 4:37-43, 1959
35. Young P.C., The instrumental variable method: a practical approach to identification and parameter estimation, the 7th IFAC/IFORS Symposium on Identification and System Parameter Estimation, York, UK, pp1–14, 1985
36. Cao, Y. Y.; Chen, X. X., A Novel Discrete ARMA-Based Model for Piezoelectric Actuator Hysteresis , IEEE/ASME Transactions on Mechatronics, https://doi.org/10.1109/TMECH.2011.2128339, 2012
37. Ibrahim R.A., Recent advances in nonlinear passive vibration isolators, Journal of Sound and Vibration 314: 371–452, 2008
38. Montague, R.; Bingham, C.; Atallah, K., Servo Control of Magnetic Gears , IEEE/ASME Transactions on Mechatronics, 17(2): 269 - 278, 2012
39. Jing X.J., Nonlinear characteristic output spectrum for nonlinear analysis and design, IEEE/ASME Transaction on Mechatronics, 19(1), 171 - 183, 2014
40. Jing X.J., Truncation Order and its Effect in a Class of Nonlinear Systems, to appear in Automatica, 2012

Chapter 6
Nonlinear Output Regulation and Its Application in Fault Detection

Abstract The identification and characterization of nonlinear behavior for the purpose of fault detection—particularly those associated with cracks or structural damage—constitutes a significant area of research in engineering. This study investigates a novel and systematic methodology for characterizing nonlinearity through nonlinear output decomposition and regulation. It is demonstrated that the proposed approach enables effective assessment of both even-order nonlinearities and those induced by cracks, which exhibit more complex behavior than simple even-order responses, even in their early stages. Specifically, the magnitude of the second-order harmonic response, which is shown to vary linearly with crack severity, serves as a reliable indicator and can be accurately quantified using the proposed technique. Theoretical derivations, illustrative examples, finite element simulations, and experimental validations are presented to substantiate the efficacy and advantages of the method in capturing nonlinear dynamic characteristics associated with incipient damage. The framework introduced in this work offers a valuable alternative for frequency-domain nonlinear signal processing, with broad applicability in the detection and evaluation of structural faults across diverse engineering domains.

Keywords Fault detection · Cracks · Nonlinear features · Nonlinear output · Output decomposition and regulation

6.1 Introduction

6.1.1 About This Chapter

This chapter is to introduce a novel theory and associated methods based on the theory in previous chapters. Nonlinearity can be deliberately manipulated such that the output response can be modulated with the introduced nonlinearity for a desired performance. This is particularly useful for different applications such as fault detection, vibration control or sensor design, since output response is expected to be nonlinearly regulated for a desired nonlinear feature to be detected or suppressed.

X. Jing, *The Bio-Inspired X-Structure/Mechanism Approach for Exploring Nonlinear Benefits in Engineering*, https://doi.org/10.1007/978-981-96-5769-8_6

Potentially, the idea of this study can present a totally new approach to these related areas. With the established theory and methods in the previous chapters and our previous work, this is now possible to do as what is demonstrated in this chapter.

This chapter places the research in the context of fault detection, although the method can be applied to many other applications. The detection and characterization of nonlinearity for the evaluation/detection of faults related to cracks/damage is a prominent topic in engineering. This study explores the potential of a new and systematic method for nonlinearity characterization through nonlinear decomposition and regulation. It is demonstrated that, by utilizing the proposed output decomposition and regulation, both the even order nonlinearity and the nonlinearity caused by a crack (which is not a simple even order nonlinearity, even at its initial stage) can be effectively assessed by the magnitude of the 2nd order harmonic response. The latter is a linear function of the severity of the crack and can be accurately estimated using the proposed method. Theoretical analysis, example studies, finite element modeling, and experimental validation are provided to showcase the benefits and effectiveness of the proposed method in characterizing nonlinear dynamics caused by an initial crack or damage. The theories and methods presented in this study offer a valuable and alternative frequency domain approach for nonlinear signal processing in the evaluation of cracks/damage, detection and characterization of nonlinearity, and can be advantageous to a wide range of engineering practices.

6.1.2 Research Background

Detecting faults related to cracks or damage is a crucial issue in various fields such as signal-based fault detection and diagnosis, nondestructive evaluation (NDE), and structure health monitoring (SHM). To address this problem, numerous methods have been developed [1–8]. In the case of vibration systems, methods for identifying damage or cracks are commonly employed [9]. These methods often involve linear analysis techniques, despite the fact that the phenomena can be highly nonlinear. For instance, crack formation can be identified by observing changes in natural frequencies and mode shapes, or by measuring energy absorption [10–12]. These methods may also rely on knowledge of crack propagation based on previously documented damage [4–7, 13, 14], and they are often not sensitive to crack initiation.

Further research findings suggest that nonlinearity or nonlinear phenomena play a crucial role in assessing the presence and severity of cracks. As a result, there has been a focus on studying nonlinear behaviors in the presence of damage [15]. One approach is to incorporate nonlinearities into crack dynamics models to study the "breathing" of cracks and associated nonlinear behaviors. This has been done through various methods, including analytical, simulation, and experimental approaches [12, 16, 17]. Typically, the changes in the nonlinear model of a system due to crack growth are used as indicators to detect crack initiation. Additionally, observing nonlinear responses during regular system operation and/or excitation is also utilized. Experimental studies have been conducted to observe changes in sub-/super-harmonic responses, which can provide estimates of the size and location of preexisting cracks [2, 17].

Despite some reported methods and results for nonlinear crack detection, accurately characterizing nonlinearity or nonlinear phenomena remains a challenging problem. In the context of vibration-based crack detection methods, there are several difficulties in detecting nonlinearity in vibration signals. Firstly, the nonlinearity caused by cracks is often very weak at the initial crack stage, making it difficult to observe obvious nonlinear responses using commonly-used vibration testing methods. Secondly, measurement data are typically corrupted by noise to some extent. The weak nonlinearity in the measured signals is greatly influenced by or buried in the noise, making it challenging to accurately characterize using existing nonlinear analysis methods. Moreover, noise-removing algorithms used for data preprocessing can also inadvertently remove the desired nonlinear signal. Importantly, many systems inherently exhibit some level of nonlinearity, which can interact with the nonlinearity caused by cracks and cause existing crack detection methods to fail.

In this study, we investigate a frequency-domain nonlinear decomposition and regulation method to address the aforementioned challenges. The nonlinear decomposition aims to break down the nonlinear system output spectrum into different order output spectra, or output response components, using a Volterra approach based solely on experimental data. This method is based on a recent nonlinear frequency-domain analysis theory [18–22], which allows for accurate decomposition of the nonlinear output spectrum and precise determination of the 2nd order output spectrum. It differs from existing methods that focus on estimating the 2nd harmonic response (i.e., the response at the 2nd harmonic frequency) [1, 2, 23]. The crack-induced harmonic response at the second harmonic frequency can be very small due to the initial state of the crack, coupling effects from inherent nonlinearity, or practical excitation magnitudes. However, the second order output spectrum or harmonic response can still be relatively large even at the initial crack state, and it is shown to be robust against potential influences from other inherent nonlinearities (as demonstrated in Proposition 1).

Additionally, our method differs from some literature ones for example the result presented in [24], which aim to estimate a nonlinear output frequency response function (NOFRF) instead of the output spectrum. On the other hand, the nonlinear regulation aspect of our method aims to control the system's output response in such a way that the dominant dynamics are restricted to a relatively "low" frequency range around the excitation frequency. It also allows for the suppression of undesirable nonlinear dynamics using the properties of nonlinear energy transfer [19]. This ensures that when a low-pass or band-pass filter is applied for data processing, the desired 2nd order harmonic response can still be accurately estimated. Similarly, when the decomposition method is applied, the system's nonlinear output response behaves in a more Volterra-like manner, enabling the effective application of the aforementioned decomposition method. These results provide innovative nonlinear signal processing techniques for fault detection and diagnosis related to cracks or damage.

The chapter is organized as follows. A simplified crack-incurred dynamic model is discussed in Sect. 6.2, where the aim is to formulate the crack-incurred nonlinear problem in a convenient way for discussion instead of studying the modeling of cracks. The decomposition and regulation methods are developed in Sect. 6.3, and demonstrated in Sect. 6.4. A conclusion is drawn in Sect. 6.5.

6.2 The Problem

For convenience in discussion, consider a typical cantilever beam with a breathing edge crack, which is a simplified model of many engineering systems [1, 2], and the equation of motion can be written with the finite element method (FEM) as

$$\begin{aligned}&\left[\overline{\mathrm{M}}\right]\{\ddot{\mathrm{x}}(\mathrm{t})\} + [\mathrm{C}]\{\dot{\mathrm{x}}(\mathrm{t})\} + \left[\overline{\mathrm{K}}\right]\{\mathrm{x}(\mathrm{t})\}\\ &\quad = \{\mathrm{F}(\mathrm{t})\} \text{ if the crack is closed}\end{aligned} \tag{6.1a}$$

$$\begin{aligned}&\left[\overline{\mathrm{M}}\right]\{\ddot{\mathrm{x}}(\mathrm{t})\} + [\mathrm{C}]\{\dot{\mathrm{x}}(\mathrm{t})\} + \left[\overline{\mathrm{K}} - \Delta\overline{\mathrm{K}}\right]\{\mathrm{x}(\mathrm{t})\}\\ &\quad = \{\mathrm{F}(\mathrm{t})\} \text{ if the crack is open}\end{aligned} \tag{6.1b}$$

where $\overline{M}$, C, $\overline{K}$ are the mass, damping and stiffness matrices respectively, $\Delta\overline{\mathrm{K}}$ represents the stiffness reduction due to the crack. Consider that the crack is closed at equilibrium and will be open due to vibration motion. Define the modal matrix as $[\emptyset] = [\emptyset_1, \emptyset_2, \ldots, \emptyset_n]$, where $\emptyset_i$ is the ith mass normalized individual mode shape of the eigenvalue problem of Eq. (6.1a). Then letting $\{\mathrm{x}(\mathrm{t})\} = [\emptyset]\{\mathrm{y}(\mathrm{t})\}$ and pre-multiplying Eqs. (6.1–6.2) with $\emptyset_1^{\mathrm{T}}$, it can be obtained for the first mode that [25]

$$\ddot{y}_1(t) + 2\zeta_1\omega_1\dot{y}_1(t) + \gamma\omega_1^2 y_1(t) = n_1(t) \tag{6.2}$$

where ζ_1 and ω_1 are corresponding modal damping and frequency of the un-cracked beam, $\mathrm{n}_1(\mathrm{t})$ is the new input after manipulation, and γ denotes the crack effect which is given by

$$\gamma = \begin{cases} 1 & \text{if the crack closed} \\ \frac{\phi_1^T\left[\overline{K} - \Delta\overline{K}\right]\phi_1}{\omega_1^2} & \text{else} \end{cases} \tag{6.3}$$

The first-modal model in (6.2–6.3) is developed with the assumption that the damping matrix is proportional [25], which can be regarded as the dominant dynamics in the forced vibration of the beam considering the frequency range of excitations and higher values of modal damping in higher modes as discussed in [26]. The parameter γ is a constant representing the severity of the crack (a larger value implies a lower severity) which can be simplified as

$$\begin{cases} \gamma = 1 & \text{if } y_1(t) < 0 \\ 0 < \gamma < 1 & else \end{cases} \tag{6.4}$$

Moreover, without loss of generality take $\omega_1 = 1$ in what follows. For convenience, (6.2) and (6.4) can be rewritten by replacing y_1, ζ_1 and n_1 with $\mathbf{q}$, ζ and u respectively as

$$\ddot{\mathbf{q}}(t) + 2\zeta\dot{\mathbf{q}}(t) + \gamma\mathbf{q}(t) = u(t) \tag{6.5a}$$

$$\begin{cases} \gamma = 1 & \text{if } q(t) < 0 \\ 0 < \gamma < 1 & else \end{cases} \tag{6.5b}$$

Equations (6.5a, b) is a nonlinear system, and actually a non-dimensional model of a typical single degree of freedom (DOF) spring-mass-damper bilinear oscillator [27]. In this study, Eqs. (6.5a, b) will be focused as a simple model of the crack-incurred nonlinear dynamics. The objective is to study how to effectively and accurately characterize the dominant nonlinear dynamics incurred by the crack of different crack severity indicated by the parameter γ in 6.5a, b.

It is known that identifying low-severity cracks in practical scenarios is challenging due to the predominance of linear dynamics under realistic excitations of limited magnitude, as well as the reasons mentioned earlier. Recent studies have shown that the harmonic response at the 2nd harmonic frequency can serve as a crucial indicator for evaluating crack severity [23]. However, accurately characterizing this harmonic response for a system subjected to excitations of normal magnitude, which may not necessarily be pure sinusoidal or have a special input form, remains a key problem to be addressed.

In this study, we propose a new systematic data-based decomposition and regulation method (to be discussed in the following section) to estimate the second order output spectrum, which corresponds to the 2nd order harmonic response, instead of focusing solely on the harmonic response at the second harmonic frequency. This approach aims to accurately characterize the harmonic response at the 2nd harmonic frequency for various types of excitations, including those of normal magnitude and non-special input forms.

6.3 The Method and Main Results

A. The Data-Based Nonlinear Output Spectrum Decomposition

The proposed method actually provides a general approach for nonlinear characterization of a large class of nonlinear systems with only experimental data and without system modeling. To present the method more clearly, consider a general nonlinear system given by the following nonlinear differential equation,

$$\sum_{m=1}^{M}\sum_{p=0}^{m}\sum_{k_1,k_{p+q}=0}^{K} c_{p,q}(k_1,\ldots,k_{p+q})\prod_{i=1}^{p}\frac{d^{k_i}\mathbf{q}(t)}{dt^{k_i}}\prod_{i=p+1}^{p+q}\frac{d^{k_i}u(t)}{dt^{k_i}}=0 \tag{6.6}$$

where $\mathbf{q}$(t) and u(t)$\in \mathbb{R}$ are system output and input respectively; M is the maximum nonlinear degree of variable $\mathbf{q}$(t) and u(t); K the largest derivative order; $p+q=m$; $\left.\frac{d^k\mathbf{q}(t)}{dt^k}\right|_{k=0}=\mathbf{q}(t)$; $\sum_{k_1,k_{p+q}=0}^{K}(\cdot)=\sum_{k_1=0}^{K}(\cdot)\cdots\sum_{k_{p+q}=0}^{K}(\cdot)$; $p+q$ is named as nonlinear degree of $c_{p,q}(\cdot)$; $c_{p,q}(\cdot)$ with p + q = 1 is referred to as linear parameters such as $c_{0,1}(.)$ and $c_{1,0}(.)$, corresponding to $\frac{d^k\mathbf{q}(t)}{dt^k}$ or $\frac{d^k u(t)}{dt^k}$, and $c_{pq}(\cdot)$ with $p+q>1$ nonlinear parameters, corresponding to nonlinear terms $\prod_{i=1}^{p}\frac{d^{k_i}\mathbf{q}(t)}{dt^{k_i}}\prod_{i=p+1}^{p+q}\frac{d^{k_i}u(t)}{dt^{k_i}}$. The input–output relationship of (6.6) can be expanded into a Volterra series up to the order N as [28]

$$y(t)=\sum_{n=1}^{N}\int_{-\infty}^{\infty}\cdots\int_{-\infty}^{\infty}h_n(\tau_1,\ldots,\tau_n)\prod_{i=1}^{n}u(t-\tau_i)d\tau_i \tag{6.7}$$

where $h_n(\tau_1,\ldots,\tau_n)$ is the nth order Volterra kernel. Volterra series has been extensively used in the literature for nonlinear system identification and filtering design etc. Define the nth order generalized frequency response function (GFRF) [28]

$$H_n(j\omega_1,\ldots,j\omega_n)=\int_{-\infty}^{\infty}\cdots\int_{-\infty}^{\infty}h_n(\tau_1,\ldots,\tau_n)$$
$$\exp(-j(\omega_1\tau_1+\cdots+\omega_n\tau_n))\mathrm{d}\tau_1\cdots\mathrm{d}\tau_n \tag{6.8}$$

The output spectrum of (6.7) can be computed by [18]

$$Y(j\omega)=\sum_{n=1}^{\infty}Y_n(j\omega) \tag{6.9}$$

which can be understood as an infinite order decomposition of output frequency response subjected to any input u(t) with Fourier transform U(jω), where the nth order component, i.e., the nth order output spectrum, given by

$$Y_n(j\omega)=\frac{1}{(2\pi)^{n-1}}\underbrace{\int_{-\infty}^{\infty}\cdots\int_{-\infty}^{\infty}}_{n-1}\Xi\,\mathrm{d}\omega_1\cdots\mathrm{d}\omega_{n-1}$$

with $U_n(j\omega,j\omega_1,\ldots,j\omega_{n-1})=U(j(\omega-\omega_1-\cdots-\omega_{n-1}))\cdot U(j\omega_1)\cdots U(j\omega_{n-1})$ and $\Xi=H_n(j\omega_1,\ldots,j\omega_{n-1},j(\omega-\omega_1-\cdots-\omega_{n-1}))\cdot U_n(j\omega,j\omega_1,\ldots j\omega_{n-1})$.

The nonlinear output spectrum (6.9) of the system (6.6) can be analytically computed with the recursive algorithms in [29–31] if the model in (6.6) is known. Because Eq. (6.9) is an explicit function of parameters $c_{p,q}(\cdot)$ [18–22], the characterization of any nonlinearity with a nonlinear degree p + q in (6.6) can be done by studying frequency properties of corresponding order output spectrum. Here the nth-order output spectrum $Y_n(j\omega)$ is targeted instead of the output response at the nth harmonic frequency $Y(jn\omega)$.

However, only some experimental data are usually available in nondestructive evaluation or fault/crack detection. The problem would be how to determine accurately the targeted nth-order output spectrum with only experimental data. The proposed data-based output spectrum decomposition algorithm without complex system modeling will be discussed in Sect. 6.3C based on (6.9), considering some practical difficulties such as selection of excitation magnitudes, inherent nonlinearity, filter design or noise problem as mentioned before.

Overall, the method proposed in this study is novel in the following (technical) points: (1) Accurate estimation of the 2nd order output spectrum with only experiment data, which will be theoretically shown the most robust to coupling influence of inherent nonlinearity (in Sect. 6.3C); (2) Construction of a virtual "regulation" output to regulate the measured output such that the desired nonlinear signal (in a specific frequency range) can be extruded or the undesired nonlinear signal can be suppressed (in Sect. 6.3BC); (3) Establishment of a systematic experimental data based algorithm or method for accurate output spectrum decomposition and thus for crack-related nonlinearity characterization and detection validated with real experiment results (in Sects. 6.3C and 6.4).

Related work can be seen in [24], which is to compute instead of directly $Y_n(j\omega)$ but the nth-order NOFRF defined by $Y_n(j\omega)/\left\{\frac{1}{(2\pi)^{n-1}}\iint U_n(\cdot)d\omega_1\cdots d\omega_{n-1}\right\}$. The method there does not focus on $Y_2(j\omega)$ (thus vulnerable to nonlinear coupling effect etc.) and cannot guarantee the accurate determination of the nth-order NOFRF [thus vulnerable to different experimental trials (see also Sect. 6.4(A))]. The work in [26] focuses on a much simplified model of (6.5a, b) and needs a known model subject to only sinusoidal inputs.

B. **The Nonlinear Output Response Regulation**

Considering the system (6.6), some information embedded in the output response $\boldsymbol{q}$(t) (given any input) could be of interest to be extracted in practice, for example the second order harmonic response as crack indicator. The desired output can be described in the frequency domain satisfying some requirements/properties (for example, all the other harmonic responses except the second order should be suppressed). To accurately and efficiently obtain the desired signal, the nonlinear regulation method can be applied with the following two methods.

The one way is to introduce nonlinear feedback to (6.6) so that the nonlinear system can be formulated with only output nonlinearities, and eventually the desired output response could be achieved [20, 32, 33]. The other method, being focused in this study, is to regard the output $\mathbf{q}$(t) of (6.6) as a "new" input to construct generally a nonlinear dynamic output equation as

$$\sum_{m=1}^{M}\sum_{p=0}^{m}\sum_{k_1,k_{p+q}=0}^{K}\overline{c}_{p,q}(k_1,\ldots,k_{p+q})\prod_{i=1}^{p}\frac{d^{k_i}\mathbf{q}(t)}{dt^{k_i}}\prod_{i=p+1}^{p+q}\frac{d^{k_i}u(t)}{dt^{k_i}}=y(t) \tag{6.10}$$

The nonlinear output spectrum of system (6.10) together with (6.6) can be computed explicitly with the recursive methods in [34] (see also Appendix 1). Obviously, the introduced nonlinearities in (6.10) with coefficients $\overline{c}_{p,q}(k_1,\ldots,k_{p+q})$ affect the output spectrum $Y(j\omega)$, and $\overline{c}_{p,q}(k_1,\ldots,k_{p+q})$ can be designed to achieve a desired output. In practice, the input $u(t)$ is not necessarily considered in (6.10) to further simplify the analysis and design. Then, (6.10) can be rewritten as

$$\sum_{m=1}^{M}\sum_{p=0}^{m}\sum_{k_1,k_p=0}^{K}\overline{c}_{p,0}(k_1,\ldots,k_p)\prod_{i=1}^{p}\frac{d^{k_i}\mathbf{q}(t)}{dt^{k_i}}=y(t) \tag{6.11}$$

Note that the new output y(t) in (6.11) can be constructed from the measured output or states of (6.6). Several methods are available for the analysis and design tasks above such as nonlinear energy transfer [19, 32, 33], nonlinear output spectrum based methods [20] and other methods related to the analytical relationship between output spectrum and model parameters [21]. Noticeably, different nonlinearities could have different effects on the output response at the same frequency. This coupling property was discussed in [19], which can be used for suppression or reinforcement of the output frequency response in a desired frequency range.

C. **Main Results**

Following Sect. 3.1–3.2, for the crack detection problem discussed in Sect. 6.2, the objective is to accurately characterize the 2nd order output spectrum which should be a function of γ and thus an indicator to the crack severity. To apply the nonlinear regulation method above to the crack detection problem, the following result can be obtained.

Proposition 1. *Consider system* (6.5a, b) *or* (6.6) *with output* $x(t) = c_1 q(t) + c_2\dot{\boldsymbol{q}}(t) + c_3\ddot{\boldsymbol{q}}(t)$ *(*c_1*,* c_2 *and* c_3 *are any constant in R). If the output is regulated (using the measured states* $\overline{x}(t)$*) as*

$$y(t) = c\cdot\overline{x}(t) + [\mu^T\cdot\overline{x}(t)]^3 \tag{6.12}$$

where $c \in R^3$ *is a constant (i.e.,* $c = [c_1, c_2, c_3]^T$*),* $\mu \in R^3$ *is a constant regulation parameter to design and* $\overline{x}(t) = [q(t), \dot{\boldsymbol{q}}(t), \ddot{\boldsymbol{q}}(t)]^T$*, then it can be obtained that,*

(1) $X_i(j\omega) = Y_i(j\omega)$ for i = 1 and 2;
(2) $Y_i(j\omega) = Y_{i,0}(j\omega) + Y_{i,1}(j\omega,\mu)$ for i = 3, 4, 5, ... where $Y_{i,0}(j\omega)$ and $Y_{i,1}(j\omega)$ are complex functions.

Proof. See Appendix 2 for the proof. □

From Proposition 1, by tuning the parameter μ, the other order output spectra could be suppressed or properly balanced due to introduced nonlinearity, while the 2nd order output spectrum is intact. Also, note that all other higher order (>2) nonlinearities have no effect on the 2nd order output spectrum (the latter is only determined by the 2nd order nonlinearity). Therefore, the 2nd order output spectrum receives the least coupling effects from other inherent nonlinearity in a system and would be the most reliable indicator to crack-induced nonlinearity, compared with the harmonic response at the 2nd harmonic frequency [23]. Proposition 1 provides an effective method for balancing coupling nonlinear effects in output response and thus facilitates the estimation or extraction of the 2nd order output spectrum $X_2(j\omega)$ from system output.

In practice, only measured output data (e.g., $\ddot{q}(t)$) are available for the purpose of non-destructive evaluation. Thus, an experiment-based output spectrum decomposition algorithm is proposed as follows. From (6.9), for a given input with Fourier transform denoted by $\rho R(j\omega)$, where ρ is an index of the relative excitation magnitude, it can be obtained that

$$\begin{aligned} Y(j\omega)_\rho &= \sum_{n=1}^{\infty} \rho^n Y_n(j\omega) = \rho Y_1(j\omega) \\ &\quad + \rho^2 Y_2(j\omega) + \rho^3 Y_3(j\omega) + \cdots \\ &= \sum_{n=1}^{N} \rho^n Y_n(j\omega) + \sigma_{[N,\rho]}(j\omega) \end{aligned} \tag{6.13}$$

where $\sigma_{[N,\rho]}(j\omega)$ is truncation error. To extract the 2nd order output spectrum from output response subjected to a given input signal, a series of experiments can be conducted with the same input signal of different excitation magnitudes. With these experiment data, an estimation method can be employed to estimate the 2nd order output spectrum $Y_2(j\omega)$ in (6.13). To this aim, the following method is adopted (Algorithm 1).

Step 1: Given a truncation order N in the Volterra series expansion (6.7), and an input signal $\rho R(j\omega)$, choose N excitation magnitudes as $\rho_1, \rho_2, \ldots, \rho_N$ satisfying

$$\begin{bmatrix} \rho_1 & \rho_1^2 & \cdots & \rho_1^N \\ \rho_2 & \rho_2^2 & \cdots & \rho_2^N \\ \vdots & \vdots & \ddots & \vdots \\ \rho_N & \rho_N^2 & \cdots & \rho_N^N \end{bmatrix} \text{ is nonsingular} \tag{6.14}$$

Step 2: Use the input $\rho_i R(j\omega)$ to excite the system for N times with different magnitude, and obtain N series of output whose FFT are denoted by $Y(j\omega)_{\rho 1}$, $Y(j\omega)_{\rho 2}$, $Y(j\omega)_{\rho 3}$, …, $Y(j\omega)_{\rho N}$;

Step 3: Then $Y_n(j\omega)$ for all $n = 1, 2, \ldots$, N can be estimated as

$$\begin{bmatrix} \hat{Y}_1(j\omega) \\ \hat{Y}_2(j\omega) \\ \vdots \\ \hat{Y}_N(j\omega) \end{bmatrix} = \begin{bmatrix} \rho_1 & \rho_1^2 & \cdots & \rho_1^N \\ \rho_2 & \rho_2^2 & \cdots & \rho_2^N \\ \vdots & \vdots & \ddots & \vdots \\ \rho_N & \rho_N^2 & \cdots & \rho_N^N \end{bmatrix}^{-1} \begin{bmatrix} Y(j\omega)_{\rho_1} \\ Y(j\omega)_{\rho_2} \\ \vdots \\ Y(j\omega)_{\rho_N} \end{bmatrix} \tag{6.15}$$

Note that it is easy to find ρ_1, ρ_2, …, ρ_N of different values so that (6.14) is satisfied. However, (6.15) can only guarantee that the estimated $\hat{Y}(j\omega)$ is sufficiently approximate to the real value $Y(j\omega)$ in an excitation range $[0, \overline{\rho}]$ ($\overline{\rho} = \max\{\rho_1, \rho_2, \ldots, \rho_N\}$) for the excited points (according to curve fitting theory), but not ensure that $\hat{Y}_n(j\omega)$ is accurate for all ρ. Actually, $\hat{Y}_n(j\omega)$ could be very biased due to truncation error $\sigma_{[N,\rho]}(j\omega)$ in (6.13). The following excitation-selection result is derived in order to accurately estimate $Y(j\omega)$ instead of fitting given points.

Definition 1. Give an input $R(j\omega)$ (i.e., the magnitude index $\rho = 1$) and a threshold ϵ, the best truncation order under the given input is denoted by $N^*(R(j\omega),\epsilon) = N^*$ satisfying

$$\left|\sigma_{[N^*,1]}(j\omega)\right| \le |Y_{N^*+1}(j\omega)| + |Y_{N^*+2}(j\omega)| + \cdots \le \varepsilon$$

Proposition 2. *For system* (6.5a, b) *or* (6.6) *with output* (6.12)*, applying Algorithm 1 and considering only the cases for* $N = 2$ *and* $N = 3$*, it can be obtained that,*

(a) if choosing $N = 2$, $\rho_1 = -\rho_2 = \rho$, then

(a1) the estimation errors for the first two orders of output spectra are

$$\Delta Y_1(j\omega) = \rho^2 Y_3(j\omega) + \rho^4 Y_5(j\omega) + \rho^6 Y_7(j\omega) + \cdots$$
$$\Delta Y_2(j\omega) = \rho^2 Y_4(j\omega) + \rho^4 Y_6(j\omega) + \rho^6 Y_8(j\omega) + \cdots$$

(a2) there exists $|\rho| \le 1$ so that $|\Delta Y_i(j\omega)| < \varepsilon$ for $i = 1, 2$ irrespectively of the value of $N^*(>1)$.

(b) if $N = 3$, $\rho_1 = -\rho_2 - \rho_3$, $\rho_2 = -\rho$, $\rho_3 = -\rho/\alpha$, $|\alpha|>1$, $0 < \rho < 1$, then

(b1) the estimation errors for the first 3 orders are

$$\Delta Y_1(j\omega) = \frac{1+\alpha}{\alpha^2}\rho^3 \cdot \left\{ Y_4(j\omega) + \frac{1+\alpha+\alpha^2}{\alpha^2}\rho^2 Y_6(j\omega) + \frac{1+\alpha}{\alpha^2}\rho^3 Y_7(j\omega) + \cdots \right\}$$

$$\Delta Y_2(j\omega) = \frac{1+\alpha+\alpha^2}{\alpha^2}\rho^2 Y_4(j\omega) + \frac{1+\alpha}{\alpha^2}\rho^3 Y_5(j\omega)$$

$$+\frac{1+2\alpha+3\alpha^2+2\alpha^3+\alpha^4}{\alpha^4}\rho^4 Y_6(j\omega)\cdots$$

$$\Delta Y_3(j\omega)=\frac{1+\alpha+\alpha^2}{\alpha^2}\rho^2 Y_5(j\omega)+\frac{1+\alpha}{\alpha^2}\rho^3 Y_6(j\omega)$$
$$+\frac{1+2\alpha+3\alpha^2+2\alpha^3+\alpha^4}{\alpha^4}\rho^4 Y_7(j\omega)\cdots$$

(b2) there exists $|\rho| \le 1$ so that $|\Delta Y_i(j\omega)| < \varepsilon$ for i = 1, 2, 3 irrespectively of the value of $N^*(>2)$.

Proof. See Appendix 3 for the proof. □

Propositions 1–2 present a general result that demonstrates the accurate determination of the 2nd order harmonic response using the nonlinear regulation method. In order to satisfy Eq. (6.14), it is necessary for the excitation magnitudes ρ_1, ρ_2, ρ_3 to be different. Proposition 2 further states that all of these magnitudes are related to a single tuning parameter ρ.

To apply the method described above, three free parameters need to be determined: the regulation coefficient μ, and the excitation magnitude parameters α and ρ. Heuristically, the parameter α has minimal impact on the performance of Algorithm 1, as it only affects the relative values of each excitation magnitude. On the other hand, the parameter ρ should be chosen carefully, as it determines the magnitudes of all the excitations involved and, consequently, affects the estimation accuracy. These will be discussed further.

6.4 Simulation Study

To demonstrate the method and results, an example is firstly given, which is to show how the method is applied for accurate estimation of the targeted 2nd order nonlinearity using experimental data. Then case study of dynamic response of the cracked beam subjected to harmonic excitation is given to show the effectiveness of the method in characterization of the 2nd order harmonic response of the crack incurred nonlinearity.

A. **An Example**

The dominant feature in crack-incurred dynamics is even-order nonlinearity. Thus, consider a nonlinear system as

$$\ddot{q}(t)+2\zeta\dot{q}(t)+q(t)+c_1 q(t)^2+c_2 q(t)^3=u(t) \tag{6.16}$$

The inclusion of the nonlinear term c_2q^3 in (6.16) is to represent potentially other inherent nonlinearity as an example for discussion. In the simulations, consider $\zeta = 0.02$ and $c_1 = 0.1$, both of which take small values, and c_2 will be given for case studies later. The input can be any signal but a sinusoidal function $u(t) = F_d{\cdot}\sin(\Omega t)$ is adopted here for convenience in discussion. The input frequency can actually be set to any value, which is taken here as $\Omega = \pi$ rad/s, a larger value than the resonance frequency of (6.16), since many practical NDE methods using ultrasonic sensors/actuators usually work at a frequency much larger than resonance frequency. The magnitude of the 2nd order harmonic response would be an indicator to the severity of the even nonlinearity in the system.

Reliable and accurate estimation with Algorithm 1 and Proposition 2. To demonstrate the effectiveness of Algorithm 1, the output is firstly taken as $y(t) = \ddot{q}(t)$, the input magnitude is set to $F_d = 10$, and only the even order nonlinearity is considered, i.e., $c_2 = 0$. Then to apply the algorithm, as an example, taking $N = 3$, $\rho = 0.001$ and $\alpha = 5$ firstly, the excitation magnitudes can be obtained by $\rho_1 = -\rho_2 - \rho_3$, $\rho_2 = -\rho$, $\rho_3 = -\rho/\alpha$ (Proposition 2), and the result is shown in Fig. 6.1a. The theoretical output spectra can also be computed with the method in Appendix 1, attached in Appendix 4. Clearly, the estimation of each order output spectrum is very accurate. Applying the same Algorithm 1 to estimate the output spectrum, instead of using Proposition 2 for selection of the excitation magnitudes, but arbitrarily taking $\rho_1 = 0.1$, $\rho_2 = 0.6$, $\rho_3 = 0.5$, and all the other parameters the same as before, the estimation is shown in Fig. 6.1b. Because the high order output spectra are very small, the estimation is also good in this case.

However, the arbitrary excitation selection method (e.g., adopted in [24]) cannot always guarantee reliable estimation in different cases. For example, when the cubic nonlinearity is considered e.g., $c_2 = 0.6$, the proposed excitation method with the same parameter setting as before can still give the same accurate estimation (see Fig. 6.2a and compare with Fig. 6.1a), while using the same arbitrary excitation magnitudes leads to obvious biased estimation (Fig. 6.2b). Note that the existence of cubic nonlinearity theoretically has no influence on the 2nd order output spectrum $Y_2(j\omega)$ but affects only higher orders [e.g., $Y_3(j\omega)$ is affected and thus varying in Figs. 6.1 and 6.2] [19–21]. Our method can give accurate estimation of $Y_2(j\omega)$ in all cases above while the arbitrary-magnitude-selection method cannot. This is the major problem not solved in [24], and also shows clearly that $Y_2(j\omega)$ is more reliable for characterization of crack-induced nonlinearity.

Moreover, the harmonic response at the 2nd harmonic frequency is $Y(j2\omega) = Y_1(j2\omega) + Y_2(j2\omega) + Y_4(j2\omega) + \ldots$. It will also be vulnerable to nonlinear coupling effect (e.g., the cubic nonlinearity has contribution to $Y_n(j\omega)$ $(n \geq 3)$ and thus affect $Y(j2\omega)$). Figure 6.3 indicates that $Y(j2\omega)$ is only about 0.016 while $Y_2(j2\omega)$ is actually about 0.07 much larger than the former value. The nonlinear coupling effect could come from system inherent nonlinearity or from measurement or data processing nonlinearity. Obviously, $Y_2(j\omega)$ is the most robust to all those coupling effect as revealed by Proposition 1.

Balance output component by tuning regulation factor μ. To show the usefulness of the regulation output given in Proposition 1, the regulated output is chosen as $y(t) = \ddot{q}(t) + \mu \cdot q(t)^3$. Consider system (6.16) with the same parameter setting as before, and $c_2 = 0.6$. When $\mu = 0$, the estimation results with Algorithm 1 can be

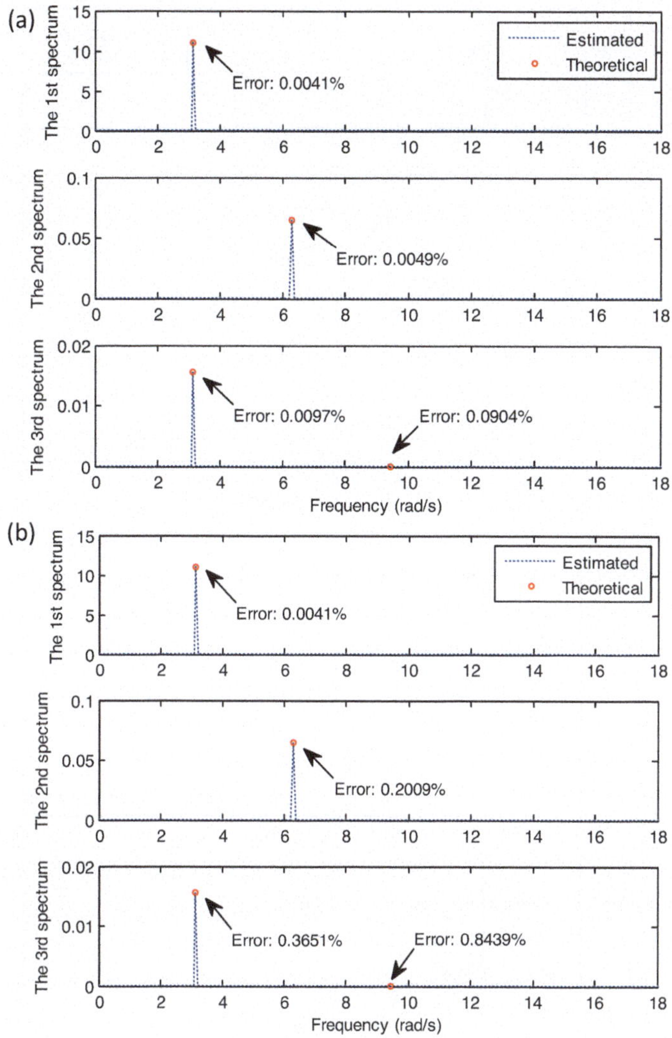

Fig. 6.1 Estimation of the first three orders of output spectra **a** with special designed excitation magnitudes and **b** with arbitrary selected excitation magnitudes. The relative errors are between the estimated and theoretical

referred to Fig. 6.2. Using the same $\rho = 0.001$ and $\alpha = 5$ as before and applying Algorithm 1, it can be seen that increasing μ to a proper value can decrease the magnitude of the higher order output spectra and thus very helpful for the accurate estimation of the first 2 orders of output spectrum components as indicated by Proposition 2. The estimation with $\mu = 0.605$ is shown in Fig. 6.4a, indicating that the third order output spectrum becomes very small in this case. Since the high orders

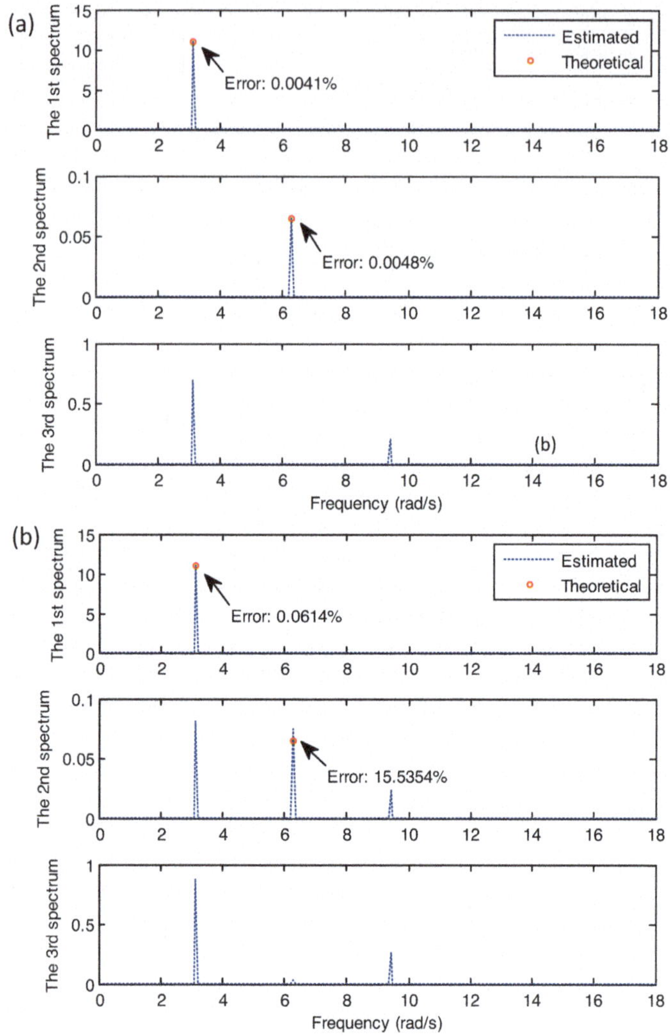

Fig. 6.2 Estimation of the first three orders of output spectra with cubic nonlinearity using **a** special designed excitation (very accurate) and **b** arbitrary selected excitation (obviously biased especially for the 2nd order)

of output spectra are very small, the Algorithm 1 with the arbitrary selected excitation magnitudes would also be effective, and the result with an arbitrary selected excitation magnitudes similar as before ($\rho_1 = 0.1, \rho_2 = 0.6, \rho_3 = 0.5$) is shown in Fig. 6.4b. Comparing with Fig. 6.2b, it can be seen that the estimation with arbitrary selected excitation magnitudes is better than before without output regulation although there are still estimation errors. This shows that the regulation method and the proposed selection method for excitation magnitudes are very useful and crucial

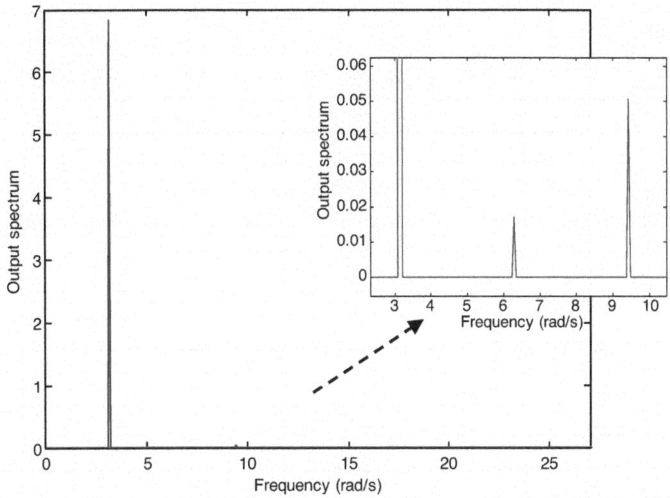

Fig. 6.3 The overall output spectrum of (6.16) without cubic nonlinearity, indicating that the output response at the 2nd harmonic frequency is affected by inherent nonlinearity and thus not as reliable as the 2nd order spectrum

for an accurate estimation. It should also be noted that the 3rd order output spectrum after regulation (Fig. 6.4a) has almost zero output at 3π rad/s and the output energy is dominantly located at the frequencies π and 2π rad/s. This property obviously facilitates application of low-pass filter for removing high frequency measurement noise but without affecting useful signal in practice.

Robust to data-cutting with Algorithm 1. The proposed method can still be accurate for noisy data if the noise effect is not very significant (e.g., the noise to signal ratio is <100%) at the concerned frequency points. This will be shown in a following section. Here it is to show that, when there is high-frequency noise or other disturbance corruption in measurement data, a digital filter can be applied to directly remove all the frequency components other than those output frequency points to be monitored with only the useful information at desired frequency points left. This can effectively suppress noise effect and does not affect the accuracy of the estimation using Algorithm 1 with the excitation magnitude selection method in Proposition 2.

To show this, consider system (6.16) with $c_1 = 0.8$, $c_2 = 1$ and output $y(t) = \ddot{q}(t)$, using the same excitation method and parameter setting as before ($N = 3$, $\rho = 0.001$ and $\alpha = 5$). The estimation result is shown in Fig. 6.5a. When cutting off all the high frequency components larger than 2.4π rad/s, the estimation result using only the rest data is shown in Fig. 6.5b. It can be seen that the estimation is very accurate in both cases.

Then, consider the measurement data above are corrupted by high frequency noise (larger than 2.5π rad/s) of signal-to-noise ratio about 2. A data series with noise is

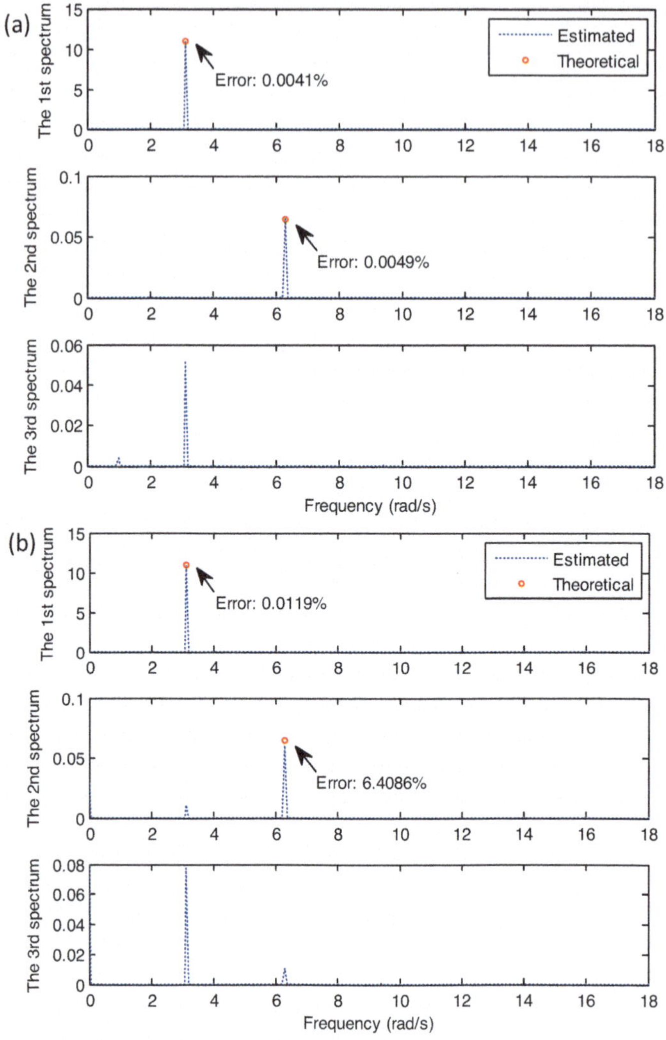

Fig. 6.4 Estimation of the first three orders of output spectrum with regulation **a** with special designed excitation magnitude and **b** with arbitrary selected excitation magnitude. The proposed method is always accurate, while the same algorithm but with arbitrary selected excitation magnitudes can be much better (compared with Fig. 6.2b) after output regulation

shown in Fig. 6.6a, and the estimation result in Fig. 6.6b, which shows that the estimation accuracy is the same as before for noise-free data.

It should also be noted that the property demonstrated here with Algorithm 1 can also be used to remove some computation burden when there are a large time series of data available for analysis. Since only some low frequency points are focused,

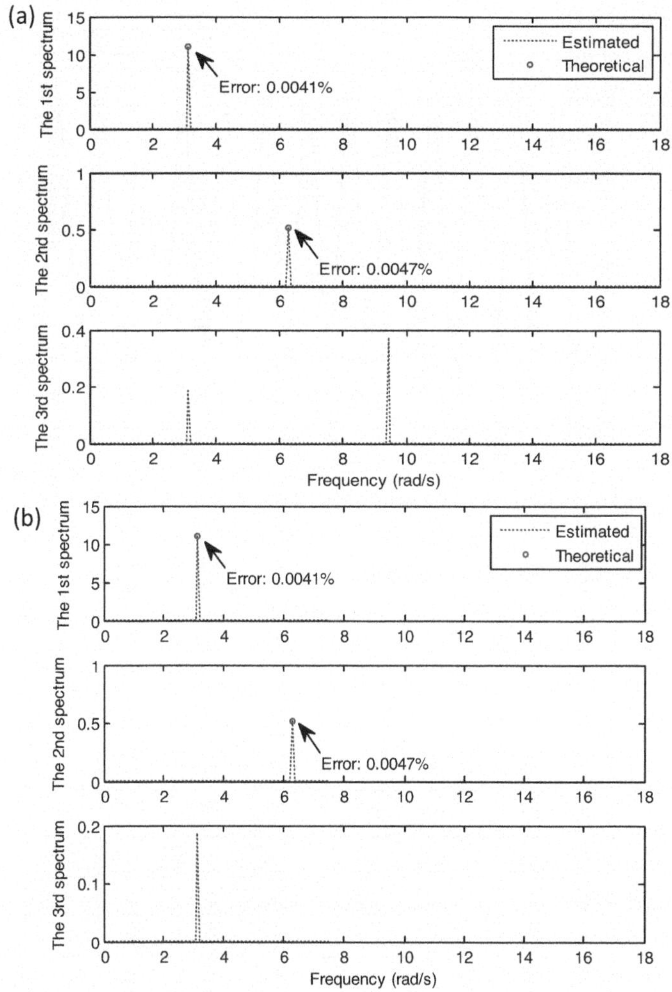

Fig. 6.5 Estimation with full data series (**a**) and with only low frequency components by cutting off all higher frequency components (**b**), showing that the estimation for noisy data can still be accurate if the noise effect is not significant at the concerned frequency points

high frequency data can directly be removed from its FFT data for the rest estimation in Algorithm 1 to speed up computation.

Selection of excitation magnitude ρ and α in Proposition 2. An important issue for applying Algorithm 1 for output component estimation is to select excitation magnitude parameters ρ and α. The parameter α is used to differentiate different excitation magnitudes and thus is not a very critical factor for the algorithm, empirically taking a suggested value around 5. The critical factor would be the parameter ρ, which represents potentially the maximum excitation level. According to Proposition 2, if

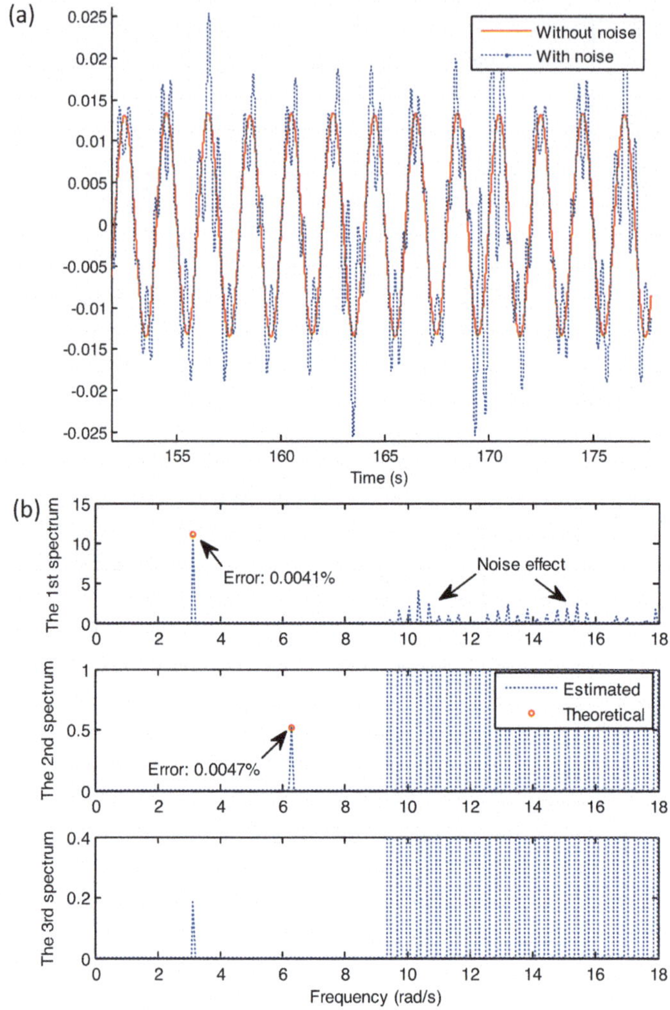

Fig. 6.6 The noise has no effect on the estimation accuracy

higher order output spectra are very small, it would be easy to choose a value for ρ. On the contrary, it would be rather difficult for an accurate estimation if higher order output spectra are very strong.

To find the best excitation parameter ρ, a ρ-selection method can be adopted. Simply speaking, ρ can start with a small number e.g., $\rho = 0.001$ and increase regularly by Δ until a larger number e.g., $\rho = 0.5$; for each ρ, run the Algorithm 1, and generate the curve $\frac{|Y_2(j\omega)_{\rho+\Delta} - Y_2(j\omega)_{\rho}|}{\Delta} \sim \rho$ or simply $|Y_2(j\omega)_{\rho}| \sim \rho$; the best value for ρ corresponds to the bottom of the resulting "U" or "J" shape curve leading to the smaller estimation error of $Y_2(j\omega)$. Usually, ρ has a large range to choose.

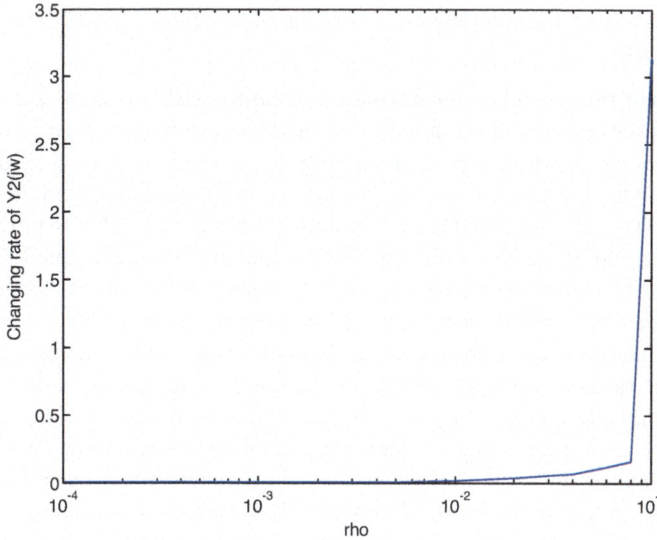

Fig. 6.7 The J-shape curve of the ρ-selection method

For the simulation case above with $c_1 = 0.8$, $c_2 = 1$, $N = 3$, $\alpha = 5$, and output $y(t) = \ddot{q}(t)$, the estimation with $\rho = 0.001$ is shown in Fig. 6.5a, and the ρ–selection curve is shown in Fig. 6.7. From the U or J-shape curve it can be seen clearly that ρ can be chosen as any value smaller than 0.01 such that the estimation would be accurate, and $\rho = 0.001$ (used before) of course results in an accurate estimation. It can be checked that, when ρ is chosen larger than 0.01, the estimation error would become larger and larger (this doesn't mean that ρ should be as small as possible in all cases). Moreover, note that the maximum input magnitude in the above analysis is $F_d * \left(\rho + \frac{\rho}{\alpha}\right) = 0.12$ for $\rho = 0.01$ and 0.012 for $\rho = 0.001$. Both cases indicate a smaller and reasonable excitation magnitude level for practical implementation.

B. **Evaluation of the Crack-Incurred Nonlinearity in (6.5a, b)**

The method and results are demonstrated in the following aspects. (A) Considering (6.5a, b), the estimation of the 2nd order output harmonic response is conducted with a crack severity parameter $\gamma = 0.98$ (a smaller value means much severer but easier to detect in practice); the selection of the parameters μ and ρ etc. follows the methods discussed before with more focus on their effects on the estimation results. (b) The relationship between the magnitude of the second order harmonic response and the severity parameter γ is studied. (c) The robustness of the proposed method with respect to measurement noise is also discussed finally. The input still takes a sinusoidal function $u(t) = F_d \cdot \sin(\Omega t)$ with $F_d = 1$ and $\Omega = \pi$ rad/s (similar results can be obtained for other input magnitude and frequency). For (6.5a, b), the regulated output is chosen as $y(t) = \ddot{q}(t) + \mu \cdot \ddot{q}(t)^3$ (i.e., $x(t) = \overline{x}(t) = \ddot{q}(t)$ in (6.12); μ can be zero). Moreover, the damping ratio of (6.5a, b) is set to $\zeta = 0.02$ in simulations if not specified.

Estimation of the second order harmonic response with $\gamma = 0.98$. The selection of an appropriate excitation magnitude ρ is crucial for estimation of the second order harmonic response, which can be done with the ρ-selection method as discussed before to find the U-shape curve. Several reasons could result in the failure to obtain a U-shape curve. The one could be that the magnitude values are chosen too large; the second one could be due to the error in FFT computation since accurate FFT results can only be obtained when the data number is an integer multiple of the underlying period in the data; and another one could be that, for given excitation range and excitation frequencies, system nonlinear dynamics behave more complicated than what a Volterra series can approximate. In the last case, the second order harmonic response is greatly corrupted by complicated nonlinear behaviours such as strong subharmonics, bifurcation and/or chaos etc., and therefore would be changed with system parameters not in a Volterra type (thus the output response cannot be exactly discomposed in a Volterra way). Moreover, when there are complicated nonlinear behaviours existing in the output response, it is also difficult to have an accurate FFT result. Therefore, to overcome this problem and to estimate the second order harmonic response accurately, the regulation method can be used such that the nonlinear output response is balanced to be more Volterra-type [19, 31, 35] and thus to protrude the second order harmonic response.

Simulation results show that, without output regulation ($\mu = 0$), the estimate of the second order harmonic response is kept decreasing with the increase of the excitation magnitude (although the decreasing rate is becoming smaller) and eventually an L-shape is formed instead of a U or J shape (see Fig. 6.8a). When checking the estimate of the 3rd order output spectrum, the output is always located at 2π and 4π rad/s only, instead of the correct positions 1π and 3π rad/s whatever excitation magnitude is used (see Fig. 6.8b). This definitely indicates more complicated nonlinear dynamics existing in the system with the given parameter setting [not just a simple even order nonlinearity as system (6.16)], and thus absolutely a biased estimation. To overcome this, the regulated output can be used, for example, taking $\mu = 1000$ (similar results can be obtained for any $\mu \geq 100$), and the results are shown in Fig. 6.9ab. Figure 6.9a gives a typical U-shape curve and the estimate when ρ takes the value at the bottom of the U-shape curve is shown in Fig. 6.9b. The estimates for each order output spectrum are reasonably discomposed, and the estimation error in the 1st order output spectrum can be checked easily as 0.11%. If ρ takes the other values, the error would be larger, which shows again that the arbitrary excitation magnitude selection method in [24] would lead to biased estimation.

The relationship between $|Y_2(j\omega)|$ and γ. To demonstrate the relationship between $|Y_2(j\omega)|$ and γ, the same parameter setting is used as before but the crack severity parameter γ is changed from 1 to 0.9 (i.e., the severity can be regarded to change from 0 to 0.1). The result is shown in Fig. 6.10. The magnitude of the 2nd order output spectrum is linearly growing with the severity $1 - \gamma$, indicating that the magnitude of $Y_2(j\omega)$ can be used as a perfect indicator to the severity of system crack-incurred

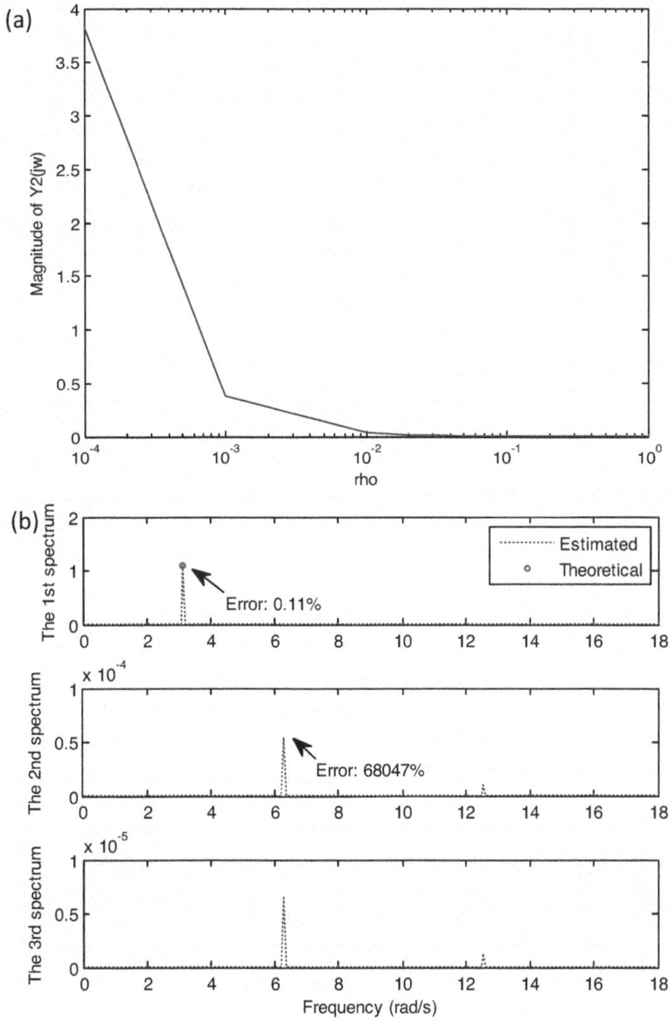

Fig. 6.8 Estimation without output regulation: **a** the estimated magnitude of $Y_2(j\omega)$ is decreasing with the increase of ρ, forming a L-shape; **b** the estimates when $\rho = 7$, indicating a wrong estimation

nonlinearity since $|Y_2(j\omega)|$ is a linear function of $1 - \gamma$. This is consistent with existing experimental and modelling results in [23, 27]. However, this study presents a completely different, but more efficient method to characterize the severity of crack-incurred nonlinearity, as discussed before including: (a) an estimation of the 2nd order harmonic response instead of harmonic response at the 2nd harmonic frequency, (b) efficient methods for the estimation with only experiment data. A simple linear approximation to this linear relationship is further shown in Fig. 6.10b with a very small root mean square error (RMSE) 1.4884e−004.

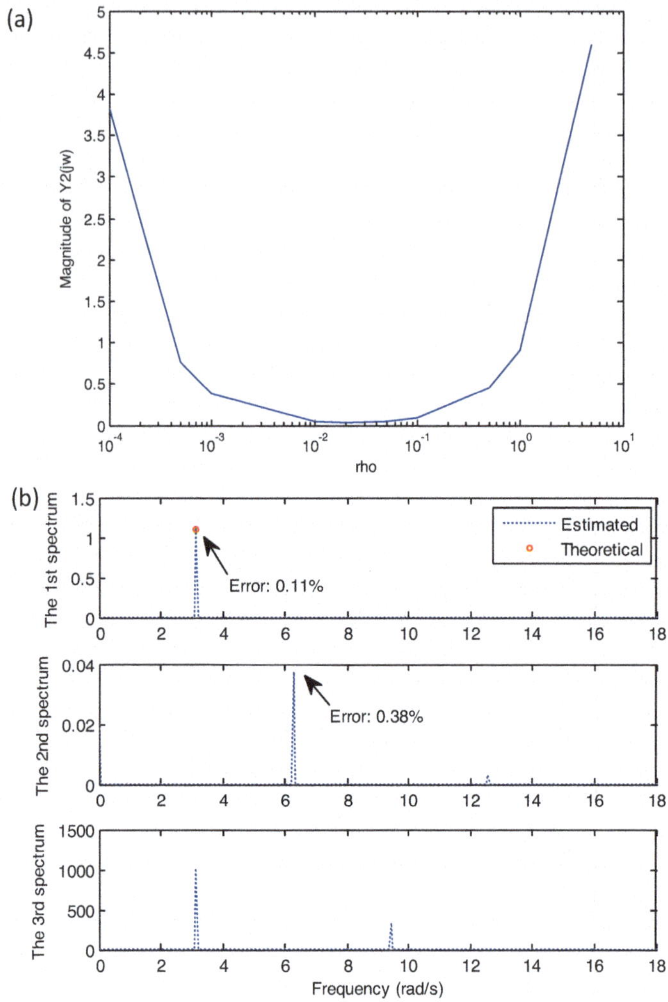

Fig. 6.9 Estimation with output regulation: **a** a typical U-shape is obtained in the $\left|Y_2(j\omega)_\rho\right| \sim \rho$ curve; **b** the estimates when $\rho = 0.02$ the bottom value of the U-shape, indicating an accurate estimation

Therefore, the higher magnitude of the 2nd order harmonic response implies the worse severity of the crack. For similar structures or systems, the severity can be compared and evaluated quantitatively due to the linear relationship above since the slope depends only on the material properties.

The robustness to measurement noise. As mentioned before, here is to show that, the estimation is robust to noise corruption. If the noise level is less than the energy at the 2nd harmonic frequency (the NTS ratio <1), the proposed method can still give satisfactory estimation. Considering the same parameter setting as before but

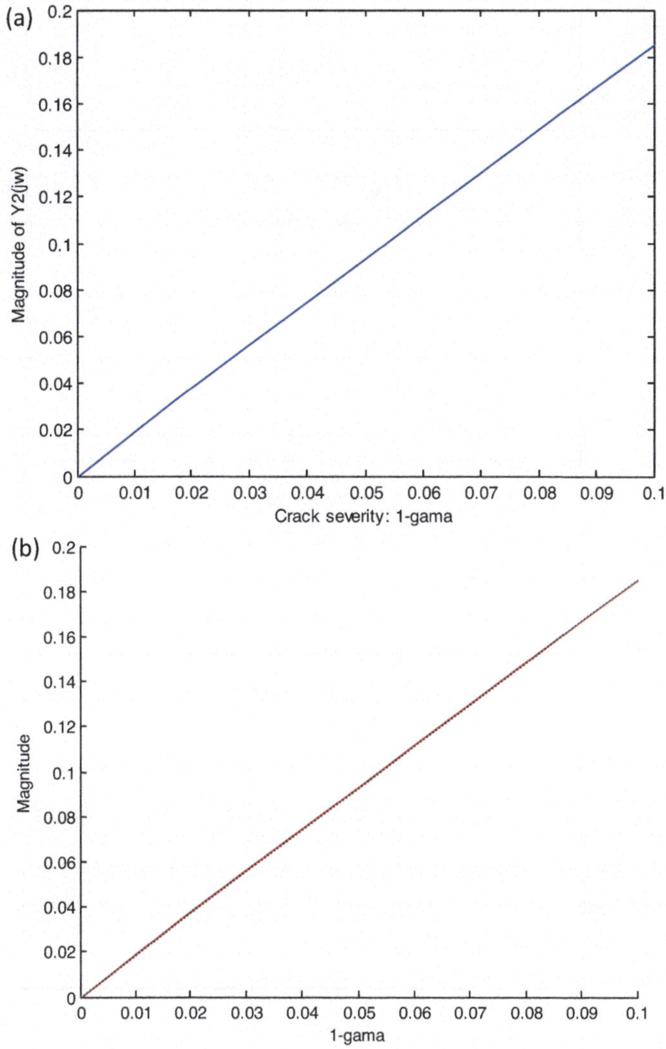

Fig. 6.10 The relationship between $|Y_2(j\omega)|$ and γ: **a** the estimation with different γ, **b** a fitting curve to the relationship is obtained as y = 1.8522× with RMSE = 1.4884e−004 (solid—original curve; dot—the fitting curve)

the data is corrupted by white measurement noise with signal to noise ratio around 20. For $\gamma = 0.9$, the estimation without noise is given in Fig. 6.11a, and the result with noise in Fig. 6.11b. Clearly, the noise effect on the estimation accuracy is very small. The error is about 0.0127% in the 1st order harmonic response, and 6% in the 2nd order.

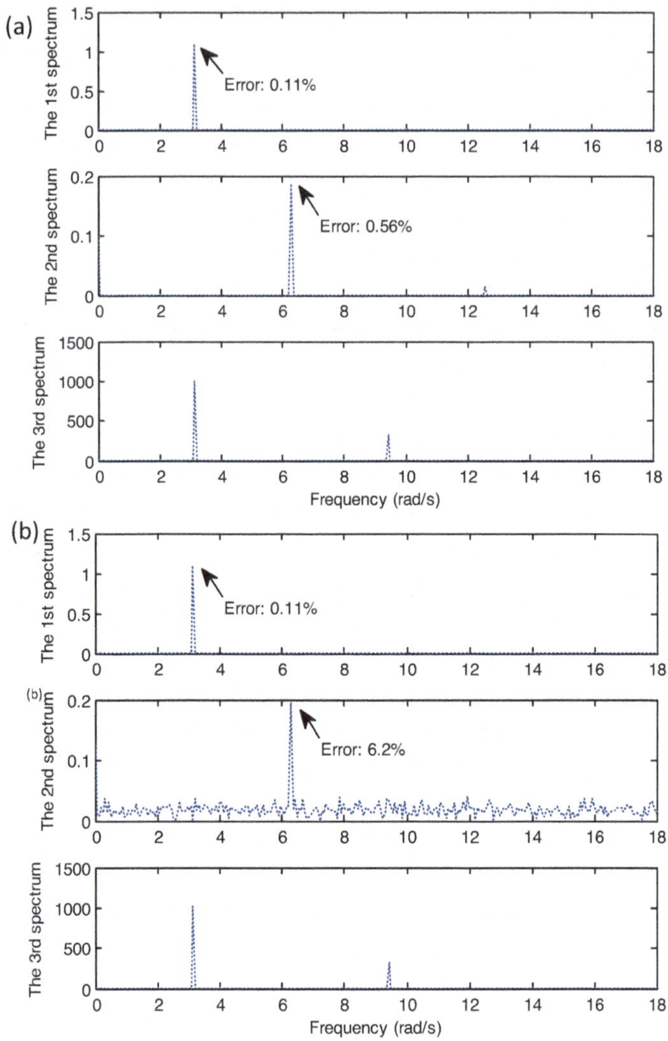

Fig. 6.11 Estimation with noise-free data (**a**) and with noisy data (**b**)

C. FEM Validation

To validate the proposed method, a Finite Element Model of a cantilever beam with a breathing crack is constructed with aluminium materials (modulus = 7e10Pa, possion ratio = 0.3, density = 2700 kg/m^3, size 1 × 0.08 × 0.04 m, damping ratio 0.1) as shown in Fig. 6.12. The force excitation is applied at the free top up and down with $R(t) = 1000 \sin(1.5\omega_0 t)$ N, where ω_0 (=33.05 Hz) is the 1st modal frequency of the

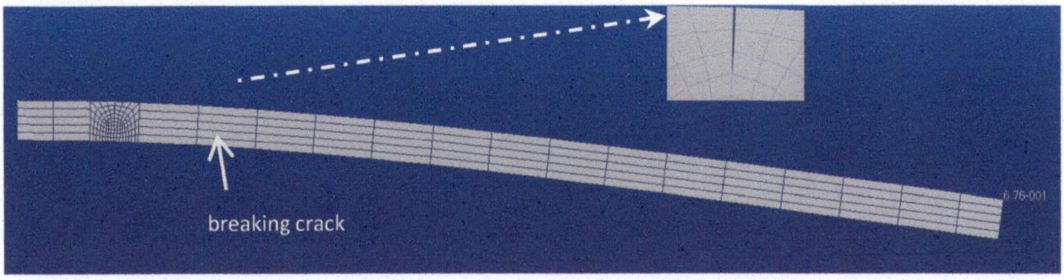

Fig. 6.12 FEM of a beam with a breathing crack at 0.1 m to the fixed root

uncracked beam. Actually, other excitation signals or frequencies can reach similar results. Acceleration signals are also taken at the free top for analysis.

With the proposed method, the 2nd order output spectra for the given beam above of different crack depth can all be estimated, which are shown in Fig. 6.13a and b. The detailed procedure can follow Procedure 1 given below with $\rho = 0.0005$, $\alpha = 5$. The output regulation adopts a similar method as the previous section for (6.5a, b). The estimation accuracy can be checked with each order output spectrum as before. Typical and easy-to-measure characteristics of accuracy are: (a) accurate estimation of the linear output response, and (b) the point (6.5b) in Procedure 1 below. The detailed discussion is omitted here. Based on our analysis, the estimation errors mainly come from FE modeling error and numerical calculation error. Note that the method in [24] does not target on the 2nd order output spectrum, takes no output regulation and apply random excitation magnitudes. These will lead to biased estimations as mentioned before, since in our case only the magnitude ρ taking values around 0.0005 gives correct estimation. For simplicity, comparisons are not given here.

It can be seen from Fig. 6.13a that the 2nd order output spectra of different crack depth roughly form a linearly-increasing trend. However, this trend increases more linearly with the 1st-modal frequency change of the cracked beam relative to the uncracked beam (Fig. 6.13b). This is an interesting, significant and also reasonable finding, since in the simplified model (6.5a, b) γ is related to system stiffness. This linear relationship with the FE modeling verifies the theoretical results above and also vindicates again that the magnitude of the 2nd order output spectrum is a reliable and measureable factor for crack severity assessment. Note also that the linear relation between the 2nd order output spectrum of the crack-incurred nonlinearity and the equivalent 1st order modal frequency has not been reported in the literature and can definitely be explored further (Table 6.1).

Overall, for a summary of the technical method proposed in this study, the following procedure can be used for a crack-related fault-detection task (Procedure 1):

(1) Take an appropriate excitation signal denoted by R(t).

    (a) It can be any signal which is convenient to use for excitation of the cracked system. The preferable signal would be single-tone sinusoidal input since it is easier for analysis and understanding in most cases.

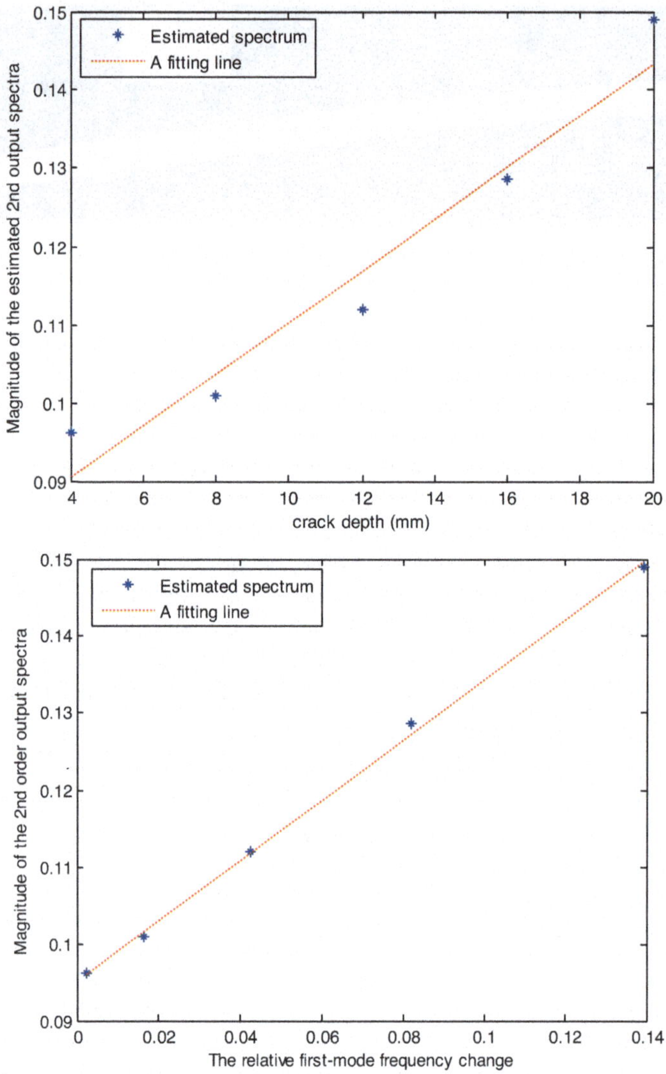

Fig. 6.13 The estimated 2nd order output spectra, increasing **a** roughly in a linear relation with crack depth, but **b** more linearly with the relative 1st modal frequency change of the cracked beam [corresponding to each crack depth in (**a**)] over the noncracked beam

(b) The magnitude of R(t) should be large enough to excite the crack-incurred dynamics.

(2) Use $\rho_1 R(t)$, $\rho_2 R(t)$ and $\rho_3 R(t)$ to excite the system and take the time domain response.

Table 6.1 Relative change of first model frequency

Crack depth (mm)	First modal frequency (Hz)	Relative frequency change (%)
0	33.05	0
4	32.974	0.23
8	32.497	1.67
12	31.643	4.26
16	30.341	8.20
20	28.449	13.92

(3) The excitation magnitude ρ_1, ρ_2 and ρ_3 should satisfy $\rho_1 = -\rho_2 - \rho_3$, $\rho_2 = -\rho$, $\rho_3 = -\rho/\alpha$, $|\alpha|>1$. α can take a vale such as 5 or −5 or else. ρ must be carefully chosen by the U-shape selection method (i.e., to find the corner (or bottom) excitation magnitude in the U-shape). If a U-shape curve cannot be obviously generated or the estimated output spectrum has obvious errors, the regulation method should be applied with an appropriate value for μ.

(4) Apply FFT to the time domain response in (6.2).

(5) Apply Algorithm 1 for output decomposition to achieve the first three order output spectra.

(5a) The first order spectrum is the linear response with input R(t), whose main frequency components are the same as R(t);

(5b) The second order spectrum is the main crack-incurred nonlinear dynamics with input R(t), whose main frequency components are 2-time multiples of the main frequency components in R(t).

D. **Experimental Validation**

Some experiments are preliminarily implemented to validate the theory and method of this study with a simple beam-like structure as shown in Fig. 6.14 (with 5 accelerometers placed). The beam is placed horizontally and the right end is closely attached to the shaker with the other end free. The shaker can excite the beam vertically (for example with a harmonic signal of a desired acceleration amplitude). The beam is consisted of two layer bolted steel panels of square size (40 cm*5 cm*0.4 cm). Breathing-like Cracks can be created in one layer at any positions for benchmark studying. Different severity of the crack can be created by using a different size or radius of steel wedge or wire inserted into the cut crack. Thus complicated cracks could be created in this way [referred to as cut-and-fill (C&F) cracks]. Three cases are considered here. The case 1 is shown in Fig. 6.14a which has no crack and the first frequency at about 95 Hz (referred to as experiment 1); the case 2 is in Fig. 6.14b with a heavier C&F crack near to the free end (experiment 2); and the case 3 with a light C&F crack near to the clamped end (experiment 3).

To apply the theoretical result in Proposition 2, the excitation is chose as a sinusoidal signal with frequency 50 Hz and TWO opposite magnitudes as shown in

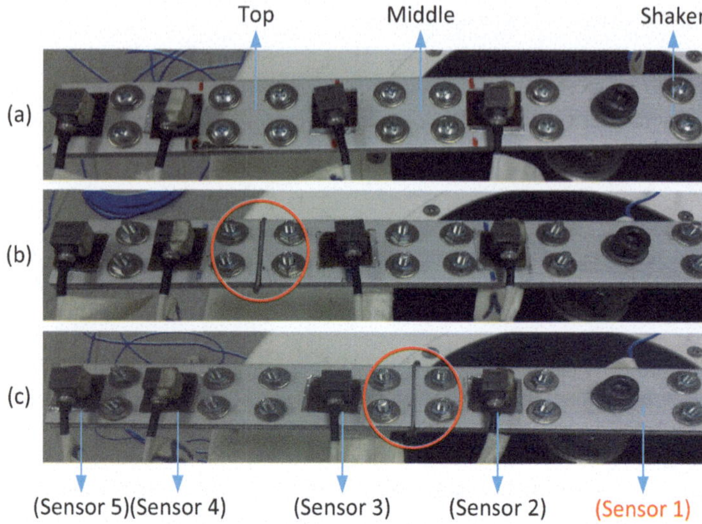

Fig. 6.14 Beam-like structures: **a** experiment 1 without crack; **b** experiment 2 with a crack near to the free end; **c** experiment 3 with a crack near to the clamped exaction point

Fig. 6.15 ($u = \pm\rho*\sin(6.28*50*t)$). The practical experiments need only one excitation with the positive magnitude, while the experimental result with the opposite excitation can be obtained by shifting the former one with a phase angle 180°. An appropriate excitation magnitude must be carefully chosen by following the ρ-selection method as discussed before. That is, (a) a series excitation magnitudes are trailed from the smaller one to bigger; (b) apply Algorithm 1 to decompose the output in the frequency domain and thus the first and second order output spectra are obtained; (c) to check the U-shape or J-shape curve and find the best excitation magnitude around the bottom. To show this, some results with real experimental data are shown in Fig. 6.16. It can be seen that the overall output spectrum of the case 1 without cracks has output at the second harmonic frequency. Thus the crack-free beam has weak inherent nonlinearity. For the 2nd order output spectrum after the decomposition, the amplitudes generally decrease with increasing the excitation magnitude until the excitation magnitude equals around 22 m/s^2, and then increase. This indicates a U-shape changing trend (Fig. 6.16), and therefore a reliable excitation magnitude can be chosen around this value. It can also be seen that the output spectra of (6.16) with this excitation magnitude have relatively smaller high frequency components (the ideal case is no higher frequency components). Similar phenomena can be seen in all the other experimental data of this beam-like structure of different cracks.

With the determined excitation magnitude, experiments with different crack cases can be done subject to similar sinusoidal inputs of the same amplitude and frequency. The data from the sensor 2–5 are used for validation of the proposed method (Propositions 1, 2, and Algorithm 1). The results are given in Tables 6.2, 6.3, and 6.4. Table 6.2 gives the magnitude results of the 2nd output spectrum obtained from each

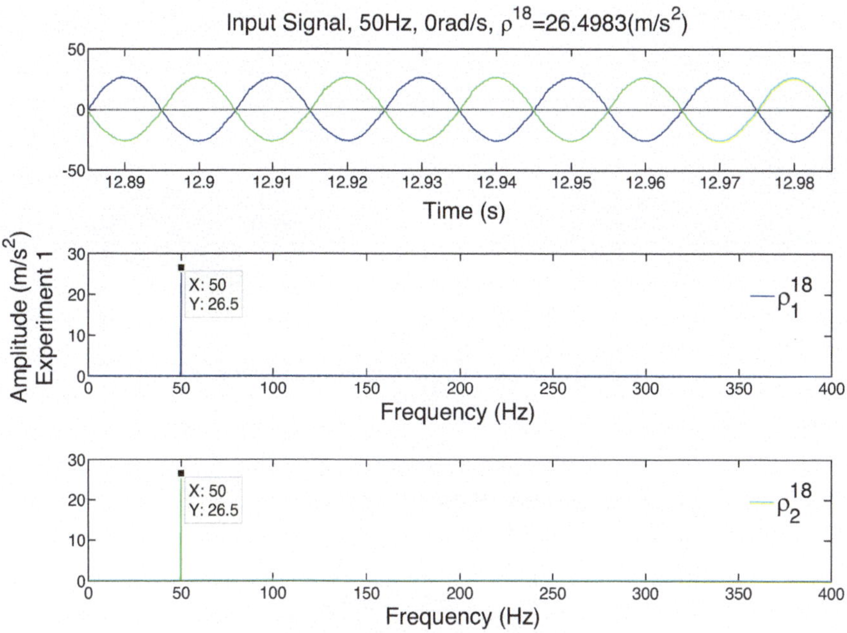

Fig. 6.15 Two opposite sinusoidal input signals in the time and frequency domain

sensor data (with Algorithm 1 using two opposite excitation (Proposition 2), and $c = [0\ 0\ 1]T$ and $\mu = 0$ in Proposition 1) for THREE cases [Experiment 1 without crack, Experiment 2 with sever crack at the middle of the beam between Sensors 3 and 4, and Experiment 3 with lighter crack at the root near to excitation point between Sensors 2 and 3 (see Fig. 6.14)].

To evaluate the results, the Experiment 1 without crack is taken as a benchmark. The 2nd order output spectra at different position of each sensor in Experiment 3 of Table 6.2 are slightly larger than those in Experiment 1 while smaller than those in Experiment 2. This indicates that the crack in Experiment 2 is worse than that in Experiment 3, which is exactly consistent with the real crack cases. Furthermore, all items of Experiment 2 or 3 in Table 6.2 can be normalized with respect to the corresponding ones of Experiment 1. This gives Table 6.3. It can be seen from Experiment 2 of Table 6.3 that the value at Sensor 3 is obviously different from the others at Sensors 2, 4 and 5. This corresponds to the crack between Sensors 3 and 4. Similarly, the value at Sensor 2 of Experiment 3 is obviously different from the others of Experiment 3, which clearly indicates the crack between Sensor 2 and Sensor 3.

To further understand these data obtained by the proposed method, the beam can be regarded as a vibration transmission path where the vibration energy is transmitted from Sensor 2, via Sensor 3, and Sensor 4, to Sensor 5. The energy ratio of the 2nd order output spectrum from each sensor following this transmission path above reflects the inherent physical relationship of successive locations. If there is a crack,

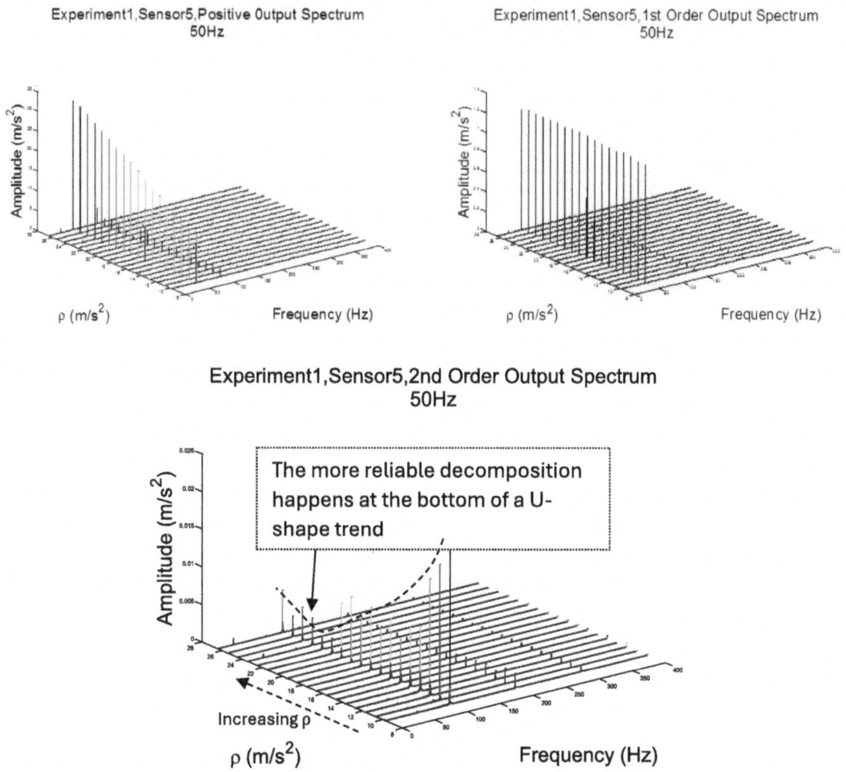

Fig. 6.16 The output spectrum and its decomposition of the case 1 with no crack

Table 6.2 The 2nd order output spectra at different positions in different cases

2nd output spectrum(m/s^2)	Sensor 2	Sensor 3	Sensor 4	Sensor 5
Experiment 1	0.000074	0.000491	0.001253	0.001696
Experiment 2	0.000250	0.001520	0.004081	0.005634
Experiment 3	0.000121	0.000531	0.001380	0.001792

Table 6.3 The relative ratio of the 2nd order output spectra of different positions with respect to Exp 1

2nd output spectrum ratio	Sensor 2	Sensor 3	Sensor 4	Sensor 5
Experiment 1	1	1	1	1
Experiment 2	3.378378	*3.095723*	3.256983	3.321934
Experiment 3	*1.635135*	1.081466	1.101357	1.056604

Table 6.4 The transfer ratio of the 2nd order output spectra from different positions in different cases

2nd output spectrum ratio	Sensor 2	Sensor 3	Sensor 4	Sensor 5
Experiment 1	1	6.635135	2.551935	1.353551
Experiment 2	1	*6.08*	*2.684868*	1.380544
Experiment 3	1	*4.38843*	2.59887	1.298551

the corresponding ratio would be changed. Table 6.3 is obtained from Table 6.2 by computing these ratios. The number of the table can be denoted by Ti(k, j), which refers to the number in Table 6.1 of Experiment k and Sensor j. For example, T2(2, 3) = 0.001520. Table 6.4 is obtained by computing T4(k, j) = T2(k, j)/T2(k, j − 1) for j = 3, 4, 5 and k = 1, 2, 3, and T4(k, 2) = 1 for k = 1, 2, 3.

From Experiment 1 of Table 6.4 it can be seen that, the energy in the 2nd order output spectrum at the position of Sensor 3 is about 6.6 times of Sensor 2, while that of Sensor 4 is about 2.5 times of Sensors 3 and 5 is about 1.3 times of Sensor 4. This trend is different in Experiment 2 at Sensors 3 and 4, while obviously different at Sensor 3 in Experiment 3. Both indicate correctly the crack between Sensor 3 and Sensor 4, and the crack between Sensors 2 and 3, respectively. The value at Sensor 3 of Experiment 3 in Table 6.4 is obviously changed. This can be understood that the crack near to the excitation point along the beam may affect the vibration energy transmission significantly than those near to the free end.

The beam-like structures are often studied in the literature as a benchmark for development of crack/damage/fault detection methods [36–38]. The fault diagnosis by measurement of the change in natural frequencies is a natural choice in practice [39–42]. However, cracks of light severity or at some uncritical positions or subject to a small excitation would bring too little change in natural frequencies to be detected in practice. The theory and methods demonstrated in this study potentially provide a new alternative approach to these issues. The beam-like structure in this section is clamped at the right end with the other end free (Fig. 6.14). By the experiment data, the first natural frequency of the case 1 without cracks is about 95 Hz, while that of the case 2 with a crack far from the excitation point is still about 95 Hz, which indicates almost no change (see Fig. 6.17). However, Table 6.2 shows clearly the obvious change of the 2nd order output spectrum of Experiment 2 which is also larger than that of Experiment 3, since the crack in Experiment 2 is sever than Experiment 3. Tables 6.3 and 6.4 also indicate potentially the location information of this crack. It shall be emphasized that our method does not require too big excitation amplitude but a reasonable small value with the ρ-selection method and also indicated by Proposition 2.

From the discussions and experiment results above, it can be concluded that,

(a) the proposed theoretical method is effectively validated with the given cracked beam-like structure;
(b) the 2nd order spectrum obtained with the method given by Algorithm 1 and Proposition 2 provides a reliable and physically-meaningful basis for diagnosis

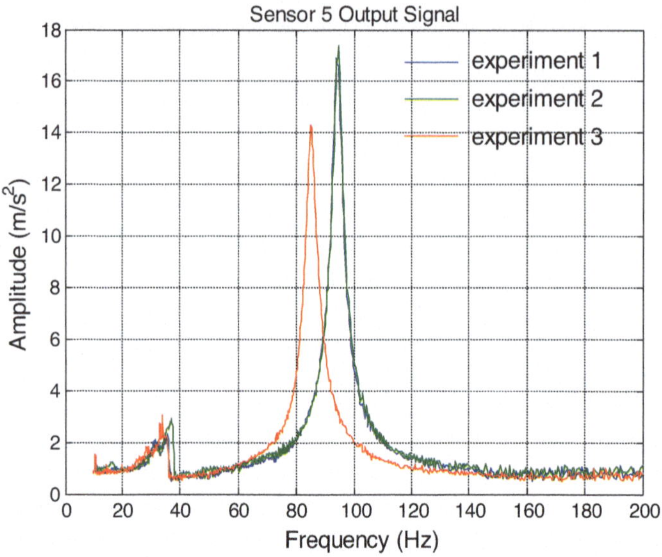

Fig. 6.17 Output response in the frequency domain of three cases

of crack severity and its location in the beam-like structure, which is sensitive to crack severity of the given beam-like structure;

(c) the relative ratio in Table 6.3 and the transmission ratio in Table 6.4 based on the 2nd order output spectrum of different position present a very special insight into the crack-incurred nonlinearity feature at different positions;

(d) only one excitation with a reasonable low amplitude is needed to compute the 2nd order output spectrum and the opposite excitation required by Proposition 2 can be obtained by shifting the former excitation result, which is very convenient to implement in practice;

(e) the noise in the data of the given experiments above brings littler effect on the proposed method, and this issue would be systematically investigated further in a future study.

6.5 Conclusions

The literature has not adequately addressed two important issues in nonlinearity characterization for crack-related fault detection. Firstly, there is a need for an effective algorithm to accurately estimate the superharmonic response, particularly during the initial stages of crack growth. Secondly, it is unclear which order of harmonic response is the most reliable for nonlinearity characterization.

This study clearly demonstrates that the second order output spectrum, rather than the extensively used second harmonic response, is the most reliable for nonlinear characterization. A systematic and feasible method for accurate estimation is proposed. The study also introduces a novel nonlinear regulation method for nonlinearity characterization in crack or damage detection. Theoretical analysis and example studies highlight the advantages and feasibility of the proposed method in characterizing nonlinear dynamics, which are further validated by experimental results.

The findings reveal that both even order nonlinearity and crack-induced nonlinearity (which is not a simple even order nonlinearity) can be effectively evaluated by examining the magnitude of the 2nd order harmonic response. Moreover, the 2nd order harmonic response is a linear function of crack severity and can be accurately estimated using the proposed method. Experimental results further confirm that the 2nd order output spectra at different positions of a beam-like structure can be effectively utilized to diagnose the location and severity of cracks.

The proposed method has the following advantages:

(a) It is robust to coupling influence incurred by system inherent nonlinearities since the 2nd order harmonic response is targeted and accurately estimated with the decomposition algorithm (i.e., Algorithm 1), instead of the harmonic response at the 2nd harmonic frequency [1, 2, 23], and non-preferable nonlinear output response can be deliberately regulated (compared with [24]);
(b) It is efficient in computation with only some experimental data, no modeling or system identification is involved (compared with [9–12, 26, 27]), and only a simple least square algorithm is conducted which can ensure accurate decomposition of output spectrum (compared with [24]);
(c) It is robust to measurement noise if the noise level is less than the output energy at the second harmonic frequency; This is not studied in e.g. [24, 26]; for noise of other properties, it would be investigated further;
(d) It is applicable to any input signal but not restrictive to sinusoidal inputs although only results with sinusoidal excitation are demonstrated. This can be seen from the output response decomposition in Sect. 6.3 (which is irrespective of any input).

The results of this study definitely provide a novel approach to nonlinearity detection and would benefit a large spectrum of engineering practice. Based on the theoretical proof of the feasibility and preliminary validation with experiment results in this chapter, further studies will focus on more experimental results and practical issues including different crack cases and different structures.

Acknowledgements The main result of this chapter can also be referred to the published work [43].

Appendices

1. **The Output Spectrum for System** (6.6, 6.10) [29, 30, 34]

The nonlinear output spectrum is determined by (6.9), i.e.,

$$Y(j\omega) = \sum_{n=1}^{N} Y_n(j\omega) \tag{6.17}$$

with $Y_n(j\omega) = \frac{1}{(2\pi)^{n-1}} \cdot \underbrace{\int_{-\infty}^{\infty} \cdots \int_{-\infty}^{\infty}}_{n-1} \Xi \;\; d\omega_1 \cdots d\omega_{n-1}$, and $U_n(j\omega, j\omega_1, \ldots, j\omega_{n-1}) =$ $U(j(\omega - \omega_1 - \cdots - \omega_{n-1})) \cdot U(j\omega_1) \cdots U(j\omega_{n-1})$,

$$\Xi = H_n^y(j\omega_1, \ldots, j\omega_{n-1}, j(\omega - \omega_1 - \cdots - \omega_{n-1})) \cdot U_n(j\omega, j\omega_1, \cdots j\omega_{n-1}).$$

The GFRFs from input u(t) to output y(t) are given by

$$\begin{aligned} H_n^y(j\omega_1, \ldots, j\omega_n) = {} & H_{n_u}^y(j\omega_1, \ldots, j\omega_n) + H_{n_\mathbf{q}}^y(j\omega_1, \ldots, j\omega_n) \\ & + H_{n_{u\mathbf{q}}}^y(j\omega_1, \ldots, j\omega_n) \end{aligned} \tag{6.18}$$

$$H_{n_u}^y(j\omega_1, \ldots, j\omega_n) = \sum_{k_1,k_n=0}^{K} \overline{c}_{0,n}(k_1, \ldots, k_n) \cdot (j\omega_1)^{k_1} \cdots (j\omega_n)^{k_n} \tag{6.19}$$

$$H_{n_\mathbf{q}}^y(j\omega_1, \ldots, j\omega_n) = \sum_{p=1}^{n} \sum_{k_1,k_p=0}^{K} \overline{c}_{p,0}(k_1, \ldots, k_p) \cdot H_{n,p}(j\omega_1, \ldots, j\omega_n) \tag{6.20}$$

$$\begin{aligned} H_{n_{u\mathbf{q}}}^y(j\omega_1, \ldots, j\omega_n) = {} & \sum_{q=1}^{n-1} \sum_{p=1}^{n-q} \sum_{k_1,k_{p+q}=0}^{K} \overline{c}_{p,q}(k_1, \ldots, k_{p+q}) \\ & \cdot H_{n-q,p}(j\omega_1, \ldots, j\omega_{n-q}) \\ & \cdot (j\omega_{n-q+1})^{k_{p+1}} \cdots (j\omega_n)^{k_{p+q}} \end{aligned} \tag{6.21}$$

$$\begin{aligned} H_{n,p}(\cdot) = {} & \sum_{i=1}^{n-p+1} H_{n-i,p-1}(j\omega_{i+1}, \ldots, j\omega_n) \\ & \cdot H_i^\mathbf{q}(j\omega_1, \ldots, j\omega_i) \cdot (j\omega_1 + \cdots + j\omega_i)^{k_p} \end{aligned} \tag{6.22}$$

$$H_{n,1}(j\omega_1, \ldots, j\omega_n) = (j\omega_1 + \cdots + j\omega_n)^{k_1} \cdot H_n^\mathbf{q}(j\omega_1, \ldots, j\omega_n) \tag{6.23}$$

$H_n^\mathbf{q}(j\omega_1, \ldots, j\omega_n)$ is the nth-order GFRF from u(t) to **q**(t) [30].

2. **Proof of Proposition** 1.

Consider **q**(t) as the input and y(t) as the output, then the GFRFs for (6.12) can be obtained by applying the results in [29, 30] as $H_n(.) = 0$ for all other orders except that $H_1(.) = c_1 + c_2(j\omega) + c_3(j\omega)^2$ and $H_3(.) = H_{3_u}(j\omega_1, \ldots, j\omega_3, \mu)$. From (6.17), the nth order output spectrum is computed by (for $n \geq 1$)

$$Y_n(j\omega) = \frac{1}{(2\pi)^{n-1}} \cdot \oiint H_n^y(j\omega_1, \ldots, j\omega_{n-1}, j(\omega - \omega_1 - \cdots - \omega_{n-1})) \cdot U_n(j\omega, j\omega_1, \ldots j\omega_{n-1}) d\omega_{1\to n-1} \tag{6.24}$$

where $\oiint \Xi d\omega_{1\to n-1}$ is actually $\underbrace{\int_{-\infty}^{\infty} \cdots \int_{-\infty}^{\infty}}_{n-1} \Xi d\omega_1 \cdots d\omega_{n-1}$. Then

$$\begin{aligned} Y(j\omega) &= H_1(j\omega)Q(j\omega) \\ &\quad + \oiint H_3(j\omega_1, j\omega_2, j(\omega - \omega_1 - \omega_2)) \cdot U_3(j\omega, j\omega_1, j\omega_2) d\omega_{1\to 2} \\ &= X(j\omega) + \oiint H_{3_u}(j\omega_1, j\omega_2, j(\omega - \omega_1 - \omega_2), \mu) \\ &\quad \cdot Q(j\omega - j\omega_1 - j\omega_2)Q(j\omega_1)Q(j\omega_2) d\omega_{1\to 2} \end{aligned}$$

where $X(j\omega) = H_1(j\omega)Q(j\omega)$. If there is nonlinear output response of any order in $Q(j\omega)$, then there will be similar output of the same order in $X(j\omega)$ due to their linear relationship. Obviously, the second term in the equation above can only produce nonlinear output of order larger than 2. Therefore,

(1) $Y_i(j\omega) = X_i(j\omega)$ for i = 1 and 2;
(2) $Y_i(j\omega)$ for i = 3, 4, 5, ... would be a function of the coefficient μ, i.e., $Y_i(j\omega) = Y_{i,0}(j\omega) + Y_{i,1}(j\omega, \mu)$

The detailed expressions for $Y_{i,0}(j\omega)$ and $Y_{i,1}(j\omega)$ can be obtained by following the method in [34].

3. **Proof of Proposition** 2.

Lemma 1. *If* $\rho_1 + \rho_2 + \cdots + \rho_N = 0$ *is satisfied, the following results can be obtained for the estimation errors:*

$$\begin{aligned} \Delta Y_1(j\omega) = (-1)^{N-1}\rho_1\rho_2\cdots\rho_N \\ \cdot \{Y_{N+1}(j\omega) + (\rho_1 + \rho_2 + \cdots + \rho_N)^{[2]} Y_{N+3}(j\omega) \\ + (\rho_1 + \rho_2 + \cdots + \rho_N)^{[3]} Y_{N+4}(j\omega) + \cdots\} \end{aligned} \tag{6.25}$$

$$\begin{aligned} \Delta Y_{N-1}(j\omega) = (\rho_1 + \rho_2 + \cdots + \rho_N)^{[2]} Y_{N+1}(j\omega) \\ + (\rho_1 + \rho_2 + \cdots + \rho_N)^{[3]} Y_{N+2}(j\omega) + \ldots \end{aligned} \tag{6.26}$$

$$\Delta Y_N(j\omega) = \left\{(\rho_1 + \rho_2 + \cdots + \rho_N)^{[2]} \cdot Y_{N+2}(j\omega) \right. \\ \left. +(\rho_1 + \rho_2 + \cdots + \rho_N)^{[3]} \cdot Y_{N+3}(j\omega) + \cdots \right\} \tag{6.27}$$

where,

$$(\mathrm{x}_1 + \mathrm{x}_2 + \cdots + \mathrm{x}_\mathrm{N})^{[\mathrm{r}]} = \sum_{\substack{\mathrm{l}_1 + \mathrm{l}_2 + \cdots + \mathrm{l}_\mathrm{N} = r \\ \mathrm{l}_1, \mathrm{l}_2, \ldots, \mathrm{l}_\mathrm{N} \in \{0,1,2,\ldots,\mathrm{r}\}}} \mathrm{x}_1^{\mathrm{l}_1} \cdot \mathrm{x}_2^{\mathrm{l}_2} \cdots \mathrm{x}_\mathrm{N}^{\mathrm{l}_\mathrm{N}}.$$

Proof. Equations (6.25) and (6.27) can be obtained by using the results in [22] considering $\rho_1 + \rho_2 + \cdots + \rho_N = 0$. Equation (6.26) is proved here. From [22], it is known that

$$\Delta Y_n(j\omega) = \sum_{j=1}^{N} \left(\frac{(-1)^{N-n} \cdot \displaystyle\sum_{\substack{m = N-n; \\ j_1 \neq \cdots \neq j_m \in \aleph\backslash\{j\}}} \rho_{j_1}\rho_{j_2}\cdots\rho_{j_m}}{\displaystyle\prod_{k\in\aleph\backslash\{j\}} (\rho_j - \rho_k)} \cdot \rho_j^N \cdot \overline{\sigma}_{[N,\rho_j]}(\cdot) \right)$$

where $\overline{\sigma}_{[N,\rho_j]}(\cdot) = \rho_j^{-N-1} \cdot \sigma_{[N,\rho_j]}(\cdot)$. It can be derived that

$$\Delta Y_{N-1}(j\omega) = \sum_{j=1}^{N} \left(\frac{(-1)^1 \displaystyle\sum_{\substack{m = 1; \\ j_1 \neq \cdots \neq j_m \in \aleph\backslash\{j\}}} \rho_{j_1}}{\displaystyle\prod_{k\in\aleph\backslash\{j\}} (\rho_j - \rho_k)} \cdot \rho_j^N \cdot \overline{\sigma}_{[N,\rho_j]}(\cdot) \right)$$

$$= \sum_{j=1}^{N} \left(\frac{(-1) \cdot \displaystyle\sum_{\substack{m = 1; \\ j_1 \in \aleph\backslash\{j\}}} \rho_{j_1}}{\displaystyle\prod_{k\in\aleph\backslash\{j\}} (\rho_j - \rho_k)} \cdot \rho_j^N \cdot \overline{\sigma}_{[N,\rho_j]}(\cdot) \right)$$

$$= \sum_{j=1}^{N} \left(\frac{(-1) \cdot (\rho_1 + \ldots + \rho_{j-1} + \rho_{j+1} + \ldots + \rho_N)}{\displaystyle\prod_{k\in\aleph\backslash\{j\}} (\rho_j - \rho_k)} \cdot \rho_j^N \cdot \overline{\sigma}_{[N,\rho_j]}(\cdot) \right)$$

$$= \sum_{j=1}^{N}\left(\frac{(-1)\cdot(-\rho_j)}{\prod\limits_{k\in\aleph\backslash\{j\}}(\rho_j-\rho_k)}\cdot\rho_j^N\cdot\overline{\sigma}_{[N,\rho_j]}(\cdot)\right)$$

$$= \sum_{j=1}^{N}\left(\frac{1}{\prod\limits_{k\in\aleph\backslash\{j\}}(\rho_j-\rho_k)}\cdot\rho_j^{N+1}\cdot\overline{\sigma}_{[N,\rho_j]}(\cdot)\right)$$

From Lemma 1 in [22], the equation above gives,

$$\begin{aligned}\Delta Y_{N-1}(j\omega) &= \sum_{j=1}^{N}\left(\frac{\rho_j^{N+1}Y_{N+1}(j\omega)}{\prod_{k\in\aleph\backslash\{j\}}(\rho_j-\rho_k)}\right)+\sum_{j=1}^{N}\left(\frac{\rho_j^{N+2}Y_{N+2}(j\omega)}{\prod_{k\in\aleph\backslash\{j\}}(\rho_j-\rho_k)}\right)+\cdots\\ &= (\rho_1+\rho_2+\cdots+\rho_N)^{[2]}Y_{N+1}(j\omega)\\ &\quad+(\rho_1+\rho_2+\cdots+\rho_N)^{[3]}Y_{N+2}(j\omega)+\cdots\end{aligned}$$

which leads to Eq. (6.26). □

Consider the two cases. For N = 2, (6.25) and (6.27) gives

$$\Delta Y_1(j\omega) = \rho^2 Y_3(j\omega)+\rho^4 Y_5(j\omega)+\rho^6 Y_7(j\omega)+\cdots \tag{6.28}$$

$$\Delta Y_2(j\omega) = \rho^2 Y_4(j\omega)+\rho^4 Y_6(j\omega)+\rho^6 Y_8(j\omega)+\cdots \tag{6.29}$$

For $N^* = 4$, then (6.28) can be written as

$$\Delta Y_1(j\omega) = \rho^2 Y_3(j\omega)+\rho^4\left[Y_5(j\omega)+\rho^2 Y_7(j\omega)+\cdots\right]$$

Therefore, if $|\rho|\le 1$ and using Definition 1,

$$\begin{aligned}|\Delta Y_1(j\omega)| &\le |\rho|^2|Y_3(j\omega)|+|\rho|^4\{|Y_5(j\omega)|+|Y_7(j\omega)|+\cdots\}\\ &< |\rho|^2|Y_3(j\omega)|+|\rho|^4\varepsilon\end{aligned}$$

Thus, it must exist $|\rho|\le 1$ so that $|\Delta Y_1(j\omega)|<\varepsilon$. Similarly, the other cases can be proved when N^* takes different values. Also, it can be proved for the boundedness of $|\Delta Y_2(j\omega)|$.

For N = 3, considering $\rho_1 = -\rho_2-\rho_3$, $\rho_2 = -\rho$, $\rho_3 = -\rho/\alpha$, the results in (b1) can be obtained from (6.25)–(6.27). For $N^* = 3$, take the estimation error in the first order as example, i.e.,

$$\begin{aligned}\Delta Y_1(j\omega) &= \frac{1+\alpha}{\alpha^2}\rho^3\\ &\quad\cdot\left\{Y_4(j\omega)+\frac{1+\alpha+\alpha^2}{\alpha^2}\rho^2 Y_6(j\omega)\right.\end{aligned}$$

$$+\frac{1+\alpha}{\alpha^2}\rho^3 Y_7(j\omega)+\cdots\Big\}$$

There must exist α and ρ so that all the coefficients in the equation above are <1. Therefore, following the same method, it can be proved that $|\Delta Y_1(j\omega)|<\varepsilon$. Similar results can be obtained for the other orders.

4. **Compute the Nth Order Output Spectrum of** (6.16)

Using the method of Appendix 1 or that in [34], for u(t) $= F_d sin\omega t$, the first 3 orders of output spectrum are

$$\begin{aligned}
Y_1(j\omega) &= H_1(j\omega)\cdot F_d\cdot (j\omega)^2\\
Y_2(j2\omega) &= \frac{c(j2\omega)^2F_d^2}{2\left[(j2\omega)^2+2\zeta(j2\omega)+1\right]}\\
&\quad\cdot H_1(j\omega)H_1(j\omega)\\
Y_3(j\omega) &= -\frac{jF_d^3}{4}(H_3(\omega,\omega,-\omega)+H_3(\omega,-\omega,\omega)\\
&\quad+H_3(-\omega,\omega,\omega))\\
Y_3(j3\omega) &= \frac{H_3(\omega,\omega,\omega)}{4}\cdot\left(jF_d^3\right)\\
H_1(j\omega) &= \frac{1}{(j\omega)^2+2\zeta(j\omega)+1},\\
H_2(j\omega_1,j\omega_2) &= (j\omega_1+j\omega_2)^2H_2^q(j\omega_1,j\omega_2),\\
H_2^q(j\omega_1,j\omega_2) &= -\frac{cH_1(j\omega_1)H_1(j\omega_2)}{(j\omega_1+j\omega_2)^2+2\zeta(j\omega_1+j\omega_2)+1}\\
H_3(j\omega_1,j\omega_2,j\omega_3) &= (j\omega_1+j\omega_2+j\omega_3)^2H_3^q(j\omega_1,j\omega_2,j\omega_3)\\
&\quad+\mu\cdot H_1(j\omega_1)H_1(j\omega_2)H_1(j\omega_3)\\
&\quad(j\omega_1)^2(j\omega_2)^2(j\omega_3)^2\\
H_3^q(j\omega_1,j\omega_2,j\omega_3) &= -\frac{c\left(H_1(j\omega_1)H_2^q(j\omega_2,j\omega_3)+H_2^q(j\omega_1,j\omega_2)H_1(j\omega_3)\right)}{(j\omega_1+j\omega_2+j\omega_3)^2+2\zeta(j\omega_1+j\omega_2+j\omega_3)+1}
\end{aligned}$$

References

1. Ou J and Li H, Structural Health Monitoring in mainland China: Review and Future Trends, Structural Health Monitoring, 9(3): 219–13, 2010
2. Thompson DO and Chimenti DE, Review of progress in quantitative non-destructive evaluation, Plenum, New York, 1993

3. Onel I Y, Benbouzid M, "Induction motor bearing failure detection and diagnosis: Park and Concordia transform approaches comparative study," IEEE Trans Mechatronics, 13(2), 257-262, 2008
4. Jay Lee, Fangji Wu, Wenyu Zhao, et al, Prognostics and health management design for rotary machinery systems—Reviews, methodology and applications, Mechanical Systems and Signal Processing, 42, Issues 1-2, January 2014, Pages 314–334
5. Yong-Hwa Kim, Young-Woo Youn, Don-Ha Hwang, etc, "High-Resolution Parameter Estimation Method to Identify Broken Rotor Bar Faults in Induction Motors ,"IEEE Trans. on Industrial Electronics, vol. 60, no. 9, pp. 4103–4117 , Sep 2013
6. Esfahani, E T, Wang S, Sundararajan V, "Multisensor wireless system for eccentricity and bearing fault detection in induction motors" IEEE Trans. on Mechatronics, vol. 19, no. 3, pp. 818-826, 2014
7. Dai X, Gao Z, From model signal to knowledge: A data-driven perspective of fault detection and diagnosis, IEEE Trans. on Industrial Informatics, vol. 9, no. 4, pp. 2226-2238, Nov. 2013
8. D.C. Swanson, J.M. Spencer, S.H. Arzoumanian, Prognostic modelling of crack growth in a tensioned steel band, Mechanical Systems and Signal Processing, 14 (2000), pp. 789–803
9. Doebling, Scott W., Farrar, C. R., and Prime, M. B., "A Summary of Vibration-Based Damage Identification Methods," Shock Vibration Digest 30, 1998, pp. 91 -105.
10. Khiem, N.T., Lien, T.V., "Multi-Crack Detection for Beam by the Natural Frequencies," Journal of Sound & Vibration 273:175-85, 2004
11. Lee, YY, Liew K. M., "Detection of Damage Locations in a Beam using the Wavelet Analysis," Int. J. of Struct. Stability and Dynamics 1, 2001, pp. 455-465.
12. Bovsunovsky, A. and Bovsunovsky, O., "Crack Detection in Beams by Means of the driving Force Parameters Variation at Non-Linear Resonance Vibrations", Key Engineering Materials, v347, Damage Assessment of Structures VII, 2007, pp. 413–420.
13. Leonard, F., Lanteigne, J., Lalonde, S., et al, "Free-Vibration of a Cracked Cantilever Beam and Crack Detection," Mechanical Systems and Signal Processing 15, 2001, p529- 548.
14. Kim, J.T., Stubbs, N., "Crack Detection in Beam-Type Structures Using Frequency Data," Journal of Sound and Vibration 259, 2003,p145-161.
15. Ding, J.L., Pazhouh, J., Lin, S.B., Burton, T.D., "Damage Characterization by Vibration Test," Scripta Metallurgica et Materialia 30, 1994, pp. 839-834.
16. Sih, G.C., Tzou, D.Y., "Mechanics of Nonlinear Crack Growth: Effects of Specimen Size and Loading Step," Martinus Nijhoff Publications, 1984, pp. 155–169.
17. Andreaus, U., Casini, P., Vestroni, F., "Nonlinear Features In The Dynamic Response of a Cracked Beam Under Harmonic Forcing," 5th International Conference on Multibody Systems, Nonlinear Dynamics, and Control, 2005, pp. 2083–2089.
18. Lang Z.Q., Billings S.A., Yue R., Li J. Output frequency response functions of nonlinear Volterra systems. Automatica 43:805-816, 2007
19. Jing X.J., Lang Z.Q., Billings S.A., Output Frequency Properties of Nonlinear Systems, International Journal of Non-Linear Mechanics, 45(7):681-690, September 2010
20. Jing X. J., Lang Z.Q. Billings S.A., Output Frequency Response Function based Analysis for Nonlinear Volterra Systems. Mechanical Systems & Signal Processing, 22:102-120, 2008
21. Jing X. J., Lang Z. Q., Billings S. A. and Tomlinson G. R., The parametric characteristic of frequency response functions for nonlinear systems. International Journal of Control, 79(12):1552–1564, 2006
22. Jing X.J., Nonlinear Characteristic Output Spectrum for Nonlinear Analysis and Design, IEEE/ ASME Transactions on Mechatronics 19 (1), 171 - 183
23. Kim JY, Jacobs LJ, and Qu J, Experimental characterization of fatigue damage in a nickel-base superalloy using nonlinear ultrasonic waves, J. Acoust. Soc. Am. 120(3), 2006, p1266-1273
24. Lang ZQ and Peng ZK, A novel approach for nonlinearity detection in vibrating systems, J. of Sound and Vibration, 314(3-5), 603-615, 2008
25. Benaroya H, Mechanical Vibration: Analysis, Uncertainties and Control, Prentice Hall, New Jersey, 1998

26. Chatterjee A, Structural damage assessment in a cantilever beam with a breathing crack using higher order frequency response functions, Journal of Sound and Vibration, 329:3325-3334, 2010
27. Friswell MI and Penny JE, Crack modelling for structural health monitoring, Structural Health Monitoring, 1(2), 2002, p139-148
28. Rugh W.J., Nonlinear System Theory: the Volterra/Wiener Approach, Baltimore, Maryland, U.S., Johns Hopkins University Press, 1981
29. Billings S.A. and Peyton-Jones J.C., Mapping nonlinear integrodifferential equation into the frequency domain, International Journal of Control, 54: 863-879, 1990
30. Jing X. J., Lang X.Q., Billings S. A., Mapping from parametric characteristics to generalized frequency response functions of nonlinear systems. International Journal of Control, 81(7):1071–1088, 2008
31. Jing XJ & Lang, Frequency domain analysis and design of nonlinear systems based on Volterra series expansion: A parametric characteristic approach, Springer International Publishing, Switzerland, 2015
32. Billings S.A.; Lang ZQ., Non-linear systems in the frequency domain: energy transfer filters, Int. J. of Control, 75(14):1066-1081,2002
33. Jing X. J., Lang Z. Q., Billings S. A., Frequency Domain Analysis for Suppression of Output Vibration from Periodic Disturbance using Nonlinearities. Journal of Sound and Vibration 314:536 -557, 2008
34. Jing X. J., Lang Z.Q. and Billings S.A., Frequency domain analysis for nonlinear Volterra systems with a general nonlinear output function. International Journal of Control, 81(2): 235-251, 2008
35. Xiao Z.L., Jing X.J.*, Cheng L., Computation of Parametric Convergence Bound and Parametric Convergence Margin for Volterra Series Expansion, the 9th ASCC 2013, Istanbul, Turkey, June23–26, 2013
36. Q. Fan, P. Qiao, Vibration-based damage identification methods: a review and comparative study, Structural Health Monitoring 10(10)(2011)83–111.
37. P. Cawley, R.D. Adams, The location of defects in structures from measurements of natural frequencies, Journal of StrainAnalysis 14(2)(1979)49–57.
38. O.S. Salawu, Detection of structural damage through changes in frequency: a review, Engineering Structures 19(9)(1997)718–723.
39. A. Morassi, M. Rollo, Identification of two cracks in a simply supported beam from minimal frequency measurements, Journal of Vibration and Control 7 (5) (2001)729–740.
40. N.T. Khiem, T.V. Lien, Multi-crack detection for beam by the natural frequencies, Journal of Sound and Vibration 273(2004)175–184.
41. F.B. Sayyad, B. Kumar, Identification of crack location and size in a simply supported beam by measurement of natural frequencies, Journal of Vibration and Control 18(2)(2012)183–190.
42. N.T. Khiem, L.K. Toan, A novel method for crack detection in beam-like structures by measurements of natural frequencies, Journal of Sound and Vibration 333(2014)4084–4103
43. Jing X.J., Li QK, A Nonlinear Decomposition and Regulation Method for Nonlinearity Characterization, Nonlinear Dynamics, 83(3): 1355-1377, 2016

Chapter 7
Case Studies: Design and Optimization of X-Structures for Energy Harvesting—The nCOS Function Based Power Generation Function

Abstract Although nonlinear energy harvesting systems have been the subject of extensive research, a clear and efficient parametric design methodology remains underdeveloped. Conventional approaches, such as harmonic balance and numerical simulations, often fail to establish explicit relationships between energy harvesting performance and system parameters, and they typically demand substantial computational resources for parametric analysis. This study introduces a novel frequency-domain framework based on the Volterra series for analyzing a representative nonlinear energy harvesting system. A concise parametric characterization technique, termed the Linear and Nonlinear Characteristic Output Spectrum (lnCOS) method, is proposed to describe both the nonlinear output frequency response and the power generation function (PGF) of the system. The lnCOS function and the associated Parametric-PGF formulation offer a direct and computationally efficient means of predicting vibration responses and power output without the need to solve complex algebraic systems. By explicitly linking power generation to key structural parameters, this method provides a transparent and analytically tractable design tool. To the best of the authors' knowledge, this is the first study to derive closed-form expressions for both vibration response and power generation in nonlinear energy harvesting systems based on structural parameters—an achievement not attainable through traditional methods.

Keyword Vibration energy harvesting · Nonlinear dynamics · Structural optimization · Performance optimization · Frequency domain

7.1 Introduction

7.1.1 About This Chapter

The nCOS function method can be well applied for analysis and design of various nonlinear systems in a straightforward way. This chapter is to present an nCOS function based method for design, analysis and optimization of X-structures in order to

X. Jing, *The Bio-Inspired X-Structure/Mechanism Approach for Exploring Nonlinear Benefits in Engineering*, https://doi.org/10.1007/978-981-96-5769-8_7

improve energy harvesting performance. It is known that there are quite a few structural parameters which can critically influence the energy harvesting performance of the X-structure based harvesters. How to optimize the structural design of a given system with respect to several critical design parameters are key issues in such a topic, and applications can be extended to many other similar structural design and optimization tasks in various engineering systems for performance improvement. The method demonstrated in this chapter is not only applicable to the structural optimization of X-structures, but also to all those nonlinear systems for performance optimization with respect to critical design parameters both from linear or nonlinear components.

Nonlinear energy harvesting systems have been extensively studied, but there is still a lack of a straightforward parametric design method for such systems. Traditional methods like harmonic balance or numerical simulation are not able to establish a clear relationship between energy harvesting performance and system parameters, and they often require significant computational resources for parametric studies. In this study, a new frequency domain method based on Volterra series is introduced for a typical nonlinear harvesting system. A concise parametric characteristic approach called the linear and nonlinear characteristic output spectrum (lnCOS) method is proposed to express the nonlinear output frequency response and power generation function (PGF) of the system. This approach, referred to as the lnCOS function and the Parametric-PGF respectively, provides a more straightforward way to obtain vibration responses and power generation without solving complex algebraic equations. The lnCOS method directly connects power outputs to critical system parameters in a concise parametric manner. This study is the first to provide explicit expressions for vibration responses and power generations based on structural parameters of a nonlinear energy harvesting system, which is not achievable using traditional methods. The obtained explicit expressions of the lnCOS function/Parametric-PGF allow for direct sensitivity analysis of vibration response and power generation with respect to structural parameters. This work offers an efficient approach to characterize both linear and nonlinear effects in a nonlinear energy harvesting system and demonstrates an efficient method for the analysis and design of such systems.

7.1.2 Research Background

The call for renewable, clean energy sources and heightened environmental conservation is on the rise. Consequently, the field of kinetic and vibration energy harvesting is gaining increased interest, leading to the development of various systems [1–6]. Notably, these energy harvesting technologies are pivotal for enabling autonomous, energy-sufficient wireless sensor networks and Internet of Things (IoT) nodes by utilizing ambient energy sources [7–9]. This is particularly crucial for advancing smart ocean [10] technologies, as well as smart homes and cities [11].

Approximately four primary mechanisms are employed for converting environmental energy into usable forms: piezoelectric [12–14], electromagnetic [15,

16], electrostatic [17], and triboelectric [18–20] transducers. To improve performance and broaden the operational bandwidth of these systems, nonlinear strategies such as variable stiffness characteristics (softening/hardening), bi-stable/multi-stable configurations, and internal resonance have been extensively researched and applied [21, 22].

Jing et al. have recently innovated X-shaped or bio-inspired limb-like structures with strategically placed spring elements [23, 24]. These designs, which draw inspiration from the skeletal structure of animal legs and limbs resembling trusses or scissors, have been shown to offer advantageous and customizable nonlinear stiffness, damping, and inertia/inerter properties [23–25]. Such structures have broad applications across various engineering challenges, including vibration isolation [26], energy absorption [27], vehicle suspensions [28], energy harvesting [29–32], and sensor development [33]. A comprehensive review of the X-structure/mechanism for advantageous nonlinear engineering design is encapsulated in Ref. [34]. This chapter delves into a typical nonlinear harvesting system, the nonlinear X-structured electromagnetic harvesting system, which will be thoroughly examined using a cutting-edge frequency domain method based on the Volterra series. It's important to note that Jing's group has previously conceptualized and crafted X-structured nonlinear energy harvesting systems (X-harvesters) as documented in Refs. [29–32]. However, these were analyzed and designed using only the harmonic balance method and numerical simulations, which did not yield explicit solutions or straightforward power functions in relation to key system parameters. Motivated by this gap, the current work not only develops a novel Volterra series-based frequency domain method but also employs the X-harvesting systems as a case study to showcase the benefits and efficacy of the proposed method.

The frequency domain method founded on the Volterra series is widely recognized as a robust mathematical tool for analyzing nonlinear systems that can be represented through Volterra series expansions [35]. This approach stands out in comparison to the harmonic balance method and other time domain techniques, offering distinct benefits [36]. For instance, the harmonic balance method requires solving a multitude of intricate nonlinear algebraic equations for various unknowns when considering the first two harmonic response orders of the X-structured system. Moreover, this method or numerical simulations only yield an implicit set of algebraic equations related to vibration responses. In contrast, the frequency domain method based on the Volterra series directly correlates model parameters with the system's output frequency response, bypassing the need to solve dynamic equations. This significantly eases the analysis and design of nonlinear systems in practical applications [35]. The Volterra series-based frequency domain method has proven effective in applications such as vibration isolation [37], detecting structural faults [38, 39], identifying systems [40], and managing active vibrations [41].

The literature offers scant examples of analytical and parametric studies that establish a direct correlation between the output power and essential design parameters of nonlinear energy harvesting systems. Diala et al. [42] explored the impact of a nonlinear cubic stiffness spring and cubic damping on a vibrational energy system through the lens of this frequency domain methodology. Furthermore, Diala et al. [43]

introduced an electromagnetic energy harvester that doubles as a vibration isolator with cubic stiffness, conducting both analysis and optimal design for the system using a method based on the output frequency domain function. Their research, however, only took into account two nonlinear factors—the cubic stiffness and damping—while presuming the linear components remain constant. Presently, there appears to be no established method for systematically designing both linear and nonlinear parameters of a nonlinear energy harvesting system in the frequency domain using a Volterra series-based approach, despite the apparent potential benefits.

Building on the insights gained, this research zeroes in on a prototypical X-structured nonlinear energy harvesting system as a case study to develop a systematic and analytical parametric methodology. This method is intended for the analysis, design, and optimization of the system's output power in relation to key design parameters, employing the Volterra series-based frequency domain approach. The X-structured system in question features a number of designable nonlinear stiffness and damping terms [29–31], which differ from those in previously reported systems. The advantageous effects of these nonlinear terms warrant further systematic exploration. The study aims to establish analytical connections between the system's output frequency responses—including vibration responses (OFRF) and the power generation function (PGF)—and the structural parameters of the X-shaped structure. Newly developed frequency domain functions will demonstrate how energy generation is influenced by the system's structural parameters in a more direct manner. A parametric method, referred to as the lnCOS-based method, will be introduced to delineate the impact of these parameters on the output frequency responses. Given that various parameters of the X-structure affect both its linear and nonlinear components, the conventional Volterra series-based frequency domain approach is not directly applicable to the current X-structured system. Therefore, new lnCOS-based functions and an innovative Parametric-PGF will be specifically formulated. These will be used for the concurrent design of the linear and nonlinear components of the X-structured system, thereby fostering a comprehensive understanding. The study will conduct a parametric analysis using both the analytical PGF and the novel lnCOS method, complemented by sensitivity analysis. The findings from the parametric and sensitivity assessments of the critical system parameters will significantly inform the practical design of the X-structured harvester.

This study's contributions are significant in several aspects:

- It includes the theoretical development of generalized frequency response functions (GFRFs) for the X-structured harvesting system's various parameters, alongside the introduction of a power generation function (PGF) to assess how structural parameters influence energy harvesting efficiency.
- A comprehensive lnCOS-based method has been formulated, along with the creation of multiple lnCOS-based functions that address vibration responses related to different structural parameters, and a pioneering Parametric-PGF. These can be utilized to concurrently design the linear and nonlinear elements of the X-structured system.

- The lnCOS-based method simplifies computations for parametric characterization. Notably, it provides the first-ever explicit expressions for vibration responses and power generation based on the X-structured harvester's key parameters, such as the assembly angle, rod length, and electrical damping. These expressions are ready for use in design and offer potential for optimizing the current harvesting system.
- The research introduces an innovative and clear-cut parametric design strategy that promises to enhance the design and optimization of structural parameters in related nonlinear energy harvesting systems.

The rest of the chapter is organized as follows. The mathematical model of the X-structured energy harvesting system is presented in Sect. 7.2. The generalized frequency response functions (GFRFs) of the system, the output frequency response function (OFRF), and the power generation function (PGF) would be obtained in Sect. 7.3, and a parametric study using the OFRF and the PGF would also be performed in Sect. 7.3. The parametric characteristic approach named lnCOS-based method and the Parametric-PGF would be proposed and introduced in Sect. 7.4. Section 7.5 would provide the lnCOS functions and Parametric-PGF-based analysis and discussions. Eventually, in Sect. 7.6, a conclusion would be drawn.

7.2 Mathematical Model of the X-Structured Harvesting System

In this section, we will delve into a representative X-structured nonlinear energy harvesting system [29–31] as a model for developing a systematic and analytical parametric methodology. This methodology is designed for the analysis and design of output power in relation to vital design parameters, utilizing a Volterra series-based frequency domain method. The mathematical model for the X-structured harvesting system will be presented.

The motion equation for the energy harvesting system is derived from an X-shaped structure [29–31], which is reminiscent of the skeletal framework of animal legs or limbs with a truss or scissor configuration, as depicted in Fig. 7.1. This structure serves as the basis for the energy harvester. Figure 7.1 details the geometric dimensions and parameters of the X-structured energy harvesting system. In modelling, L and θ denote the length and the assembly angle of the connecting rods within the X-structure, respectively; N signifies the total number of layers in the X-structure; k is the stiffness of the horizontal spring element; Crucially, cs represents the equivalent structural damping of the X-structure, while ce is the equivalent electrical damping from the generator unit. It is important to note that the m-layered X-structure incorporates only a linear stiffness component and equivalent damping components, which include both structural and electrical damping, both of which are considered as viscous damping.

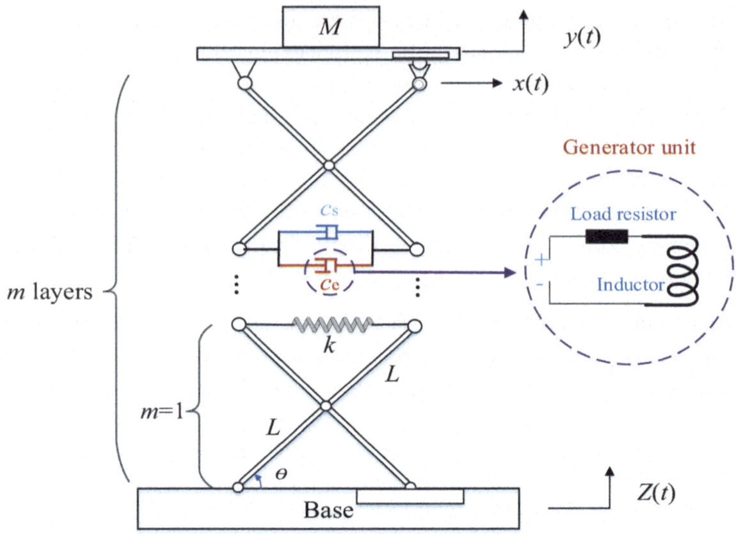

Fig. 7.1 The proposed X-structured energy harvesting system

The kinetic energy and the potential energy of the X-shaped structure can be given as

$$T = \frac{1}{2}M\dot{y}^2, \quad U = \frac{1}{2}kx^2, \tag{7.1}$$

where M is the mass on the X-shaped structure. x is the horizontal motion of the connecting joint in the bottom right of the X-shaped structure (also the deformation of the horizontal spring and damper), which can be expressed as

$$x = 2L\cos\theta - 2\sqrt{L^2 - \left(L\sin\theta + \left(\frac{\hat{y}}{2m}\right)\right)^2}.$$

The Lagrange principle is utilized to establish the dynamic equation of motion for the X-shaped structure, which can be written as

$$\frac{d}{dt}\left(\frac{\partial}{\partial\dot{y}}(T-U)\right) - \frac{\partial}{\partial y}(T-U) = Q, \tag{7.2}$$

where $Q = -c\dot{x}\cdot\frac{dx}{dy}$, and $c = c_s + c_e$ represents the total damping of the energy harvesting system.

The dynamical equation of the X-structured energy harvesting system can be obtained as

$$M\ddot{\hat{y}} + kx\frac{dx}{d\hat{y}} + c\left(\frac{dx}{d\hat{y}}\right)^2\dot{\hat{y}} = -M\ddot{Z}, \tag{7.3}$$

where $\hat{y} = y - Z$ is the relative motion between base and the mass on the platform.

By using the Taylor series expansion for the terms of $\frac{dx}{d\hat{y}}$ and $\left(\frac{dx}{d\hat{y}}\right)^2$, the governing equation of the proposed X-structured energy harvesting system can be further described as

$$\begin{aligned} M\ddot{\hat{y}}(t) + k\delta_0\hat{y}(t) + c\delta_0\dot{\hat{y}}(t) + \left(\delta_1\hat{y}^2(t) + \delta_2\hat{y}^3(t)\right) \\ + \left(\delta_3\hat{y}(t)\dot{\hat{y}}(t) + \delta_4\hat{y}^2(t)\dot{\hat{y}}(t)\right) = -Mu(t), \end{aligned} \tag{7.4}$$

where $u(t) = \ddot{Z}(t) = A_0\cos(\Omega t)$, $\delta_0 = \frac{\tan^2\theta}{m^2}$, $\delta_1 = k\frac{3\sin\theta}{4m^3\cos^4\theta}\frac{1}{L}$, $\delta_2 = k\frac{5-4\cos^2\theta}{8m^4\cos^6\theta}\frac{1}{L^2}$, $\delta_3 = c\frac{\sin^2\theta}{m^2\cos^4\theta}\frac{1}{L}$, and $\delta_4 = c\frac{4-3\cos^3\theta}{4m^4\cos^6\theta}\frac{1}{L^2}$. The derivation process can also be referred to as Refs. [29, 30], and [44].

It should be note that the linear component δ_0 is determined by θ and m, and the nonlinear components δ_1, δ_2, δ_3, and δ_4 are determined by θ, m, and L (the structural parameters of the X-structure).

7.3 Output Frequency Response Function (OFRF) and Power Generation Function (PGF)

This section will outline the principles of the generalized frequency response functions (GFRFs) and the output frequency response function (OFRF), as well as introduce the notion of the power generation function (PGF). Following this, a theoretical analysis utilizing the established OFRF and PGF will be conducted for various case studies.

7.3.1 Generalized Frequency Response Functions (GFRFs)

The analysis of nonlinear systems within the frequency domain can be executed using the theory of Volterra series expansion. It is understood that in the vicinity of the zero equilibrium, nonlinear systems can be represented by an approximation using a Volterra series to the highest order N [45, 46]. For the above nonlinear system, in the time domain, it can be expressed as

$$\hat{y}(t) = \sum_{n=1}^{N}\int_{-\infty}^{\infty}\cdots\int_{-\infty}^{\infty} h_n(\tau_1, \ldots, \tau_n)\prod_{i=1}^{n} u(t-\tau_i)d\tau_i, \tag{7.5}$$

where $h_n(\tau_1, \ldots, \tau_n)$ is the nth order impulse function or the nth order Volterra kernel of the system. The multidimensional Fourier transforms of the nth order impulse response functions would yield the nth order transfer functions or the generalized frequency response functions (GFRFs) as

$$H_n^{\hat{y}}(j\omega_1, \ldots, j\omega_n) = \int_{-\infty}^{\infty} \cdots \int_{-\infty}^{\infty} h_n(\tau_1, \ldots, \tau_n) e^{-(j\omega_1 + \cdots + j\omega_n)} d\tau_1 \cdots d\tau_n, \tag{7.6}$$

where $H_n^{\hat{y}}(j\omega_1, \ldots, j\omega_n)$ means the nth order GFRF of the system. It should be noted the GFRFs are also called higher-order frequency response functions in some other publications such as Ref. [47].

Employing a recursive technique [48], one can derive the first-order generalized frequency response function (GFRF), which is synonymous with the linear transfer function, as

$$H_1^{\hat{y}}(j\omega_1) = \frac{M}{-\left(M(j\omega_1)^2 + c\delta_0(j\omega_1) + k\delta_0\right)}. \tag{7.7}$$

Next, the second order GFRF can be obtained as

$$\begin{aligned} H_2^{\hat{y}}(j\omega_1, j\omega_2) = {} & L^{-1}(j\omega_1 + j\omega_2)(\delta_1 + \delta_3(j\omega_1)) \\ & H_1(j\omega_1)H_1(j\omega_2), \end{aligned} \tag{7.8}$$

where $L(\cdot) = -\left(M(\cdot)^2 + c\delta_0(\cdot) + k\delta_0\right)$ and $L^{-1}(\cdot) = 1/L(\cdot)$. Herein, the expression of $L(\cdot)$ can be considered as an operator and will be further used in the following parts.

The third and fourth order GFRFs can be written as

$$\begin{aligned} H_3^{\hat{y}}(j\omega_1, j\omega_2, j\omega_3) = {} & L^{-1}(j\omega_1 + j\omega_2 + j\omega_3) \\ & \times \left(\begin{array}{l} \left(2\delta_1 + \delta_3 \sum_{i}^{3} (j\omega_i)\right)(H_1(j\omega_1)H_2(j\omega_2, j\omega_3)) \\ +(\delta_2 + \delta_4(j\omega_1))H_1(j\omega_1)H_1(j\omega_2)H_1(j\omega_3) \end{array} \right), \end{aligned} \tag{7.9}$$

and

$$H_4^{\hat{y}}(j\omega_1, j\omega_2, j\omega_3, j\omega_4) = L^{-1}(4)$$

$$\times\left(\begin{array}{l}\left(2\delta_1+\delta_3\sum_i^4(j\omega_i)\right)H_1(j\omega_1)H_3(j\omega_2,j\omega_3,j\omega_4)\\+(\delta_1+\delta_3(j\omega_1+j\omega_2))H_2(j\omega_1,j\omega_2)H_2(j\omega_3,j\omega_4)\\+\left(3\delta_2+\delta_4\sum_i^4(j\omega_i)\right)H_1(j\omega_1)H_1(j\omega_2)H_2(j\omega_3,j\omega_4)\end{array}\right), \tag{7.10}$$

where $L^{-1}(4)=L^{-1}(j\omega_1+j\omega_2+j\omega_3+j\omega_4)$.

Similarly, the expressions of the higher order GFRFs, namely $H_5^{\hat{y}}(j\omega_1,j\omega_2,j\omega_3,j\omega_4,j\omega_5)$, $H_6^{\hat{y}}(j\omega_1,j\omega_2,j\omega_3,j\omega_4,j\omega_5,j\omega_6)$ and $H_7^{\hat{y}}(j\omega_1,j\omega_2,j\omega_3,j\omega_4,j\omega_5,j\omega_6,j\omega_7)$, can also be obtained by using the recursive method, which are shown in detail in Appendix 1 due to the lengthy expression. It should be mentioned here, the orders higher than seven would be omitted as their contributions on the output spectrum are negligible.

7.3.2 *Output Frequency Response Function*

If the system input is a multi-tone function such as $u(t)=\sum_{k=1}^{K}|F_k|\cos(\omega_k t+\angle F_k)$, the system output spectrum or the output frequency response function (OFRF) of model Eq. (7.4) can be expressed as [49]

$$Y(j\omega)=\begin{cases}\sum_{n=1}^{N}\frac{1}{2^n}\sum_{\omega=\omega_{k_1}+\omega_{k_2}+\cdots+\omega_{k_n}}H_n(j\omega_{k_1},\ldots,j\omega_{k_n})F(\omega_{k_1})\cdots F(\omega_{k_n}), & \omega>0\\ \sum_{n=1}^{N}\frac{1}{2^{n+1}}\sum_{\omega_{k_1}+\omega_{k_2}+\cdots+\omega_{k_n}=0}H_n(j\omega_{k_1},\ldots,j\omega_{k_n})F(\omega_{k_1})\cdots F(\omega_{k_n}), & \omega=0\end{cases}, \tag{7.11}$$

where

$$F(\omega)=\begin{cases}|F_k|e^{j\angle F_k\cdot \mathrm{sgn}(k)} & \text{if } \omega\in\{\omega_k,k\in\pm1,\ldots,\pm K\}\\ 0 & \text{else}\end{cases},$$
$$\mathrm{sgn}(k)=\begin{cases}1 & k\geq 0\\ -1 & k<0\end{cases}. \tag{7.12}$$

The methodologies outlined in the previous subsections are applicable for the analysis of the suggested nonlinear system. It's important to note that for certain nonlinear systems, the output frequency response function (OFRF) might not converge within specific parameter ranges. Nevertheless, this research will employ simulation results to confirm the validity of the OFRF-based analysis, ensuring its convergence. While the convergence of Volterra series-based nonlinear systems is a topic worth exploring, it falls outside the scope of this current study.

For a single-tone input in this study, by substituting the GFRFs into Eq. (7.11), the expressions of the output spectrums or the OFRFs can be obtained as

$$\begin{aligned} Y^{\hat{y}}(j\omega) &= Y_1^{\hat{y}}(j\omega) + Y_3^{\hat{y}}(j\omega) + Y_5^{\hat{y}}(j\omega) + \cdots \\ &= \frac{1}{2} H_1^{\hat{y}}(j\omega_1)\Big|_{\omega_1=\omega} F(\omega_1)|_{\omega_1=\omega} \\ &+ \frac{1}{2^3} \sum_{\omega=\omega_1+\omega_2+\omega_3} H_3^{\hat{y}}(j\omega_1, j\omega_2, j\omega_3) F(\omega_1)F(\omega_2)F(\omega_3) \\ &+ \frac{1}{2^5} \sum_{\omega=\omega_1+\omega_2+\omega_3+\omega_4+\omega_5} H_5^{\hat{y}}(j\omega_1, j\omega_2, j\omega_3, j\omega_4, j\omega_5) \\ &\quad F(\omega_1)F(\omega_2)F(\omega_3)F(\omega_4)F(\omega_5) + \cdots, \end{aligned} \tag{7.13a}$$

$$\begin{aligned} Y^{\hat{y}}(j0) &= Y_2^{\hat{y}}(j0) + Y_4^{\hat{y}}(j0) + \cdots \\ &= \frac{1}{2^3} \sum_{\omega_1+\omega_2=0} H_2^{\hat{y}}(j\omega_1, j\omega_2) F(\omega_1)F(\omega_2) \\ &+ \frac{1}{2^5} \sum_{\omega_1+\omega_2+\omega_3+\omega_4=0} H_4^{\hat{y}}(j\omega_1, j\omega_2, j\omega_3, j\omega_4) \\ &\quad F(\omega_1)F(\omega_2)F(\omega_3)F(\omega_4) + \cdots, \end{aligned} \tag{7.13b}$$

$$\begin{aligned} Y^{\hat{y}}(j2\omega) &= Y_2^{\hat{y}}(j2\omega) + Y_4^{\hat{y}}(j2\omega) + \cdots \\ &= \frac{1}{2^2} \sum_{\omega_1+\omega_2=2\omega} H_2^{\hat{y}}(j\omega_1, j\omega_2) F(\omega_1)F(\omega_2) \\ &+ \frac{1}{2^4} \sum_{\omega_1+\omega_2+\omega_3+\omega_4=2\omega} H_4^{\hat{y}}(j\omega_1, j\omega_2, j\omega_3, j\omega_4) \\ &\quad F(\omega_1)F(\omega_2)F(\omega_3)F(\omega_4) + \cdots \end{aligned} \tag{7.13c}$$

where $Y^{\hat{y}}(j\omega)$, $Y^{\hat{y}}(j0)$, and $Y^{\hat{y}}(j2\omega)$ represent the first order harmonic output, the 0th order harmonic output (the DC component) and the second order harmonic output respectively. If $n = 1$ only, $Y_1^{\hat{y}}(j\omega)$ would be the output spectrum of the corresponding linear system. It is interesting to find that, only odd-ordered components would contribute to $Y^{\hat{y}}(j\omega)$ and only even-ordered components would contribute to $Y^{\hat{y}}(j0)$, and $Y^{\hat{y}}(j2\omega)$(i.e., $Y_2^{\hat{y}}(j\omega) = Y_4^{\hat{y}}(j\omega) = 0$, $Y_1^{\hat{y}}(j0) = Y_3^{\hat{y}}(j0) = 0$, and $Y_1^{\hat{y}}(j2\omega) = Y_3^{\hat{y}}(j2\omega) = 0$). This is consistent with the results and descriptions in Ref. [35].

When $Y^{\hat{y}}(j\omega)$, $Y^{\hat{y}}(j0)$, and $Y^{\hat{y}}(j2\omega)$ calculated, $|2Y^{\hat{y}}(j\omega)|$, $|Y^{\hat{y}}(j0)|$ and $|2Y^{\hat{y}}(j2\omega)|$ would be readily obtained, which represent the corresponding physical magnitudes of the system outputs of the vibration response $\hat{y}$.

7.3.3 *The Power Generation Function*

The instantaneous power generated by the equivalent electrical damper is the product of the force on the damper and its relative velocity at that moment, which could be given by

$$P_{\text{ins}}(t) = \left(c_{\text{e}}\delta_0 + \tilde{\delta}_3\hat{y}(t) + \tilde{\delta}_4\hat{y}^2(t)\right)\dot{\hat{y}}^2(t), \tag{7.14}$$

where $\tilde{\delta}_3 = c_{\text{e}}\frac{\sin^2\theta}{m^2\cos^4\theta}\frac{1}{L}$ and $\tilde{\delta}_4 = c_{\text{e}}\frac{4-3\cos^3\theta}{m^4\cos^6\theta}\frac{1}{L^2}$.

By considering $\hat{y}(t) = -\left|Y^{\hat{y}}(j0)\right| + \left|2Y^{\hat{y}}(j\omega)\right|\cos(\omega t + \varphi_\omega) + \left|2Y^{\hat{y}}(j2\omega)\right|\cos(2\omega t + \varphi_{2\omega})$, the average power generated by the generator unit or the equivalent electrical damper in one cycle could then be given by

$$|P_{\text{ave}}(j\omega)| = \frac{\omega^2}{8}\left(\begin{array}{l} c_{\text{e}}\delta_0\left(4\left|2Y^{\hat{y}}(j\omega)\right|^2 + 16\left|2Y^{\hat{y}}(j2\omega)\right|^2\right) + \tilde{\delta}_3\left(\begin{array}{l} -4\left|Y^{\hat{y}}(j0)\right|\left|2Y^{\hat{y}}(j\omega)\right|^2 \\ -16\left|Y^{\hat{y}}(j0)\right|\left|2Y^{\hat{y}}(j2\omega)\right|^2 \\ +6\left|2Y^{\hat{y}}(j\omega)\right|^2\left|2Y^{\hat{y}}(j2\omega)\right| \end{array}\right) \\ +\tilde{\delta}_4\left(\begin{array}{l} 4\left|Y^{\hat{y}}(j0)\right|^2\left|2Y^{\hat{y}}(j\omega)\right|^2 + \left|2Y^{\hat{y}}(j\omega)\right|^4 - 12\left|Y^{\hat{y}}(j0)\right|\left|2Y^{\hat{y}}(j\omega)\right|^2\left|2Y^{\hat{y}}(j2\omega)\right| \\ +16\left|Y^{\hat{y}}(j0)\right|^2\left|2Y^{\hat{y}}(j2\omega)\right|^2 + 10\left|2Y^{\hat{y}}(j\omega)\right|^2\left|2Y^{\hat{y}}(j2\omega)\right|^2 + 4\left|2Y^{\hat{y}}(j2\omega)\right|^4 \end{array}\right) \end{array}\right), \tag{7.15}$$

in which $\left|2Y^{\hat{y}}(j\omega)\right|$, $\left|Y^{\hat{y}}(j0)\right|$ and $\left|2Y^{\hat{y}}(j2\omega)\right|$ can be readily obtained from Eqs. (7.13a–7.13c).

It is important to recognize that Eq. (7.15) can be considered the power generation function (PGF) for the proposed X-structured energy harvesting system. Since the structural parameters are incorporated into $Y^{\hat{y}}(j0)$, $Y^{\hat{y}}(j\omega)$, and $Y^{\hat{y}}(j2\omega)$, integrating them into Eq. (7.15) means that the PGF $|P_{\text{ave}}(j\omega)|$ will also encompass these structural parameters in addition to the excitation frequency. Consequently, this establishes a derived and connected analytical relationship between the PGF and the structural parameters of the X-structure.

7.3.4 *Theoretical Analysis Using OFRF and PGF*

7.3.4.1 A Case Study

As an example, the structural parameters and the base excitation are set as: $M = 2$ kg, $k = 4000$ N/m, $L = 0.1$ m, $\theta = \pi/6$, $m = 2$, and $u(t) = 3\cos(\omega t)$. In this chapter, a two layer X-structured harvester would be considered, as shown in Fig. 7.2a.

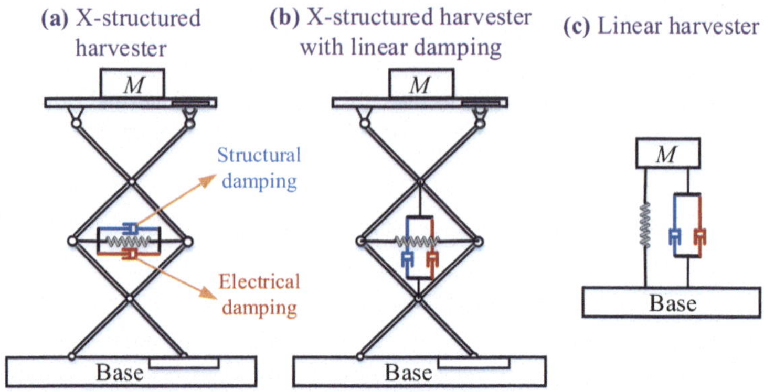

Fig. 7.2 Different configurations: **a** The X-structured harvester; **b** The X-structured harvester with linear damping; **c** The traditional linear harvester (The blue one represents structural damping and the red one represents electrical damping)

Utilizing Eqs. (7.13a–7.13c) derived earlier, we can calculate the vibration response's 0th, 1st, and 2nd order harmonic outputs, as illustrated in Fig. 7.3a–d. To assess the method's validity and precision, simulations using ode45 with FFT were performed. The outcomes demonstrate a strong concordance between the theoretical frequency domain approach and the simulation results, particularly for $N = 5$ and $N = 7$. Moreover, the results for $N = 3$, $N = 5$, and $N = 7$, as seen in Fig. 7.3, are quite similar, with $N = 5$ and $N = 7$ being nearly identical. This suggests that an N value of 3 is sufficiently precise, while $N = 5$ or higher offers ample accuracy for this case study.

Crucially, when comparing to the baseline case of $N = 1$, which neglects higher-order terms, the first harmonic output below the resonant frequency is notably higher for $N = 3$, $N = 5$, and $N = 7$, as shown in Fig. 7.3a, b. For the 0th and 2nd harmonic outputs, the output is zero at $N = 1$. Given that a greater amplitude correlates with increased power generation, the scenarios incorporating nonlinear damping and stiffness ($N = 3$, $N = 5$, and $N = 7$) are evidently more advantageous than the linear case of $N = 1$. This advantage is a distinctive feature of the X-structured harvester, stemming from its intrinsic geometrical nonlinear stiffness and damping characteristics.

Additionally, to complement the study, the relative discrepancies between the theoretical predictions and the simulation outcomes are detailed in Tables 7.1, 7.2 and 7.3. From Table 7.1, it's evident that the relative errors for the first order harmonic output generally range from 0.18 to 2.70% for N values of 3, 5, and 7. With N set to 1, the relative errors hover around 5% within the lower frequency spectrum (0.4–1.6 Hz). This aligns with the findings illustrated in Fig. 7.3a, b, indicating that an N value of 3 is sufficiently precise, while N values of 5 or higher are deemed highly accurate for this particular analysis. Moreover, the relative errors for the DC component (0th order harmonic) and the second order output are predominantly under or about 10%. It's important to highlight that the first harmonic output, which is the

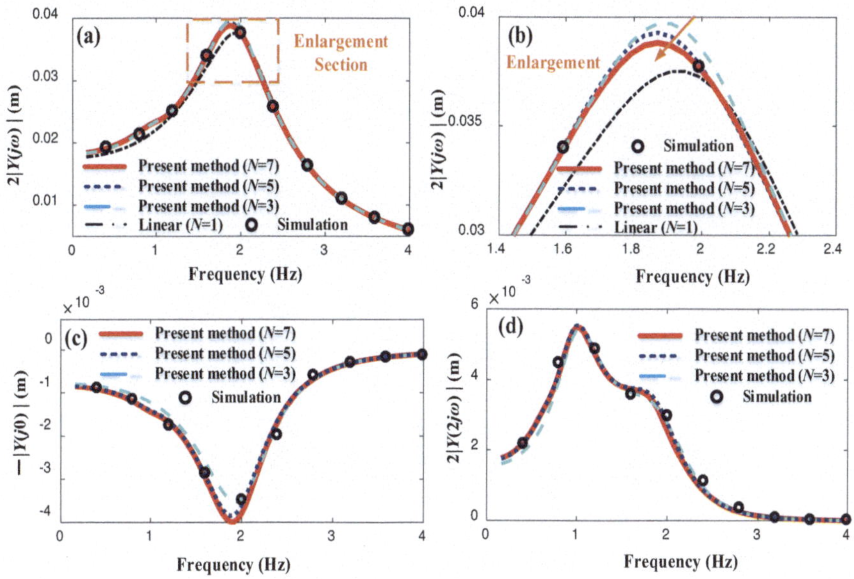

Fig. 7.3 Vibration responses (amplitudes of the output spectrum) versus the frequency and comparisons between the present method and simulation results: **a** The first order harmonic output (the fundamental harmonic output); **b** Enlarged view of the first order harmonic output; **c** The DC component (the 0th order harmonic output); **d** The second order harmonic output

primary contributor to power generation, exhibits relative errors that are considered acceptable for the purposes of this study.

For additional comparison, the scenario of an X-structured harvester equipped solely with linear damping is examined. In this setup, the damping elements are vertically integrated within the structure, as shown in Fig. 7.2b. Utilizing Eq. (7.15), one can calculate the average power output across various frequencies, which is depicted in Fig. 7.4. It is noticeable that the nonlinear X-structured harvester is capable of producing more power than its linear-damping-only counterpart (refer to Fig. 7.2b) within the lower frequency spectrum. This characteristic is often highly sought after in practical applications. Moreover, it is apparent that a conventional linear harvester, as illustrated in Fig. 7.2c, would exhibit its resonant peak at a higher frequency, resulting in considerably lower power output. This observation underscores the superiority of the current X-structured design in harvesting energy at lower frequencies.

7.3.4.2 Effects of Excitation Amplitude

In order to further investigate the influence of the excitation amplitude on the accuracy of the proposed theoretical OFRF and PGF, more cases with different excitation amplitudes are considered in this subsection. Herein, the cases of $u(t) = 3.5\cos(\omega t)$,

Table 7.1 Relative errors between the present method and simulation results: the first order harmonic output (the fundamental harmonic output)

First order harmonic output	Frequency (Hz)								
	0.4	0.8	1.2	1.6	2.0	2.4	2.8	3.2	3.6
Simulation (m)	0.01932	0.02147	0.02513	0.03406	0.03775	0.02578	0.01645	0.01115	0.008056
Linear (m) ($N = 1$)	0.01824	0.02027	0.02451	0.03217	0.03730	0.02663	0.01658	0.01111	0.007999
Relative errors	5.59%	5.59%	2.47%	5.55%	1.19%	3.30%	0.79%	0.36%	0.71%
Present method (m) ($N = 3$)	0.01902	0.02166	0.02626	0.03396	0.03877	0.02600	0.01641	0.01107	0.007987
Relative errors	1.55%	0.88%	4.50%	0.29%	2.70%	0.85%	0.24%	0.72%	0.86%
Present method (m) ($N = 5$)	0.01911	0.02177	0.02522	0.03406	0.03793	0.02598	0.01642	0.01107	0.007987
Relative errors	1.09%	1.40%	0.36%	0%	0.48%	0.78%	0.18%	0.72%	0.86%
Present method (m) ($N = 7$)	0.01913	0.02178	0.02521	0.03396	0.03777	0.02600	0.01642	0.01107	0.007987
Relative errors	0.98%	1.44%	0.32%	0.29%	0.05%	0.85%	0.18%	0.72%	0.86%

Table 7.2 Relative errors between the present method and simulation results: the DC component (the 0th order harmonic output)

0th harmonic output	Frequency (Hz)								
	0.4	0.8	1.2	1.6	2.0	2.4	2.8	3.2	3.6
Simulation (-10^{-3} m)	0.8756	1.154	1.735	2.844	3.470	1.950	0.6225	0.2750	0.1604
Present method (-10^{-3} m) ($N = 3$)	0.8318	1.027	1.502	2.588	3.479	1.772	0.6872	0.3084	0.1600
Relative errors	5.00%	11.0%	13.4%	9.0%	2.59%	9.13%	10.4%	12.1%	0.25%
Present method (-10^{-3} m) ($N = 5$)	0.8970	1.174	1.638	2.878	3.635	1.671	0.6710	0.3057	0.1594
Relative errors	2.44%	1.73%	5.59%	1.20%	4.76%	14.3%	7.79%	11.2%	0.62%
Present method (-10^{-3} m) ($N = 7$)	0.9073	1.189	1.667	2.925	3.658	1.884	0.6725	0.3058	0.1594
Relative errors	3.62%	3.03%	3.92%	2.85%	5.42%	3.38%	8.03%	11.2%	0.62%

$u(t) = 4\cos(\omega t)$, $u(t) = 4.5\cos(\omega t)$, and $u(t) = 5\cos(\omega t)$ are taken into consideration. The obtained results for different excitation amplitudes are demonstrated in Fig. 7.5a–d. It should be noted that, only the first order harmonic output (the fundamental harmonic output) is analyzed, as it is dominant in vibration response and power generation, which has been mentioned above.

In Fig. 7.5a–d, it can be observed that for the cases of $u(t) = 3.5\cos(\omega t)$ and $u(t) = 4\cos(\omega t)$, the results of the present method (including $N = 5$ and $N = 7$) are very close to the simulation results. However, when with $u(t) = 4.5\cos(\omega t)$ and $u(t) = 5\cos(\omega t)$, the curves for $N = 7$ (red ones) would experience a downward shift at around the resonant regions, although good agreements can still be observed for non-resonant regions. This means that, by increasing the excitation amplitude, the obtained results of the truncation order $N = 7$ would become not accurate or even biased at about the resonant regions. This may be due to the Volterra series would be not convergent in this narrow region of the resonant peak. It is interesting that, this downward shift phenomenon has also been observed in the case study in Ref. [46]. In this case, according to the comparisons with simulation results, one can still use the truncation order $N = 5$ as the maximum order, which is relatively reliable and acceptable.

Table 7.3 Relative errors between the present method and simulation results: the second order harmonic output

Second order harmonic output	Frequency (Hz)								
	0.4	0.8	1.2	1.6	2.0	2.4	2.8	3.2	3.6
Simulation (10^{-3} m)	2.188	4.490	4.886	3.602	2.971	1.131	0.3052	0.1139	0.0432
Present method (10^{-3} m) ($N = 3$)	2.018	4.213	4.842	3.643	2.824	0.9757	0.2816	0.0999	0.0427
Relative errors	7.77%	6.17%	0.90%	1.14%	4.95%	13.7%	7.74%	12.3%	1.16%
Present method (10^{-3} m) ($N = 5$)	2.154	4.450	4.791	3.765	2.841	1.002	0.2705	0.0984	0.0423
Relative errors	1.55%	0.89%	1.94%	4.53%	4.38%	11.4%	11.4%	13.6%	2.08%
Present method (10^{-3} m) ($N = 7$)	2.194	4.435	4.748	3.740	2.630	0.9829	0.2709	0.0983	0.0424
Relative errors	0.27%	1.22%	2.82%	3.83%	11.5%	13.1%	11.2%	13.7%	1.85%

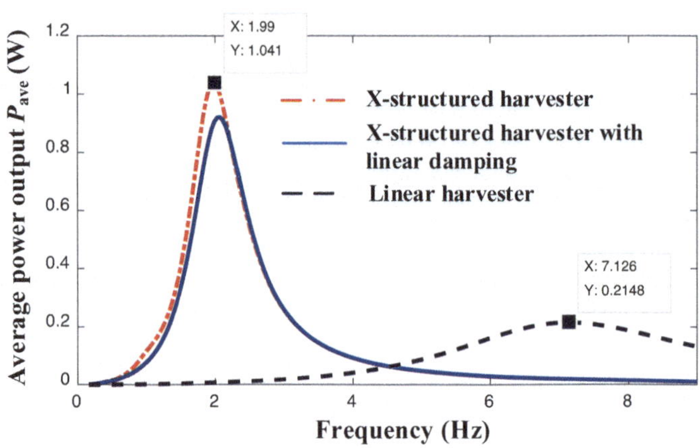

Fig. 7.4 The average power output: the X-structured systems versus the corresponding linear system

While the convergence of the Volterra series-based method is not the central concern of this study, its validity remains an important consideration. It is advisable to monitor the outcome trends as the series order escalates, or to benchmark them against simulations or alternative approaches [46]. The insights provided in this subsection should serve as a precursor to the lnCOS function, which will be detailed

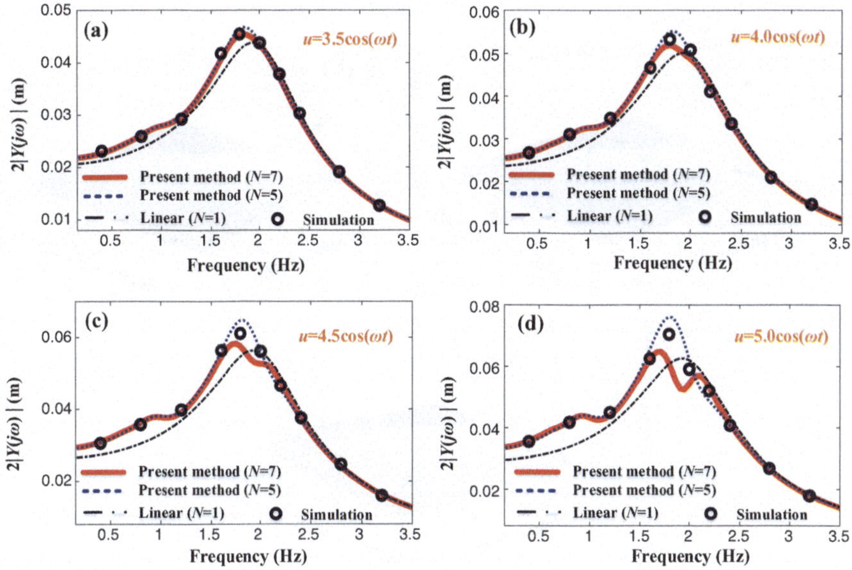

Fig. 7.5 Influence of excitation amplitude on the first order harmonic output (the fundamental harmonic output): **a** $u = 3.5\cos(\omega t)$; **b** $u = 4.0\cos(\omega t)$; **c** $u = 4.5\cos(\omega t)$; **d** $u = 5.0\cos(\omega t)$

in Sects. 7.4 and 7.5. For a more comprehensive understanding of the convergence limits of Volterra series-based nonlinear systems, one may consult Ref. [50].

7.3.4.3 Parametric Study

In order to explore the effects of the structural parameters on the performance of the X-structured harvester, the average power generations versus different structural parameters and different excitation frequencies have been obtained by using the theoretically derived power generation function (PGF), which are demonstrated in Figs. 7.6, 7.7, and 7.8. It should be mentioned that the excitation amplitude is kept at 3 m/s^2 ($u(t) = 3\cos(\omega t)$). It should be mentioned that, the two structural parameters (L and θ) would be restricted by the structural design and the loading capacity of the X-structure [31, 44]. Based on our experimental practice [31, 44], the working ranges $\theta = \left[\pi/6, \pi/3\right]$ and $L = [0.06\ \text{m},\ 0.36\ \text{m}]$ are chosen in this study.

From Fig. 7.6, it can be observed that, for all the three cases with the rod lengths of $L = 0.06$ m, $L = 0.10$ m, and $L = 0.14$ m, the ridges of the surfaces of the average power locate obliquely from around (2 Hz, 0.5 rad (π/6)) to around (6 Hz, 1 rad (π/3)). This is because the assembly angle would determine the resonant frequency of the X-structured system directly. With the increase of the assembly angle of the X-structure from π/6 to π/3, the resonant frequency rise from around 2–6 Hz. In addition, for the three cases, their peak values are located at (1.83 Hz, 0.52 rad), (1.99 Hz, 0.52 rad), and (1.99 Hz, 0.52 rad) respectively. This indicates that, the rod

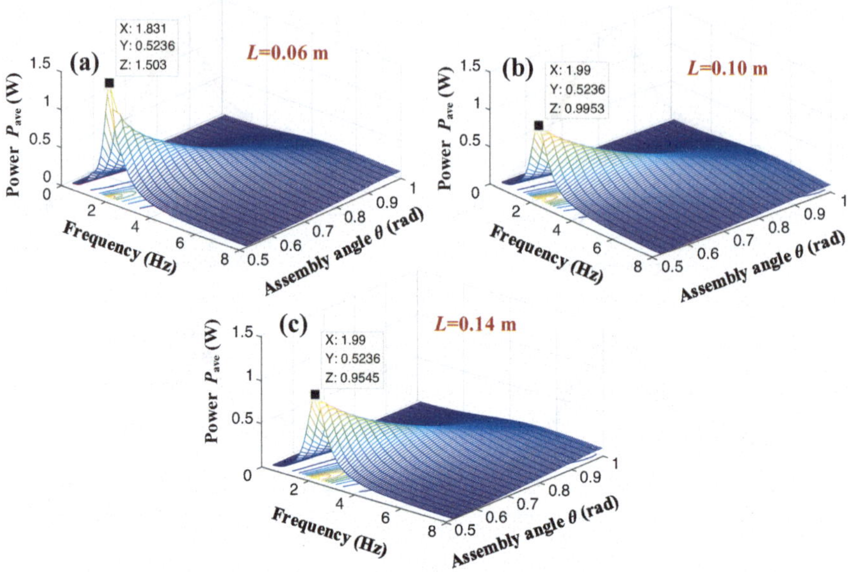

Fig. 7.6 The average power output of the X-structured harvester with different assembly angles θ and excitation frequencies: **a** $L = 0.06$ m; **b** $L = 0.10$ m; **c** $L = 0.14$ m

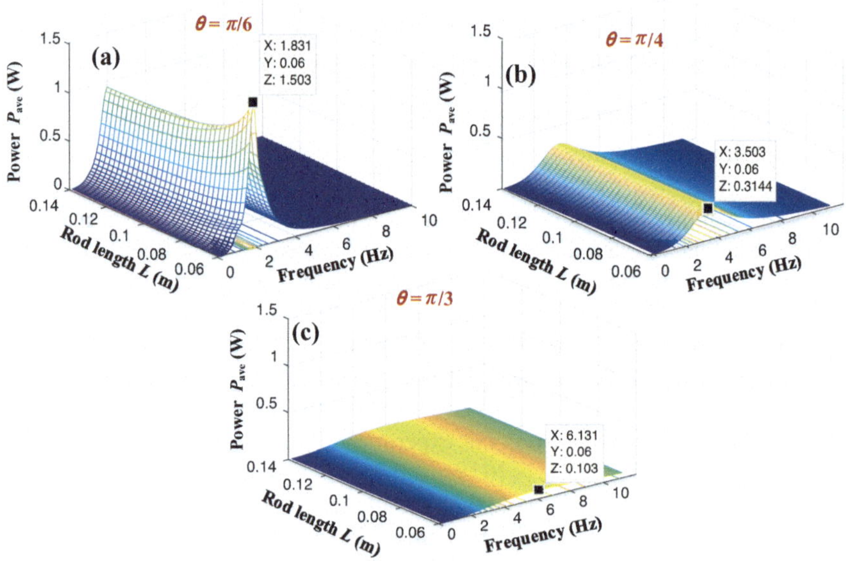

Fig. 7.7 The average power output of the X-structured harvester with different rod lengths L and excitation frequencies: **a** $\theta = \pi/6$; **b** $\theta = \pi/4$; **c** $\theta = \pi/3$

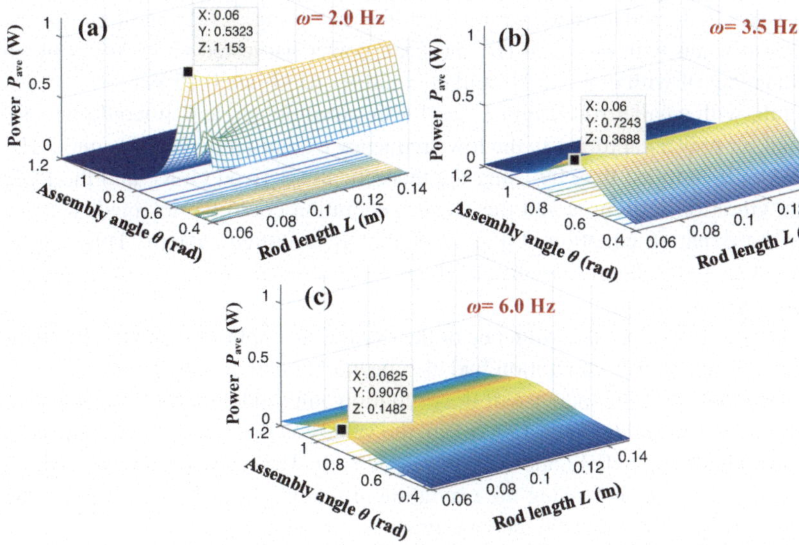

Fig. 7.8 The average power output of the X-structured harvester with different assembly angles θ and rod lengths L: **a** $\omega = 2.0$ Hz; **b** $\omega = 3.5$ Hz; **c** $\omega = 6.0$ Hz

length L would contribute less to the resonant frequency of the system, which could be also observed from Eq. (7.4). Actually, the rod length L would only contribute to the nonlinear components of the proposed X-structured system. With a smaller rod length L, the nonlinear components would be larger and the nonlinear effect would become obvious. More importantly, with $L = 0.06$ m, the power harvested would be much larger than those with $L = 0.10$ m and $L = 0.14$ m, especially when at (1.83 Hz, 0.52 rad), where a peak value is achieved. This indicates that, a smaller rod length L would be helpful for generating more power.

From Fig. 7.7, it can be seen that, for the three cases with the assembly angles of $\theta = \pi/6$, $\theta = \pi/4$, and $\theta = \pi/3$, the ridges of the surfaces of the average power locate at about 1.83, 3.50, and 6.13 Hz respectively. This indicates again that the assembly angle would determine the resonant frequency of the X-structured system directly. For the cases of $\theta = \pi/4$ and $\theta = \pi/3$, the peak value would only change slightly along their ridges with the increase of the rod length L. However, for the case of $\theta = \pi/6$, with the increase of the rod length L, the peak values on the ridge would sharply decrease and then become nearly unchanged. This means again that, with smaller rod length L, much higher would be captured, especially for the case with lower assembly angle θ. In addition, for the three cases, when with $\theta = \pi/6$, the peak power harvested (1.50 W) is much larger than that of with $\theta = \pi/4$ and $\theta = \pi/3$ (0.31 W and 0.10 W).

In Fig. 7.8, for the three cases with different excitation frequencies of $\omega = 2.0$ Hz, $\omega = 3.5$ Hz, and $\omega = 6.0$ Hz, the ridges of the surfaces of the average power locate at about 0.53, 0.72, and 0.9 rad respectively. For the effects of the rod length L, it demonstrates that similar trends would exist like that in Fig. 7.7. In addition, for the three cases, when with $\omega = 2.0$ Hz, the peak power harvested (1.15 W) is much larger than that of with $\omega = 3.5$ Hz and $\omega = 6.0$ Hz (0.37 and 0.15 W).

Overall, with smaller rod length L and lower assembly angle θ, it would be very beneficial for power capturing in the low-frequency excitation range at around 2 Hz. It should be mentioned that, by using the theoretically derived OFRF and the PGF, both the vibration responses and the power generation could be obtained, without solving the dynamic equation (Eq. 7.4) of the X-structured system. This can be considered as a very helpful feature of the present Volterra series-based frequency domain approach.

In addition, to explore the influence of the excitation amplitudes and the frequencies. The average power generation had also been obtained as shown in Fig. 7.9. It can be seen that, with the increase of the excitation amplitude A_0, for the cases with $\theta = \pi/4$, $L = 0.1$ m and $\theta = \pi/3$, $L = 0.1$ m, the values on their ridges of the surfaces increase gradually up to 0.42 and 0.14 W respectively. However, for the case with $\theta = \pi/6$, $L = 0.1$ m, by increasing the excitation amplitude from 1.5 to 3.5 m/s^2, the value on its ridge of the surface increase dramatically up to 1.43 W.

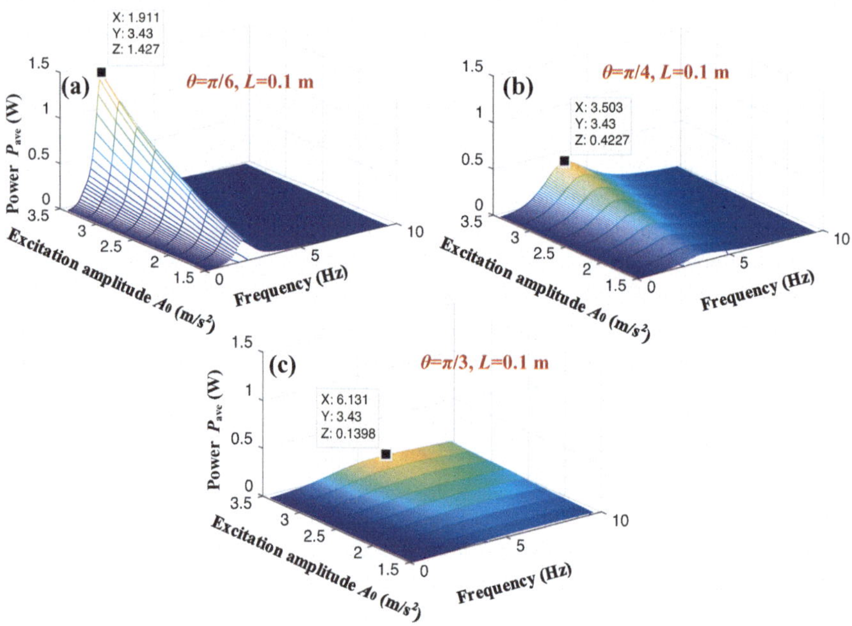

Fig. 7.9 The average power output of the X-structured harvester with different excitation amplitudes A_0 and excitation frequencies: **a** $\theta = \pi/6$, $L = 0.10$ m; **b** $\theta = \pi/4$, $L = 0.10$ m; **c** $\theta = \pi/3$, $L = 0.10$ m

7.4 Parametric Characteristic Approach and the Parametric-PGF

In this section, a novel Linear-Nonlinear characteristic output spectrum (lnCOS) based method is currently developed, characterizing both linear and nonlinear parameters. Based on the proposed lnCOS-based method, the lnCOS functions and the Parametric-PGF of the representative X-structured harvesting system would be obtained.

7.4.1 Novel lnCOS-Based Method

7.4.1.1 Parametric Characterization of Linear Parameters-A General Version

This study aims to elucidate the connection between the model parameters and the energy output of the X-structured energy harvesting system by formulating and examining a novel method that integrates both linear and nonlinear characteristics, known as the hybrid linear and nonlinear characteristic output spectrum (lnCOS)-based method, along with the parametric power generation function (Parametric-PGF). This new approach offers a more streamlined and direct expression of the generalized frequency response functions (GFRFs) and the output frequency response functions (OFRFs) compared to the traditional recursive algorithm used for calculating GFRFs, OFRFs, or output spectrums [35, 51].

Previously, Jing [52] comprehensively introduced and reviewed a standard parametric characteristic method called the nonlinear characteristic output spectrum (nCOS) function-based method. Building on this foundation, Xiao and Jing [53] later innovated a unique characteristic parameter technique tailored for analyzing and designing the linear elements within nonlinear systems. According to [53], the linear parameters included in $L(\cdot)$ could also be separated into the two parts as $c_{1,0}(l) = \hat{c}_{1,0}(l) \oplus \tilde{c}_{1,0}(l)$, where $\hat{c}_{1,0}(l)$ and $\tilde{c}_{1,0}(l)$ denote the linear components of no interest (or known in advance) and the linear parameters of interest (or to be analyzed and designed) respectively. In addition, $L(\cdot)$ could be separated into the two parts as $L(\cdot) = \hat{L}(\cdot) + \tilde{L}(\cdot)$. In the present study, a novel lnCOS-based function concerning both the linear and nonlinear parameters would be further developed and discussed as follows.

If $\left|\hat{L}^{-1}(\omega_n)\tilde{L}(\omega_n)\right| < 1$, the following equations hold: $L^{-1}(\omega_n) = C_{\mathrm{L}}\varphi(\omega_n)$ and $H_1(\omega_n) = C_L\phi(\omega_n)$, where C_{L} represents the linear parameters to be analyzed and design [53, 54]. $\varphi(\omega_n)$ and $\phi(\omega_n)$ only involve output linear parameters of no interest and frequency variables ω_n. It should be mentioned that, in this study, the linear parameter of the input is known in advance and only one output linear parameter $\tilde{c}_{1,0}$ would be considered. In this case, the following expressions could be obtained as

$$C_{\mathrm{L}} = [1, \tilde{c}_{1,0}, \tilde{c}_{1,0}^{2}, \ldots], \tag{7.16a}$$

$$\begin{aligned} L^{-1}(\omega_n) &\approx \varphi_1(\omega_n) + \tilde{c}_{1,0}\varphi_2(\omega_n) + \tilde{c}_{1,0}^{2}\varphi_3(\omega_n), \\ H_1(\omega_n) &\approx \phi_1(\omega_n) + \tilde{c}_{1,0}\phi_2(\omega_n) + \tilde{c}_{1,0}^{2}\phi_3(\omega_n). \end{aligned} \tag{7.16b}$$

However, if $\left|\hat{L}^{-1}(\omega_n)\tilde{L}(\omega_n)\right| > 1$, the above structure of the expansion in Refs. [53, 54] would not hold, and then the detailed expressions would be very different as

$$C_{\mathrm{L}} = \left[\tilde{c}_{1,0}^{-1}, \tilde{c}_{1,0}^{-2}, \tilde{c}_{1,0}^{-3}, \ldots\right], \tag{7.17a}$$

$$\begin{aligned} L^{-1}(\omega_n) &\approx \tilde{c}_{1,0}^{-1}\varphi_1(\omega_n) + \tilde{c}_{1,0}^{-2}\varphi_2(\omega_n) + \tilde{c}_{1,0}^{-3}\varphi_3(\omega_n), \\ H_1(\omega_n) &\approx \tilde{c}_{1,0}^{-1}\phi_1(\omega_n) + \tilde{c}_{1,0}^{-2}\phi_2(\omega_n) + \tilde{c}_{1,0}^{-3}\phi_3(\omega_n). \end{aligned} \tag{7.17b}$$

The proof of Eqs. (7.17a, 7.17b) is presented in Appendix 2. It should be noted that, in [54, 55], the case of Eqs. (7.17a, 7.17b) has not yet been considered, which would be critical in the linear characteristic analysis. In this way, the determination of the expressions of $L^{-1}(\omega_n)$ and $H_1(\omega_n)$ would depend on the relative ratio between the linear parameter to be designed and the linear parameters of no interest. The above results could be regarded as a more comprehensive and more general version of the previous linear characteristics analysis in Ref. [53].

7.4.1.2 New lnCOS Function

In this study, the new lnCOS function is for the first time proposed. By taking both the linear and nonlinear parameters to be designed into consideration, the new lnCOS function could be given as

$$\begin{aligned} Y(j\omega) = \Bigg(\bigoplus_{n=1}^{N} \big[&\mathrm{CE_N}(H_n(j\omega_1, j\omega_2, \cdots, j\omega_n)) \\ &\otimes \mathrm{CE_L}(H_n(j\omega_1, j\omega_2, \cdots, j\omega_n)) \big] \Bigg) \boldsymbol{\Psi}(j\omega)^{\mathrm{T}}, \end{aligned} \tag{7.18}$$

where $\mathrm{CE_N}(\cdot)$ and $\mathrm{CE_L}(\cdot)$ represent the nonlinear coefficients extractor and the linear coefficients extractor respectively; $\oplus$ and $\otimes$ represents the reduced vectorized sum and the reduced Kronecker product operators. More rigorous mathematic definitions of these operators can be referred to [51, 52]. $\boldsymbol{\Psi}(j\omega)$ is a complex-valued vector that only involves the parameters of no interest (known in advance) and frequency variables ω. It should be mentioned that, the proposed lnCOS function can be used for calculating $Y^{\hat{y}}(j0)$, $Y^{\hat{y}}(j\omega)$, and $Y^{\hat{y}}(j2\omega)$ respectively.

The concept of the nonlinear coefficients extractor $\mathrm{CE_N}(\cdot)$ was proposed by Jing et al. [51], which is a powerful tool and can be extended to the case of both linear and nonlinear characteristics extraction. According to [51], the expressions of the nonlinear parameter characteristics could be

$$\mathrm{CE_N}(H_n(\cdot)) = \sum_{p=2}^{n} C_{p,0} \otimes \mathrm{CE_N}(H_{n-p+1}(\cdot)),$$
$$\mathrm{CE_N}(H_1(\cdot)) = 1, \tag{7.19a}$$

where p and $C_{p,0}$ represent the order of the nonlinear parameters and the corresponding nonlinear parameter included.

Importantly, concerning the linear parameter characteristic extractor $\mathrm{CE_L}(\cdot)$, this study proposed the expressions of the linear parameter characteristics as

$$\mathrm{CE_L}(H_n(\cdot)) = [C_L]^\beta = \begin{cases} [1, \tilde{c}_{1,0}, \tilde{c}_{1,0}^2, \ldots]^\beta, & \left|\hat{L}^{-1}(\omega_n)\tilde{L}(\omega_n)\right| < 1 \\ [\tilde{c}_{1,0}^{-1}, \tilde{c}_{1,0}^{-2}, \tilde{c}_{1,0}^{-3}, \ldots]^\beta, & \left|\hat{L}^{-1}(\omega_n)\tilde{L}(\omega_n)\right| > 1 \end{cases}, \tag{7.19b}$$

where $\beta = (2n-1) - \alpha$, α is the maximum order of the cubic nonlinear coefficient that appeared in parameter series of the corresponding $\mathrm{CE_N}(H_n(\cdot))$, which would be further illustrated in the following parts in detail with respect to different cases.

It should be noted that, the case of $\left|\hat{L}^{-1}(\omega_n)\tilde{L}(\omega_n)\right| > 1$ is currently developed, and this new form given in Eq. (7.19b) can thus be regarded as a more general form of previous reports in [53–55]. This is significant for facilitating parametric analysis and design of those related nonlinear systems.

Generally, the process of utilization of the lnCOS function would be as follows: (1) The first step is the determination of the parameter series by using the nonlinear and the linear coefficient extractors; (2) The second step is to calculate $\Psi(j\omega)$ by using the least square method with the required simulation results [35, 56]; (3) By combining the them together, the parametric lnCOS functions could be obtained. The procedures of the lnCOS-based parametric approach are presented in Fig. 7.10 to clearly show the principle of the proposed method.

To further show the utilization of the proposed lnCOS function, an example with respect to characterizing the damping c has been given in Appendix 3. The developed new lnCOS functions could be utilized for analyzing the proposed nonlinear energy harvesting system, which would be demonstrated in detail as follows.

7.4.2 New lnCOS Functions and the Parametric-PGF

7.4.2.1 New lnCOS Function with Respect to θ

In this subsection, the assembly angle θ of the X-structure would be considered and other structural parameters are of no interest.

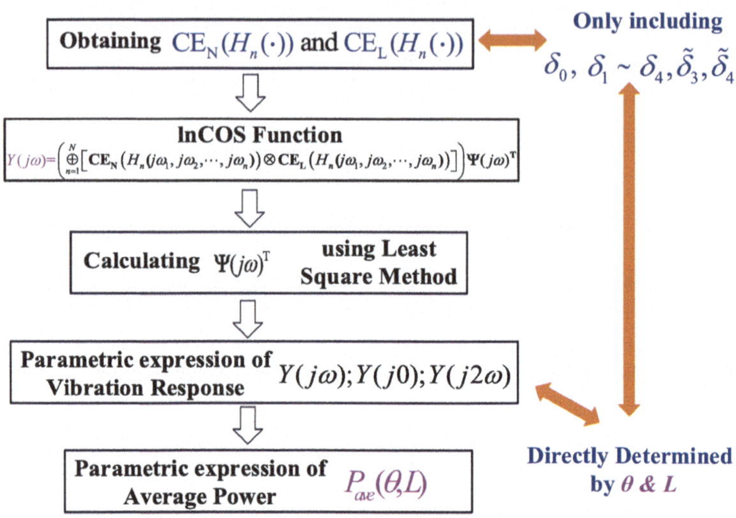

Fig. 7.10 Procedures of the InCOS-based parametric approach

It should be noted that, based on previous theoretical analysis, it has been known that applying $N = 5$ would be accurate enough and even applying $N = 3$ could be acceptably accurate. Base on this, the case of $N = 5$ would be considered in the following parts. Also, as in the theoretical analysis it has been found that only odd-ordered components would contribute to $Y^{\hat{y}}(j\omega)$ and only even-ordered components would contribute to $Y^{\hat{y}}(j0)$, and $Y^{\hat{y}}(j2\omega)$, the unnecessary terms that equals to be zero have been ignored in the following parts. This case would be also utilized for InCOS function with respect to L and c_e in the next subsections.

The new InCOS function of $Y^{\hat{y}}(j\omega)$, $Y^{\hat{y}}(j0)$ concerning θ could be given as

$$Y^{\hat{y}}(j\omega;\theta) = Y_1^{\hat{y}}(j\omega;\theta) + Y_3^{\hat{y}}(j\omega;\theta) + Y_5^{\hat{y}}(j\omega;\theta) + \cdots, \tag{7.20a}$$

$$\begin{cases} Y_1^{\hat{y}}(j\omega\,;\theta) = \left[\delta_0^{-1}, \delta_0^{-2}, \delta_0^{-3}\right]\boldsymbol{\psi}_1(j\omega)^{\mathrm{T}},\ Y_3^{\hat{y}}(j\omega\,;\theta) \\ \quad = \begin{bmatrix} \left[\delta_1^2, \delta_1\delta_3, \delta_3^2\right] \otimes \left[\delta_0^{-1}, \delta_0^{-2}, \delta_0^{-3}\right]^5 \\ \oplus[\delta_2, \delta_4] \otimes \left[\delta_0^{-1}, \delta_0^{-2}, \delta_0^{-3}\right]^4 \end{bmatrix} \boldsymbol{\psi}_3(j\omega)^{\mathrm{T}} \\ Y_5^{\hat{y}}(j\omega\,;\theta) \\ \quad = \begin{bmatrix} \left[\delta_1^4, \delta_1^3\delta_3, \delta_1^2\delta_3^2, \delta_1\delta_3^3, \delta_3^4\right] \otimes \left[\delta_0^{-1}, \delta_0^{-2}, \delta_0^{-3}\right]^9 \\ \oplus\left[\delta_1^2\delta_2, \delta_1^2\delta_4, \delta_1\delta_2\delta_3, \delta_1\delta_3\delta_4, \delta_2\delta_3^2, \delta_3^2\delta_4\right] \otimes \left[\delta_0^{-1}, \delta_0^{-2}, \delta_0^{-3}\right]^8 \\ \oplus\left[\delta_2^2, \delta_2\delta_4, \delta_4^2\right] \otimes \left[\delta_0^{-1}, \delta_0^{-2}, \delta_0^{-3}\right]^7 \end{bmatrix} \boldsymbol{\psi}_5(j\omega)^{\mathrm{T}} \end{cases}, \tag{7.20b}$$

$$Y^{\hat{y}}(j0;\theta) = Y_2^{\hat{y}}(j0;\theta) + Y_4^{\hat{y}}(j0;\theta) + \cdots. \tag{7.21a}$$

$$\begin{cases} Y_2^{\hat{y}}(j0\,;\theta) = \left[[\delta_1, \delta_3] \otimes \left[\delta_0^{-1}, \delta_0^{-2}, \delta_0^{-3}\right]^3\right]\psi_2(j\omega)^T \\ Y_4^{\hat{y}}(j0\,;\theta) = \begin{bmatrix} [\delta_1^3, \delta_1^2\delta_3, \delta_1\delta_3^2, \delta_3^3] \otimes \left[\delta_0^{-1}, \delta_0^{-2}, \delta_0^{-3}\right]^7 \\ \oplus[\delta_1\delta_2, \delta_1\delta_4, \delta_2\delta_3, \delta_3\delta_4] \otimes \left[\delta_0^{-1}, \delta_0^{-2}, \delta_0^{-3}\right]^6 \end{bmatrix}\psi_4(j\omega)^T \end{cases}. \tag{7.21b}$$

It can be observed that, the vibration response in the frequency domain of the X-structured system can be expressed as a complex-valued function in terms of the assembly angle θ. In addition, the assembly angle θ would influence the linear and nonlinear parts of the X-structured harvesting system simultaneously. It should be mentioned that, the structure of the lnCOS function of $Y^{\hat{y}}(j2\omega\,;\theta)$ is the same as that of $Y^{\hat{y}}(j0\,;\theta)$, which is not listed due to space limitation. Similarly, in the following cases with respect to L and c_e, the function of $Y^{\hat{y}}(j2\omega)$ would not be listed but be taken into consideration to calculate the parametric power generation function (Parametric-PGF).

7.4.2.2 New lnCOS Function with Respect to *L*

The rod length L of the X-structure would be considered and other structural parameters are of no interest. The new lnCOS function of $Y^{\hat{y}}(j\omega)$, $Y^{\hat{y}}(j0)$ with respect to L could be given as

$$Y^{\hat{y}}(j\omega; L) = Y_1^{\hat{y}}(j\omega; L) + Y_3^{\hat{y}}(j\omega; L) + Y_5^{\hat{y}}(j\omega; L) + \cdots, \tag{7.22a}$$

$$\begin{cases} Y_1^{\hat{y}}(j\omega; L) = \boldsymbol{\vartheta}_1(j\omega);\ Y_3^{\hat{y}}(j\omega\,; L) = \left[\delta_1^2, \delta_1\delta_3, \delta_3^2, \delta_2, \delta_4\right]\boldsymbol{\vartheta}_3(j\omega)^{\mathrm{T}} \\ Y_5^{\hat{y}}(j\omega; L) = \begin{bmatrix} \delta_1^2\delta_2, \delta_1^2\delta_4, \delta_1^4, \delta_1^3\delta_3, \delta_1^2\delta_3^2, \delta_1\delta_2\delta_3, \delta_1\delta_3\delta_4, \\ \delta_1\delta_3^3, \delta_2\delta_3^2, \delta_3^2\delta_4, \delta_3^4, \delta_2^2, \delta_2\delta_4, \delta_4^2 \end{bmatrix}\boldsymbol{\vartheta}_5(j\omega)^{\mathrm{T}} \end{cases}, \tag{7.22b}$$

$$Y^{\hat{y}}(j0; L) = Y_2^{\hat{y}}(j0; L) + Y_4^{\hat{y}}(j0; L) + \cdots, \tag{7.23a}$$

$$\begin{cases} Y_2^{\hat{y}}(j0; L) = [\delta_1, \delta_3]\boldsymbol{\vartheta}_2(j\omega)^{\mathrm{T}} \\ Y_4^{\hat{y}}(j0; L) = [\delta_1\delta_2, \delta_1\delta_4, \delta_1^3, \delta_1^2\delta_3, \delta_1\delta_3^2, \delta_2\delta_3, \delta_3\delta_4, \delta_3^3]\boldsymbol{\vartheta}_4(j\omega)^{\mathrm{T}} \end{cases}. \tag{7.23b}$$

It can be seen that, the rod length L would only influence the nonlinear parts of the X-structured harvesting system. It should be noted that, the structure of the lnCOS function of $Y^{\hat{y}}(j2\omega\,; L)$ is the same as that of $Y^{\hat{y}}(j0; L)$.

7.4.2.3 New lnCOS Function with Respect to c_e

In this subsection, the electrical damping coefficient c_e of the X-structured harvesting system would be considered and other structural parameters are of no interest. The new lnCOS function of $Y^{\hat{y}}(j\omega)$, $Y^{\hat{y}}(j0)$ with respect to c_e could be given as

$$Y^{\hat{y}}(j\omega; c_e) = Y_1^{\hat{y}}(j\omega; c_e) + Y_3^{\hat{y}}(j\omega; c_e) + Y_5^{\hat{y}}(j\omega; c_e) + \cdots, \tag{7.24a}$$

$$\begin{cases} Y_1^{\hat{y}}(j\omega; c_e) = \left[1,\ c_e,\ c_e^2\right]\zeta_1(j\omega)^{\mathrm{T}} \\ Y_3^{\hat{y}}(j\omega; c_e) = \left[\delta_3^2 \otimes \left[1, c_e, c_e^2\right]^5 \oplus \delta_4 \otimes \left[1, c_e, c_e^2\right]^4\right]\zeta_3(j\omega)^{\mathrm{T}} \\ Y_5^{\hat{y}}(j\omega; c_e) = \left[\delta_3^4 \otimes \left[1, c_e, c_e^2\right]^9 \oplus \delta_3^2\delta_4 \otimes \left[1, c_e, c_e^2\right]^8 \oplus \delta_4^2 \otimes \left[1, c_e, c_e^2\right]^7\right]\zeta_5(j\omega)^{\mathrm{T}} \end{cases}, \tag{7.24b}$$

$$Y^{\hat{y}}(j0; c_e) = Y_2^{\hat{y}}(j\omega; c_e) + Y_4^{\hat{y}}(j\omega; c_e) + \cdots, \tag{7.25a}$$

$$\begin{cases} Y_2^{\hat{y}}(j\omega; c_e) = \left[\delta_3 \otimes \left[1, c_e, c_e^2\right]^3\right]\zeta_1(j\omega)^{\mathrm{T}} \\ Y_4^{\hat{y}}(j\omega; c_e) = \left[\delta_3^3 \otimes \left[1, c_e, c_e^2\right]^7 \oplus \delta_3\delta_4 \otimes \left[1, c_e, c_e^2\right]^6\right]\zeta_4(j\omega)^{\mathrm{T}} \end{cases}. \tag{7.25b}$$

The equivalent electrical damping coefficient c_e would contribute to both the linear and nonlinear damping components of the X-structured harvesting system. This is the inherent structural damping characteristic of the X-structured harvester. Similarly, the structure of the InCOS function of $Y^{\hat{y}}(j2\omega; c_e)$ is the same as that of $Y^{\hat{y}}(j0; c_e)$.

7.4.2.4 The Parametric-PGF

By substituting Eqs. (7.20–7.21), (7.22–7.23) or (7.24–7.25) into the previous PGF (Eq. 7.15), the parametric power generation function (Parametric-PGF) could be given as

$$|P_{\mathrm{ave}}(j\omega; J)| = P(J),\ J = \theta,\ L,\ c_e. \tag{7.26}$$

It should be noted that, the Parametric-PGF is a more concise expression with regard to the assembly angle θ, the rod length L and/or the equivalent electrical damping c_e.

7.5 The InCOS Functions and Parametric-PGF Based Analysis

In this section, based on the obtained InCOS functions and the Parametric-PGF of the representative X-structured harvesting system, parametric analysis and design would be carried out. Sensitive analysis with respect to critical structural parameters would be conducted. In addition, detailed discussions and comparisons with the harmonic balance method would be provided.

7.5.1 Parametric Analysis and Design Using lnCOS Functions and Parametric-PGF

By using Eqs. (7.20–7.25), the lnCOS functions with respect to θ, L, or c_e, can be further simplified, which could be applied for the following analysis. For example, the lnCOS functions with respect to L and c_e can be expressed as

$$\begin{cases} Y^{\hat{y}}(j\omega; L) = \left[1, L^{-2}, L^{-4}, \ldots\right]\tilde{\vartheta}(j\omega)^{\mathrm{T}} \\ Y^{\hat{y}}(j0; L) = \left[L^{-1}, L^{-3}, \ldots\right]\overline{\vartheta}(j\omega)^{\mathrm{T}} \\ Y^{\hat{y}}(j2\omega; L) = \left[L^{-1}, L^{-3}, \ldots\right]\widehat{\vartheta}(j\omega)^{\mathrm{T}} \end{cases}, \quad \begin{cases} Y^{\hat{y}}(j\omega; c_e) = \left[1, c_e, c_e^2, \ldots\right]\tilde{\xi}(j\omega)^{\mathrm{T}} \\ Y^{\hat{y}}(j0; c_e) = \left[c_e, c_e^2, c_e^2, \ldots\right]\overline{\xi}(j\omega)^{\mathrm{T}} \\ Y^{\hat{y}}(j2\omega; c_e) = \left[c_e, c_e^2, c_e^2, \ldots\right]\widehat{\xi}(j\omega)^{\mathrm{T}} \end{cases}. \tag{7.27}$$

Equation (7.27) reveals that the output spectrum for the vibration responses is presented in a straightforward and unambiguous manner, taking the form of polynomials in terms of L^{-1} and c_e. It's noteworthy that this study is the inaugural one to articulate the intricate and direct correlations between the vibration responses, power generation, and the system parameters of the X-structured harvesting system. Such insights were not attainable with the harmonic balance method or simulation techniques previously referenced in [29, 30]. The scenario involving θ, is more intricate than those involving L and c_e, and is detailed in Appendix 4. By integrating the streamlined lnCOS functions into the PGF, a simplified Parametric-PGF is derived, which can then be applied to analyze the power generation performance in relation to the system parameters of interest. This process enables the parametric characterization of the essential structural parameters within the X-structured harvesting system.

In this part, the structural parameters and the base excitation are set as: $M = 2$ kg, $k = 4000$ N/m, $m = 2$ and $u(t) = 3\cos(1.8 \cdot 2\pi t)$. It should be noted that, the excitation frequency is set as 1.8 Hz, which is regarded as the interested external frequency. Under this case or this aiming frequency, the structural parameters of the X-structured harvester could be designed to achieve the potential highest power generation. This is the idea of the present lnCOS function and the Parametric-PGF-based analysis and design.

By utilizing the lnCOS functions (Eqs. 7.27, 7.37–7.39) and the Parametric-PGF (Eq. 7.26) with respect to θ, L, or c_e, the vibration responses and the average power generated of the X-structured could be calculated, as shown in Figs. 7.11, 7.12 and 7.13. It could be observed that, the lnCOS function and the Parametric-PGF-based representations of the system output spectrum have been validated by the simulated results with good agreements. As a supplement, the relative errors between

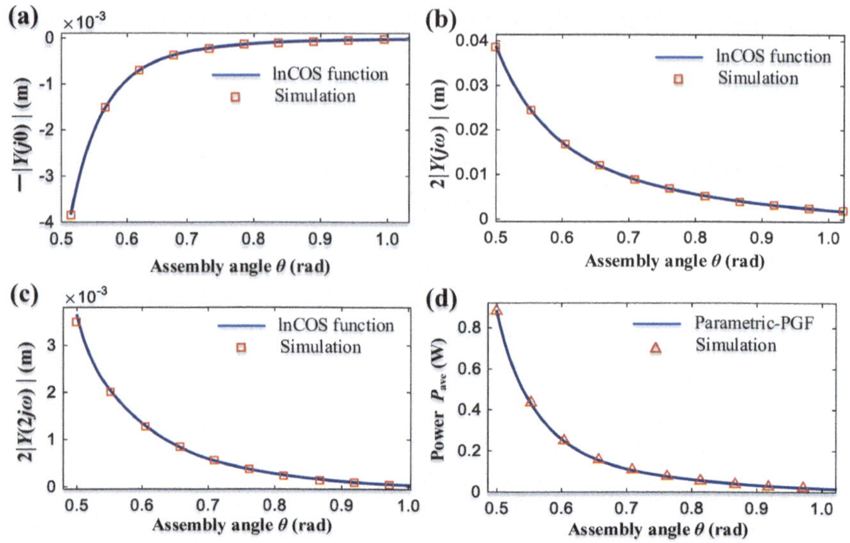

Fig. 7.11 The amplitudes of the output spectrum of the vibration response and the average power output of the X-structured harvester with different assembly angle θ: **a** The 0th order harmonic output; **b** The first order harmonic output; **c** The second order harmonic output; **d** The average power output ($L = 0.10$ m)

the InCOS-based method and the simulation results (the Parametric-PGF and the simulations) are also presented in Tables 7.4, 7.5 and 7.6. It could be seen that the relative errors are mostly lower than 2% (in between 0.01 and 2.07%), which means the InCOS-based approach is accurate and reliable in this study.

In Fig. 7.11, by increasing the assembly angle θ from around 0.5–1.0 rad ($\theta = [\pi/6, \pi/3]$), the amplitudes of the output spectrum (including the 0th, first, and second harmonic outputs) of the vibration response would dramatically decrease. Importantly, the average power output would also dramatically decrease from about 0.9 W to a very low value around 0. This indicates that the assembly angle θ is a very critical system parameter that influences the performance of the power generation, which is consistent with the theoretical results in Sect. 7.5.

Figure 7.12 illustrates that extending the rod length (L) from ~0.05 to 0.35 m results in a significant reduction in the amplitudes of both the 0th and second harmonic outputs of the vibration response. On the other hand, the amplitude of the first order harmonic output decreases more modestly, from about 0.046 m to roughly 0.037 m. Notably, once the rod length (L) exceeds about 0.15 m, the amplitude of the first harmonic output stabilizes, while the amplitudes of the 0th and second harmonics remain minimal. This trend suggests that as the rod length increases, the system begins to behave more linearly, with the nonlinear components diminishing, which is also reflected in Eq. (7.4). Similarly, the average power output drops from around 1.4–0.8 W before leveling off when the rod length is >0.15 m. The rod length (L)

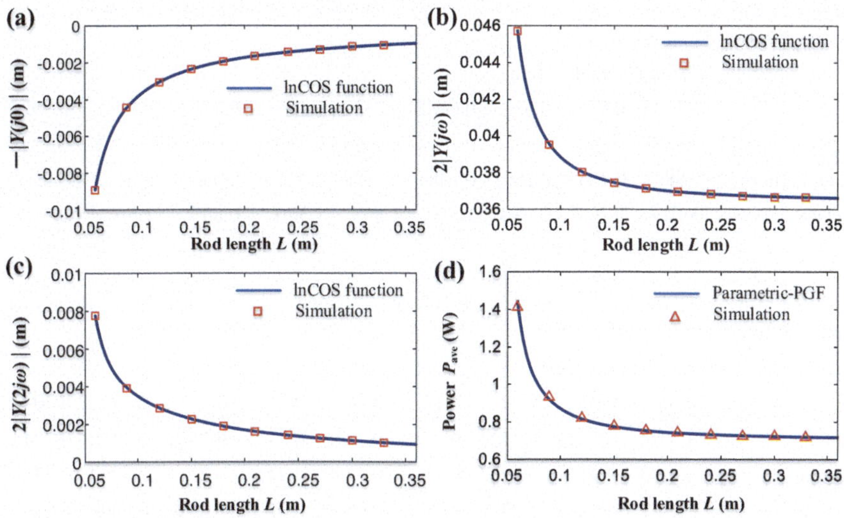

Fig. 7.12 The amplitudes of the output spectrum of the vibration response and the average power output of the X-structured harvester with different rod length L: **a** The 0th order harmonic output; **b** The first order harmonic output; **c** The second order harmonic output; **d** The average power output ($\theta = \pi/6$)

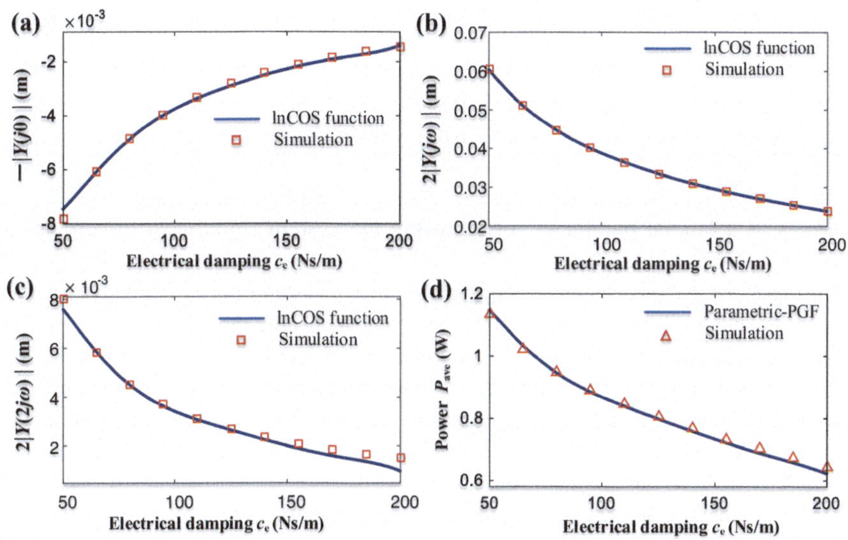

Fig. 7.13 The amplitudes of the output spectrum of the vibration response and the average power output of the X-structured harvester with different equivalent electrical damping: **a** The 0th order harmonic output; **b** The first order harmonic output; **c** The second order harmonic output; **d** The average power output ($\theta = \pi/6$ and $L = 0.10$ m)

Table 7.4 Relative errors of the average power output between the parametric-PGF and simulation results: different assembly angle θ

	Assembly angle θ (rad)								
	0.5236	0.5760	0.6283	0.6807	0.7330	0.7854	0.8378	0.8901	0.9425
Simulation	0.8831	0.4384	0.2526	0.1614	0.1099	0.07787	0.05677	0.04222	0.0318
Parametric-PGF	0.8822	0.4351	0.2501	0.1584	0.1082	0.07686	0.05672	0.04229	0.0316
Relative errors	0.10%	0.75%	0.99%	1.86%	1.55%	1.30%	0.88%	1.66%	0.63%

Table 7.5 Relative errors of the average power output between the parametric-PGF and simulation results: different rod length L

	Rod length L (m)								
	0.06	0.09	0.12	0.15	0.18	0.21	0.24	0.27	0.30
Simulation	1.4160	0.9343	0.8229	0.7783	0.7556	0.7424	0.7341	0.7284	0.7244
Parametric-PGF	1.4055	0.9259	0.8202	0.7766	0.7549	0.7418	0.7337	0.7282	0.7243
Relative errors	0.74%	0.90%	0.33%	0.22%	0.09%	0.08%	0.05%	0.03%	0.01%

Table 7.6 Relative errors of the average power output between the parametric-PGF and simulation results: different equivalent electrical damping c_e

	Equivalent electrical damping c_e (Ns/m)								
	50	65	80	95	110	125	140	155	170
Simulation	1.136	1.024	0.9482	0.8918	0.8455	0.8050	0.7682	0.7344	0.7029
Parametric-PGF	1.149	1.032	0.9475	0.8861	0.8400	0.7982	0.7590	0.7217	0.6884
Relative errors	1.14%	0.78%	0.74%	0.64%	0.65%	0.84%	1.20%	1.73%	2.06%

emerges as a crucial parameter for adjusting the system's nonlinear effects when necessary.

In Fig. 7.13, by increasing the equivalent electrical damping c_e from around 50–200 Ns m, the amplitudes of the output spectrum (including the 0th, first, and second harmonic outputs) of the vibration response would gradually decrease. Similarly, the average power output would gradually decrease from about 1.15–0.65 W. Overall, the equivalent electrical damping c_e would be determined by the generator component, which would contribute to the total damping (together with the frictional damping) of the X-structured harvesting system.

In addition, by considering both the structural parameters θ and L, the vibration responses (including the 0th, first, and second harmonic outputs) and the average power generated by the X-structured could be obtained, as shown in Fig. 7.14. Similarly, the lnCOS function and the Parametric-PGF-based analysis of the system output spectrum have been compared by the simulation results with good agreements.

It could be observed that, in Fig. 7.14, the optimal structural parameters are $\theta = \pi/6$ and $L = 0.06$ m (within the range of $\theta = \left[\pi/6, \pi/3\right]$ and $L = [0.06\ \text{m},\ 0.36\ \text{m}]$) with the average power output of 1.4 W. Overall, the rod length L could make

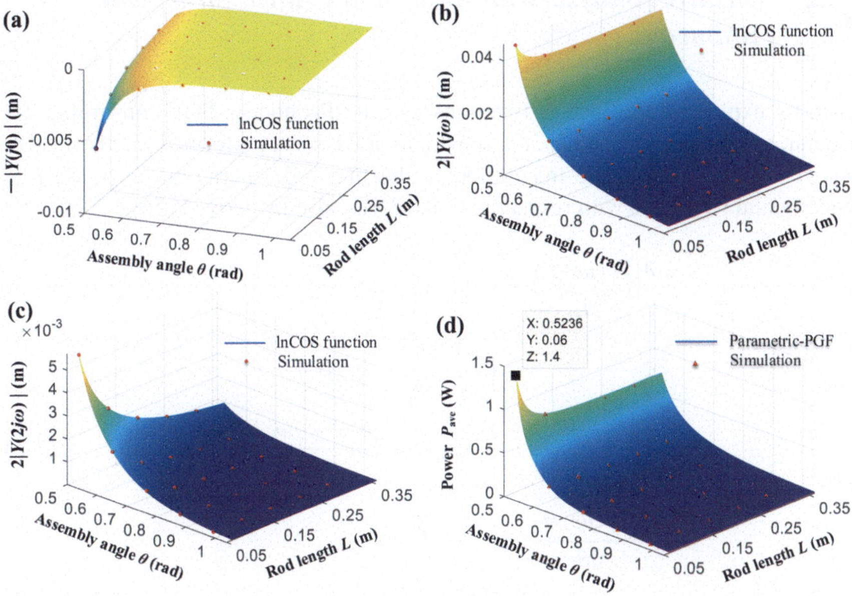

Fig. 7.14 The amplitudes of the output spectrum of the vibration response and the average power output of the X-structured harvester with different equivalent assembly angle θ and rod length L: **a** The 0th order harmonic output; **b** The first order harmonic output; **c** The second order harmonic output; **d** The average power output

the nonlinear effect much more obvious when it is smaller than 0.15 m, while the assembly angle θ would greatly influence the power generation performance almost in the whole range.

In summary, by tuning the rod length L, the nonlinear effects could be designed without interfering with the linear components, while by adjusting the assembly angle θ and the electrical damping c_e, the linear and nonlinear components could be designed simultaneously. Compared with previous results by utilizing the harmonic balance method or the ode45 simulation in [29, 30], this work for the first time, achieves parametric characterization and obtains straightforward expressions of the vibration responses/power output concerning the critical system parameters of the X-structured nonlinear system, by exploiting the proposed lnCOS-based functions and the Parametric-PGF.

7.5.2 Sensitive Analysis with Respect to Critical Structural Parameters

Also, by exploiting the InCOS functions and the Parametric-PGF concerning the structural parameters (θ and L), the sensitivity analysis is performed in this subsection. The sensitivity of the magnitude of the first order harmonic response (the fundamental response) with respect to θ and L can also be derived [56]

$$\begin{aligned}\frac{\partial\left|Y^{\hat{y}}(j\omega;L)\right|}{\partial L}=&\frac{1}{2\left|Y^{\hat{y}}(j\omega;L)\right|}\\&\left(\frac{\partial Y^{\hat{y}}(j\omega;L)}{\partial L}Y^{\hat{y}}(-j\omega;L)\right.\\&\left.+Y^{\hat{y}}(j\omega;L)\frac{\partial Y^{\hat{y}}(-j\omega;L)}{\partial L}\right),\end{aligned}\tag{7.28a}$$

$$\begin{aligned}\frac{\partial\left|Y^{\hat{y}}(j\omega;\theta)\right|}{\partial\theta}=&\frac{1}{2\left|Y^{\hat{y}}(j\omega;\theta)\right|}\\&\left(\frac{\partial Y^{\hat{y}}(j\omega;\theta)}{\partial L}Y^{\hat{y}}(-j\omega;\theta)\right.\\&\left.+Y^{\hat{y}}(j\omega;\theta)\frac{\partial Y^{\hat{y}}(-j\omega;\theta)}{\partial L}\right).\end{aligned}\tag{7.28b}$$

It should be noted that, for the DC component and the second order harmonic response, similar sensitivity functions exist and they are not listed here due to space limitation. Based on the sensitivity functions of the vibration responses, the sensitivity function of the magnitude of the power generation could also be obtained. By using the derived sensitivity functions, the sensitivity analysis can be carried out as shown in Figs. 7.15 and 7.16.

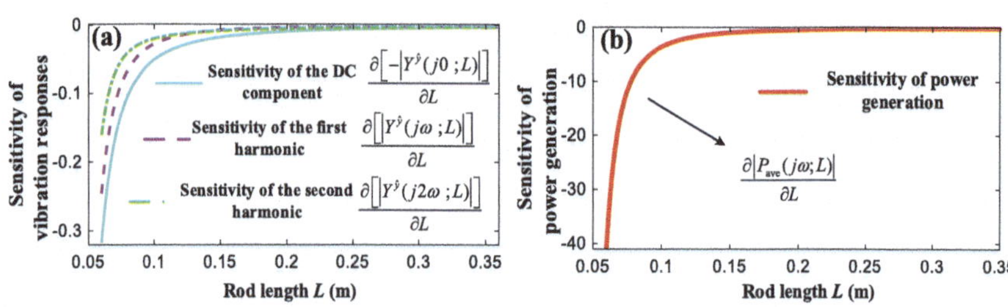

Fig. 7.15 The sensitivity of the vibration responses/power generation with respect to the rod length L of the X-structure: **a** Sensitivity of the DC component, the first order harmonic, and the second harmonic; **b** Sensitivity of the power generation

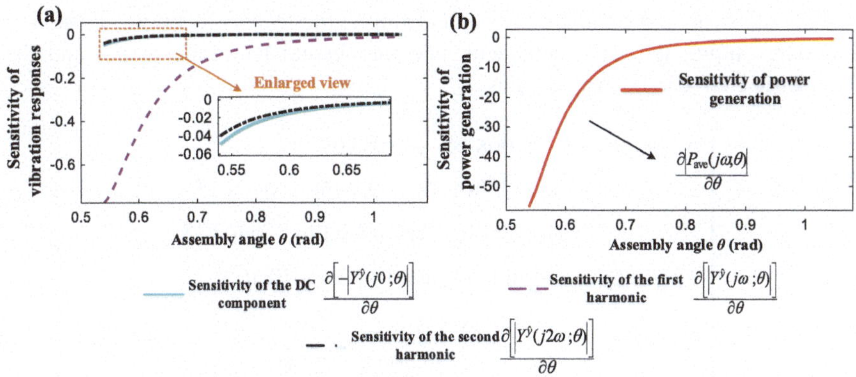

Fig. 7.16 Sensitivity of the vibration responses/power generation with respect to the assembly angle θ of the X-structure: **a** Sensitivity of the DC component, the first order harmonic, and the second harmonic; **b** Sensitivity of the power generation

Figures 7.15 and 7.16 demonstrate that when the rod length (L) is <0.10–0.15 m and the assembly angle (θ) is below 0.6–0.8 radians, there is a heightened sensitivity in the vibration responses and power generation relative to these structural parameters. It's noted that the influence of the rod length (L) on power generation becomes minimal once it exceeds 0.15 m, aligning with the observations from Fig. 7.12. Intriguingly, Fig. 7.15a reveals that the sensitivity of the DC component and the first and second harmonic responses to the rod length are quite comparable. However, Fig. 7.16a shows a marked difference in the sensitivity of the first harmonic response compared to the DC component and the second harmonic response. This discrepancy arises because the rod length (L) primarily affects the nonlinear aspects of the vibration responses, whereas the assembly angle (θ) impacts both the linear and nonlinear responses. This analysis offers valuable insights into how structural parameters influence system behavior.

7.5.3 Discussions and Comparisons with the Harmonic Balance Method

This study is pioneering in providing a comprehensive and explicit analysis of how the vibration responses and power generation correlate with the system parameters of the X-structured harvesting system. Unlike previous research that relied on the harmonic balance method or ode45 simulations referenced in [29, 30], this work uniquely identifies the distinct impact of various structural parameters on the vibration responses of the X-structured harvester, including the DC component and the first and second harmonic responses, as well as on its power generation capabilities. This is achieved through the use of the newly introduced lnCOS-based functions and the Parametric-PGF, which offer clear expressions related to L, θ, and c_e.

Take the structural parameter L as an example, the InCOS functions of the first harmonic output, the DC component, and the second harmonic output and the Parametric-PGF can be expressed as

$$\begin{cases} Y^{\hat{y}}(j\omega; L) = (-0.008789 + 0.01597i) \\ \qquad +(1.77e-05 + 2.225e-05i)L^{-2} \\ \qquad +(4.414e-08 + 9.217e-09i)L^{-4} \\ Y^{\hat{y}}(j0; L) = (-0.0003323 + 1.402e-20i)L^{-1} \\ \qquad +(4.019e-07 - 3.798e-22i)L^{-3} \\ \qquad +(-1.047e-09 + 1.253e-09i)L^{-5} \\ Y^{\hat{y}}(j2\omega; L) = (5.685e-05 - 0.0001611i)L^{-1} \\ \qquad +(-4.522e-07 - 1.904e-07i)L^{-3} \\ \qquad +(-1.057e-09 + 4.645e-10i)L^{-5} \\ |P_{\text{ave}}(L)| = 0.7083818053 + 1.485436505e-03L^{-2} \\ \qquad +1.134556957e-06L^{-4} \\ \qquad +3.104364612e-09L^{-6} + \ldots \end{cases}, \tag{7.29}$$

which cannot be achieved by using the previous harmonic balance method or simulation approach in Refs. [29, 30].

The expressions of the InCOS functions are quite explicit and straightforward to analyze and design the critical structural parameters that of interests. This can be considered as the most important advantage over the harmonic balance method. Based on the InCOS functions and the Parametric-PGF, analysis and design of the critical parameter analysis can be carried out (Sect. 7.5.1) and sensitivity analysis could also be performed (Sect. 7.5.2).

In addition, the procedures of the InCOS-based method and the harmonic balance method are listed in Table 7.7. For the harmonic balance method, if the DC component, the first and the second harmonic outputs are taken into consideration ($\hat{y} = Y_0 + Y_1\cos(\Omega\tau + \phi) + Y_2\cos(2\Omega\tau + \varphi)$), the set of harmonic balance equations including five unknowns would be generated as $f(Y_0, Y_1, Y_2, \phi, \varphi) = 0$, which can be solved by using the Newton–Raphson method [29, 30]. The authors have already utilized the harmonic balance method for the X-structured harvesters in Ref. [29, 30]. For some strongly nonlinear systems with multi-valued solutions, the harmonic balance method can capture the jump phenomenon, which is the merits of the harmonic balance method [46].

To further make further comparisons between the proposed InCOS-based method and the harmonic balance (HB) method, the procedures in Table 7.7 are conducted and the obtained results are shown in Fig. 7.17. It can be observed that, the results of the InCOS-based method, the second order HB method, and the simulation have good agreements with each other. The first order HB method slightly overestimates the solution when with a smaller rod length L.

Table 7.7 Procedures of the lnCOS-based method and the harmonic balance method

	lnCOS-based method	The harmonic balance method
Step 1	Determinations of the parameter series $\bigoplus_{n=1}^{N}(\mathrm{CE_N}(H_n(\cdot)) \otimes \mathrm{CE_L}(H_n(\cdot)))$	Assuming the expression of the solution $\hat{y} = Y_0 + Y_1\cos(\Omega\tau + \phi) + Y_2\cos(2\Omega\tau + \varphi)$
Step 2	Calculation of $\boldsymbol{\Psi}(j\omega)$ by using the least square method	Generation and solving of the harmonic balance equations $f(Y_0, Y_1, Y_2, \phi, \varphi) = 0$
Step 3	Forming the lnCOS functions	Forming the expression of the solution

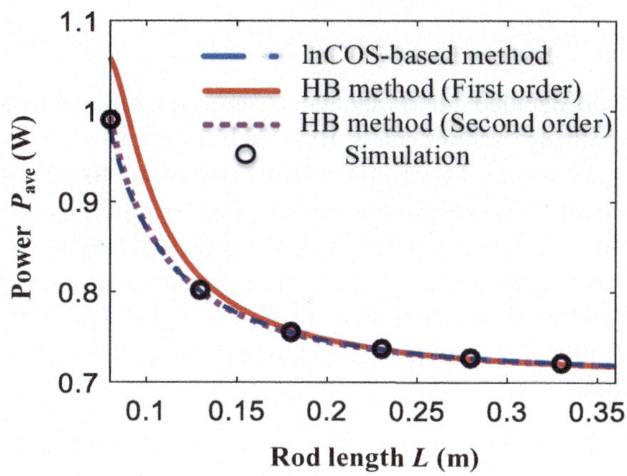

Fig. 7.17 Comparisons between the proposed lnCOS-based method and the harmonic balance method: an example with respect to the rod length L

Importantly, for the new lnCOS-based method, the system outputs can be obtained in a more straightforward way without solving complex algebraic balance equations. The issue would be to precisely determine the parameter series of interest by using $\oplus_{n=1}^{N}(\mathrm{CE_N}(H_n(\cdot)) \otimes \mathrm{CE_L}(H_n(\cdot)))$ (Sect. 7.4.2) and to accurately calculate the required $\boldsymbol{\Psi}(j\omega)$ that in complex form (Sect. 7.5.1). The proposed lnCOS-based method does not involve any equation solution procedure or any numeric iteration, which is therefore considerably straightforward and convenient for computation (especially for parametric study). This can be regarded as another important advantage of the present method. Furthermore, the advantages and drawbacks are summarized in Table 7.8 to clearly show the comparisons between the novel lnCOS-based method and the HB method. It should be also mentioned that, the other important advantage of the volterra-series based parametric method is that it can tackle the cases with general inputs such as multi-tone input and random input [35], which would be further investigated in a future study.

It should be mentioned that, although the diverse merits of the lnCOS-based method (the Volterra series-based method), the validity issue or the convergence issue of the Volterra series-based approach should be considered when using the present method. It is a good practice to check the behavior of the results as the order is increased [46], or to compare with results from simulation or the other methods.

Table 7.8 Advantages and drawbacks of the lnCOS-based method and the HB method

	lnCOS-based method	The harmonic balance method
Advantages	(1) Explicit solution; (2) Straightforward power functions for critical parameters; (3) Easy to optimize; (4) Less computational costs (5) Suitable for general inputs (Ref. [35])	(1) Can capture the jump phenomenon; (2) Fewer convergence issues; (3) Easy to tackle strong nonlinear system
Drawbacks	(1) Facing convergence issues; (2) Not easy to tackle strong nonlinear stiffness	(1) Involving complex algebraic balance equations; (2) Implicit solutions

More detailed reports about the convergence bound issues of the Volterra series-based nonlinear systems can be referred to Ref. [50].

Moreover, based on explicit expressions like Eq. (7.29), a future optimization would be more straightforward. For the optimization of the vibration responses/ power generation performance, if a desired magnitude of vibration responses Y^* or power generation P^*_{ave} is given, optimal structural parameters of θ^* and L^* could be found by utilizing $\min\left(\left|Y^{\hat{y}}(j\omega)\right| - Y^*\right)$ and $\min\left(\left|P_{ave}(j\omega)\right| - P^*_{ave}\right)$. A systematic method for this purpose is yet to be developed, which could be discussed in the future study.

7.6 Conclusions

The frequency domain method based on the Volterra series is applied to systematically analyze and optimize a novel bio-inspired nonlinear X-structured energy harvesting system. The generalized frequency response functions (GFRFs), output frequency response function (OFRF), and power generation function (PGF) are derived theoretically through a recursive algorithm to evaluate the energy harvesting efficiency. Numerical simulations have corroborated the effectiveness of the proposed method. Crucially, the parametric approach introduces the innovative lnCOS function and Parametric-PGF, which succinctly and directly correlate the vibration responses and power generation to the key parameters of the nonlinear X-structured system. The key findings include:

- The OFRF and PGF derived from theory can be directly applied to compute the vibration response and power generation. The lnCOS-based method, a product of the parametric approach, allows for the determination of system outputs without the need to solve complex nonlinear algebraic equations, such as those required by the harmonic balance method. This makes it a straightforward and convenient computational tool, particularly for the parametric analysis of the X-structured system.

- The lnCOS-based method renders the vibration response and power generation as concise and explicit functions concerning both linear and nonlinear parameters that are under design consideration for the system (including L, θ, and c_e). This method holds promise for systematic analysis and optimization of nonlinear energy harvesting systems.
- Power generation results indicate that the X-structured energy harvester can produce more power than a traditional linear harvester within the low-frequency range. Regarding system parameters, the rod length (L) can be adjusted to enhance the nonlinear stiffness and damping effects, particularly when it is <0.15 m. The assembly angle (θ) significantly affects power generation performance across nearly the entire range. An increase in the equivalent electrical damping (c_e) will cause a gradual decrease in power generation within a certain range.

In conclusion, the PGF and lnCOS-based frequency domain approach derived from theory offer a clear method to characterize the advantageous nonlinearities introduced by critical system parameters. This research contributes a novel and straightforward parametric design method that will benefit the analysis and design of structural parameters in related energy harvesting systems. Future work will concentrate on optimizing energy harvesting performance through more systematic structural design.

Acknowledgements The main modelling and analysis results of this study give credit to a previous PhD student, Li and Jing [57].

Appendix 1

The expressions of the higher order GFRFs are given as follows:

$$H_5^{\hat{y}}(j\omega_1, j\omega_2, j\omega_3, j\omega_4, j\omega_5) = L^{-1}(5) \times \left(\begin{array}{l} \left(2\delta_1 + \delta_3 \sum_i^5 (j\omega_i)\right) \left(\begin{array}{l} H_1(j\omega_1) H_4(j\omega_2, j\omega_3, j\omega_4, j\omega_5) \\ + H_2(j\omega_1, j\omega_2) H_3(j\omega_3, j\omega_4, j\omega_5) \end{array} \right) \\ + \left(3\delta_2 + \delta_4 \sum_i^5 (j\omega_i)\right) \left(\begin{array}{l} H_1(j\omega_1) H_1(j\omega_2) H_3(j\omega_3, j\omega_4, j\omega_5) \\ + H_1(j\omega_1) H_2(j\omega_2, j\omega_3) H_2(j\omega_4, j\omega_5) \end{array} \right) \end{array} \right), \tag{7.30}$$

$$H_6^{\hat{y}}(j\omega_1, j\omega_2, j\omega_3, j\omega_4, j\omega_5, j\omega_6) = L^{-1}(6)$$

$$\times\left(\begin{array}{l}\left(2\delta_1+\delta_3\sum_i^6(j\omega_i)\right)\left(\begin{array}{l}H_1(j\omega_1)H_5(j\omega_2,j\omega_3,j\omega_4,j\omega_5,j\omega_6)\\+H_2(j\omega_1,j\omega_2)H_4(j\omega_3,j\omega_4,j\omega_5,j\omega_6)\end{array}\right)\\+(\delta_1+\delta_3(j\omega_4+j\omega_5+j\omega_6))H_3(j\omega_1,j\omega_2,j\omega_3)H_3(j\omega_4,j\omega_5,j\omega_6)\\+\left(3\delta_2+\delta_4\sum_i^6(j\omega_i)\right)\left(\begin{array}{l}H_1(j\omega_1)H_1(j\omega_2)H_4(j\omega_3,j\omega_4,j\omega_5,j\omega_6)\\+H_1(j\omega_1)H_2(j\omega_2,j\omega_3)H_3(j\omega_4,j\omega_5,j\omega_6)\end{array}\right)\\+(\delta_2+\delta_4)H_2(j\omega_1,j\omega_2)H_2(j\omega_3,j\omega_4)H_2(j\omega_5,j\omega_6)\end{array}\right), \tag{7.31}$$

$$H_7^{\hat{y}}(j\omega_1,j\omega_2,j\omega_3,j\omega_4,j\omega_5,j\omega_6,j\omega_7)=L^{-1}(7)$$
$$\times\left(\begin{array}{l}\left(2\delta_1+\delta_3\sum_i^7(j\omega_i)\right)\left(\begin{array}{l}H_1(j\omega_1)H_6(j\omega_2,j\omega_3,j\omega_4,j\omega_5,j\omega_6,j\omega_7)\\+H_2(j\omega_1,j\omega_2)H_5(j\omega_3,j\omega_4,j\omega_5,j\omega_6,j\omega_7)\\+H_3(j\omega_1,j\omega_2,j\omega_3)H_4(j\omega_4,j\omega_5,j\omega_6,j\omega_7)\end{array}\right)\\+\left(3\delta_2+\delta_4\sum_i^7(j\omega_i)\right)\left(\begin{array}{l}H_1(j\omega_1)H_1(j\omega_2)H_5(j\omega_3,j\omega_4,j\omega_5,j\omega_6,j\omega_7)\\+H_1(j\omega_1)H_2(j\omega_2,j\omega_3)H_4(j\omega_4,j\omega_5,j\omega_6,j\omega_7)\\+H_1(j\omega_1)H_3(j\omega_2,j\omega_3,j\omega_4)H_3(j\omega_5,j\omega_6,j\omega_7)\\+H_2(j\omega_1,j\omega_2)H_2(j\omega_3,j\omega_4)H_3(j\omega_5,j\omega_6,j\omega_7)\end{array}\right)\end{array}\right), \tag{7.32}$$

where $L^{-1}(5)=L^{-1}(j\omega_1+j\omega_2+j\omega_3+j\omega_4+j\omega_5)$,$L^{-1}(6)=L^{-1}(j\omega_1+j\omega_2+j\omega_3+j\omega_4+j\omega_5+j\omega_6)$, and $L^{-1}(7)=L^{-1}(j\omega_1+j\omega_2+j\omega_3+j\omega_4+j\omega_5+j\omega_6+j\omega_7)$.

Appendix 2

The proof of Eqs. (7.17a, 7.17b) is given as follows. If $\left|\hat{L}^{-1}(\boldsymbol{\omega}_n)\tilde{L}(\boldsymbol{\omega}_n)\right|>1$, namely $\left|\hat{L}(\boldsymbol{\omega}_n)\tilde{L}^{-1}(\boldsymbol{\omega}_n)\right|<1$, then

$$\begin{aligned}L^{-1}(\boldsymbol{\omega}_n)&=-\hat{L}^{-1}(\boldsymbol{\omega}_n)\left[-1+\frac{1}{1-\hat{L}(\boldsymbol{\omega}_n)\tilde{L}^{-1}(\boldsymbol{\omega}_n)}\right]\\&=-\hat{L}^{-1}(\boldsymbol{\omega}_n)\left[\sum_{i=1}^{\infty}\left(\hat{L}(\boldsymbol{\omega}_n)\tilde{L}^{-1}(\boldsymbol{\omega}_n)\right)^i\right]\\&\approx-\hat{L}^{-1}(\boldsymbol{\omega}_n)\left[\sum_{i=1}^{3}\left(\hat{L}(\boldsymbol{\omega}_n)\tilde{L}^{-1}(\boldsymbol{\omega}_n)\right)^i\right].\end{aligned} \tag{7.33}$$

By expanding Eq. (7.33) and extracting the linear parameters to be designed, Eqs. (7.17a, 7.17b) could be easily obtained. This completes the proof.

Appendix 3

Herein, an example exploiting the lnCOS method with respect to characterizing the damping c has been given. Namely, assuming that only the damping c is of interest (to be designed) and other parameters are known in advance. By using Eq. (7.19a), the parameter series of the nonlinear coefficients δ_3 and δ_4 ($\delta_3 = c\frac{\sin^2\theta}{m^2\cos^4\theta}\frac{1}{L}$ and $\delta_4 = c\frac{4-3\cos^3\theta}{4m^4\cos^6\theta}\frac{1}{L^2}$) could be determined as

$$\begin{aligned}
\mathrm{CE_N}(H_1(j\omega\,;c)) &= 1,\\
\mathrm{CE_N}(H_2(j\omega\,;c)) &= C_{2,0} = \delta_3\\
\mathrm{CE_N}(H_3(j\omega\,;c)) &= C_{3,0}+C_{2,0}\otimes C_{2,0}\\
&= [\delta_4,\delta_3^2]\\
\mathrm{CE_N}(H_4(j\omega\,;c)) &= C_{3,0}\otimes C_{2,0}\\
&= [\delta_4,\delta_3^2]\otimes\delta_3\\
&= [\delta_3\delta_4,\,\delta_3^3]\\
\mathrm{CE_N}(H_5(j\omega\,;c)) &= C_{3,0}\otimes C_{3,0}\\
&= [\delta_4,\delta_3^2]\otimes[\delta_4,\delta_3^2]\\
&= [\delta_3^4,\,\delta_3^2\delta_4,\,\delta_4^2].
\end{aligned}\tag{7.34}$$

Then, using Eq. (7.19b), the parameter series of the linear coefficient c could be determined as

$$\begin{aligned}
&\mathrm{CE_L}(H_1(j\omega;c)) = [1,c,\,c^2],\quad \mathrm{CE_L}(H_2(j\omega;c)) = [1,c,\,c^2]^3\\
&\mathrm{CE_L}(H_3(j\omega;c)) = \begin{cases}[1,c,\,c^2]^5, & \text{for } \delta_3^2\\ [1,c,\,c^2]^4, & for\ \delta_4\end{cases}\\
&\mathrm{CE_L}(H_4(j\omega;c)) = \begin{cases}[1,c,\,c^2]^7, & \text{for } \delta_3^3\\ [1,c,\,c^2]^6, & for\ \delta_3\delta_4\end{cases}\\
&\mathrm{CE_L}(H_5(j\omega;c)) = \begin{cases}[1,c,\,c^2]^9, & \text{for } \delta_3^4\\ [1,c,\,c^2]^8, & for\ \delta_3^2\delta_4\ .\\ [1,c,\,c^2]^7, & for\ \delta_4^2\end{cases}
\end{aligned}\tag{7.35}$$

Afterwards, by using Eq. (7.18), the lnCOS functions with respect to the damping c could be expressed as

$$\begin{aligned}
Y(j\omega) &= \left(\bigoplus_{n=1}^{5}\Big[\mathrm{CE_N}(H_n(j\omega_1,j\omega_2,\ldots,j\omega_n))\right.\\
&\quad\left.\otimes\mathrm{CE_L}(H_n(j\omega_1,j\omega_2,\ldots,j\omega_n))\Big]\right)\boldsymbol{\Psi}(j\omega)^{\mathrm{T}}\\
&= [1,c,\,c^2]\boldsymbol{\Psi}_1(j\omega)^{\mathrm{T}}
\end{aligned}$$

$$
\begin{aligned}
&+\left[\delta_3 \otimes [1, c,\ c^2]^3\right]\boldsymbol{\Psi}_2(j\omega)^{\mathrm{T}}\\
&+\left[\delta_3^2 \otimes [1, c,\ c^2]^5 \oplus \delta_4\right.\\
&\left.\otimes[1, c,\ c^2]^4\right]\boldsymbol{\Psi}_3(j\omega)^{\mathrm{T}}\\
&+\left[\delta_3^3 \otimes [1, c,\ c^2]^7 \oplus \delta_3\delta_4\right.\\
&\left.\otimes[1, c,\ c^2]^6\right]\boldsymbol{\Psi}_4(j\omega)^{\mathrm{T}}\\
&+\left[\delta_3^4 \otimes [1, c,\ c^2]^9 \oplus \delta_3^2\delta_4 \otimes [1, c,\ c^2]^8\right.\\
&\left.\oplus\delta_4^2 \otimes [1, c,\ c^2]^7\right]\boldsymbol{\Psi}_5(j\omega)^{\mathrm{T}}. \qquad (7.36)
\end{aligned}
$$

It should be mentioned that, the above lnCOS function can be utilized for calculating $Y^{\hat{y}}(j0;\, c)$, $Y^{\hat{y}}(j\omega;\, c)$, and $Y^{\hat{y}}(j2\omega;\, c)$ respectively.

Appendix 4

The lnCOS functions with respect to θ can be expressed as

$$
Y^{\hat{y}}(j\omega;\theta) = \begin{bmatrix} \left[\left(\tan^2\theta\right)^{-1}, \left(\tan^2\theta\right)^{-2}, \left(\tan^2\theta\right)^{-3}, \ldots\right] \\ \oplus\left[\dfrac{\tan^2\theta}{\cos^6\theta}, \dfrac{\tan^3\theta}{\cos^5\theta}, \dfrac{\tan^4\theta}{\cos^4\theta}, \ldots\right] \\ \otimes\left[\left(\tan^2\theta\right)^{-5}, \left(\tan^2\theta\right)^{-6}, \left(\tan^2\theta\right)^{-7}, \ldots\right] \end{bmatrix} \tilde{\boldsymbol{\psi}}(j\omega)^{\mathrm{T}}, \qquad (7.37)
$$

$$
\begin{aligned}
Y^{\hat{y}}(j0;\theta) = &\left[\frac{\tan\theta}{\cos^3\theta}, \frac{\tan^2\theta}{\cos^2\theta}, \ldots\right]\\
&\otimes\left[\left(\tan^2\theta\right)^{-3}, \left(\tan^2\theta\right)^{-4}, \left(\tan^2\theta\right)^{-5}, \ldots\right]\overline{\boldsymbol{\psi}}(j\omega)^{\mathrm{T}}, \qquad (7.38)
\end{aligned}
$$

$$
\begin{aligned}
Y^{\hat{y}}(j2\omega;\theta) = &\left[\frac{\tan\theta}{\cos^3\theta}, \frac{\tan^2\theta}{\cos^2\theta}, \ldots\right]\\
&\otimes\left[\left(\tan^2\theta\right)^{-3}, \left(\tan^2\theta\right)^{-4}, \left(\tan^2\theta\right)^{-5}, \ldots\right]\widehat{\boldsymbol{\psi}}(j\omega)^{\mathrm{T}}. \qquad (7.39)
\end{aligned}
$$

References

1. Fu H, Mei X, Yurchenko D, Zhou S, Theodossiades S, Nakano K, Yeatman EM. Rotational energy harvesting for self-powered sensing. Joule 2021, 5(5), 1074-1118.
2. Liu H, Zhong J, Lee C, Lee SW, Lin L. A comprehensive review on piezoelectric energy harvesting technology: Materials, mechanisms, and applications. Applied Physics Reviews 2018;5(4):041306.

3. Zhou M, Al-Furjan MS, Zou J, Liu W. A review on heat and mechanical energy harvesting from human–Principles, prototypes and perspectives. Renewable and Sustainable Energy Reviews 2018;82:3582-609.
4. Wang J, Geng L, Ding L, Zhu H, Yurchenko D. The state-of-the-art review on energy harvesting from flow-induced vibrations. Applied Energy 2020;267:114902.
5. Zou H, Zhao L, Gao Q, Zuo L, Liu F, Tan T, et al. Mechanical modulations for enhancing energy harvesting: Principles, methods and applications. Applied Energy 2019;255:113871.
6. Liu H, Fu H, Sun L, Lee C, Yeatman EM. Hybrid energy harvesting technology: From materials, structural design, system integration to applications. Renewable and Sustainable Energy Reviews 2020;110473.
7. Li X, Teng L, Tang H, Chen J, Wang H, Liu Y, et al. ViPSN: A vibration-powered IoT platform. IEEE Internet of Things Journal. 2020 Aug 17.
8. Zeadally S, Shaikh FK, Talpur A, Sheng QZ. Design architectures for energy harvesting in the Internet of Things. Renewable and Sustainable Energy Reviews 2020;128:109901.
9. Newell D, Duffy M. Review of power conversion and energy management for low-power, low-voltage energy harvesting powered wireless sensors. IEEE Transactions on Power Electronics 2019;34(10):9794-805.
10. Orekan T, Zhang P. Underwater Wireless Power Transfer: Smart Ocean Energy Converters. Springer International Publishing; 2019.
11. Zhong Y, Zhao H, Guo Y, Rui P, Shi S, Zhang W, et al. An Easily Assembled Electromagnetic-Triboelectric Hybrid Nanogenerator Driven by Magnetic Coupling for Fluid Energy Harvesting and Self-Powered Flow Monitoring in a Smart Home/City. Advanced Materials Technologies 2019;4(12):1900741.
12. Yang Z, Zhou S, Zu J, Inman D. High-performance piezoelectric energy harvesters and their applications. Joule 2018; 2(4):642-97
13. Safaei M, Sodano HA, Anton SR. A review of energy harvesting using piezoelectric materials: state-of-the-art a decade later (2008–2018). Smart Materials and Structures 2019;28(11):113001.
14. Sezer N, Koç M. A Comprehensive Review on the State-of-the-Art of Piezoelectric Energy Harvesting. Nano Energy 2020:105567.
15. Zuo L, Tang X. Large-scale vibration energy harvesting. Journal of Intelligent Material Systems and Structures 2013;24(11):1405-30.
16. Carneiro P, dos Santos MP, Rodrigues A, Ferreira JA, Simões JA, Marques AT, et al. Electromagnetic energy harvesting using magnetic levitation architectures: A review. Applied Energy 2020;260:114191.
17. Khan FU, Qadir MU. State-of-the-art in vibration-based electrostatic energy harvesting. Journal of Micromechanics and Microengineering 2016;26(10):103001.
18. Wang Z, Jiang T, Xu L. Toward the blue energy dream by triboelectric nanogenerator networks. Nano Energy 2017;39:9-23.
19. Chen J, Wang Z. Reviving vibration energy harvesting and self-powered sensing by a triboelectric nanogenerator. Joule 2017;1(3):480-521.
20. Ahmed A, Hassan I, El-Kady MF, Radhi A, Jeong CK, Selvaganapathy PR, et al. Integrated Triboelectric Nanogenerators in the Era of the Internet of Things. Advanced Science 2019;6(24):1802230.
21. Tran N, Ghayesh MH, Arjomandi M. Ambient vibration energy harvesters: A review on nonlinear techniques for performance enhancement. International Journal of Engineering Science 2018;127:162-85.
22. Yang T, Zhou S, Fang S, Qin W, Inman DJ. Nonlinear vibration energy harvesting and vibration suppression technologies: Designs, analysis, and applications. Applied Physics Reviews, 2021;8(3), 031317.
23. Sun X, Jing X, Xu J, Cheng L. Vibration isolation via a scissor-like structured platform. Journal of Sound and Vibration 2014;333(9):2404-20.
24. Wu Z, Jing X, Bian J, Li F, Allen R. Vibration isolation by exploring bio-inspired structural nonlinearity. Bioinspiration & biomimetics 2015;10(5):056015.

25. Feng X, Jing X, Guo Y. Vibration isolation with passive linkage mechanisms. Nonlinear Dynamics, 2021; 106(3): 1891-1927.
26. Chai Y, Jing X, Chao X. X-shaped mechanism based enhanced tunable QZS property for passive vibration isolation. International Journal of Mechanical Sciences, 2022;218: 107077.
27. Bian J, Jing X. A nonlinear X-shaped structure based tuned mass damper with multi-variable optimization (X-absorber). Communications in Nonlinear Science and Numerical Simulation, 2021;99: 105829.
28. Zhang M, Jing X, Wang G. Bioinspired Nonlinear Dynamics-Based Adaptive Neural Network Control for Vehicle Suspension Systems with Uncertain/Unknown Dynamics and Input Delay. IEEE Transactions on Industrial Electronics. 2020;68(12): 12646-12656.
29. Liu C, Jing X. Vibration energy harvesting with a nonlinear structure. Nonlinear dynamics 2016;84(4):2079-98.
30. Wei C, Jing X. Vibrational energy harvesting by exploring structural benefits and nonlinear characteristics. Communications in Nonlinear Science and Numerical Simulation. 2017; 48:288-306.
31. Li M, Jing X. Novel tunable broadband piezoelectric harvesters for ultralow-frequency bridge vibration energy harvesting. Applied Energy 2019;255:113829.
32. Li M, Jing X. A bistable X-structured electromagnetic wave energy converter with a novel mechanical-motion-rectifier: Design, analysis, and experimental tests. Energy Conversion and Management, 2021;244: 114466.
33. Li Z, Jing X, Yu J. Fault detection based on a bio-inspired vibration sensor system. IEEE Access 2018; 6:10867-77.
34. Jing X. The X-structure/mechanism approach to beneficial nonlinear design in engineering. Applied Mathematics and Mechanics, 2022: 1–22.
35. Jing X, Lang Z. Frequency domain analysis and design of nonlinear systems based on Volterra series expansion: A Parametric Characteristic Approach. 2015; Springer.
36. Cheng CM, Peng ZK, Zhang WM, Meng G. Volterra-series-based nonlinear system modeling and its engineering applications: A state-of-the-art review. Mechanical Systems and Signal Processing 2017;87:340-64.
37. Ho C, Lang Z, Billings SA. Design of vibration isolators by exploiting the beneficial effects of stiffness and damping nonlinearities. Journal of Sound and Vibration 2014;333(12):2489-504.
38. Li Q, Jing X. A systematic second-order output spectrum based method for fault diagnosis with a local tuning approach. Journal of Sound and Vibration 2020:115283.
39. Liu Y, Li J, Feng K, Zhao Y, Yan X, Ma H. A novel fault diagnosis method for rotor rub-impact based on nonlinear output frequency response functions and stochastic resonance. Journal of Sound and Vibration 2020:115421.
40. He F, Wei H, Billings SA. Identification and frequency domain analysis of non-stationary and nonlinear systems using time-varying NARMAX models. International Journal of Systems Science 2015;46(11):2087-100.
41. Xiao Z, Jing X. Frequency-domain analysis and design of linear feedback of nonlinear systems and applications in vehicle suspensions. IEEE/ASME transactions on mechatronics 2015;21(1):506-17.
42. Diala U, Gunawardena R, Zhu Y, Lang Z. Nonlinear design and optimisation of a vibration energy harvester. In 2018 UKACC 12th International Conference on Control (CONTROL). 2018; (pp.180–185).
43. Diala U, Mofidian SM, Lang Z, Bardaweel H. Analysis and optimal design of a vibration isolation system combined with electromagnetic energy harvester. Journal of Intelligent Material Systems and Structures 2019;30(16):2382-95.
44. Li M, Zhou J, Jing X. Improving low-frequency piezoelectric energy harvesting performance with novel X-structured harvesters. Nonlinear Dynamics. 2018; 94(2):1409-28.
45. Boyd S, Chua LO. Fading memory and the problem of approximating nonlinear operators with Volterra series. IEEE Transactions on Circuits Systems 1985;32(11):1150-1161.
46. Jones JP, Yaser KS. Recent advances and comparisons between harmonic balance and Volterra-based nonlinear frequency response analysis methods. Nonlinear Dynamics 2018;91(1):131-45.

47. Lee GM. Estimation of non-linear system parameters using higher-order frequency response functions. Mechanical Systems and Signal Processing. 1997; 11(2):219-28.
48. Billings SA, Jones JP. Mapping nonlinear integro-differential equation into the frequency domain. International Journal of Control 1990;54:863-879.
49. Jing X, Lang Z, Billings SA, Tomlinson GR. Frequency domain analysis for suppression of output vibration from periodic disturbance using nonlinearities. Journal of Sound and Vibration 2008;314(3-5):536-557.
50. Jing X, Lang Z, Billings SA. Magnitude bounds of generalized frequency response functions for nonlinear Volterra systems described by NARX model. Automatica. 2008; 44(3):838-45.
51. Jing X, Lang Z, Billings SA, Tomlinson GR. The parametric characteristic of frequency response functions for nonlinear systems. International Journal of Control 2006;79(12):1552-64.
52. Jing X. Nonlinear characteristic output spectrum for nonlinear analysis and design. IEEE/ASME Transactions on Mechatronics 2014;19(1):171-183.
53. Xiao Z, Jing X. A novel characteristic parameter approach for analysis and design of linear components in nonlinear systems. IEEE transactions on signal processing 2016;64(10):2528-40.
54. Li J, Jing X, Li Z, Huang X. A Novel Linear and Nonlinear Characteristic Parameter Approach for Analysis and Design of Nonlinear Systems. 2020 IEEE Conference on Industrial Cyberphysical Systems (ICPS) 2020 Jun 10 (Vol. 1, pp. 185–190). IEEE.
55. Zhu Y, Lang Z. The effects of linear and nonlinear characteristic parameters on the output frequency responses of nonlinear systems: The associated output frequency response function. Automatica 2018;93:422-7.
56. Jing X, Lang Z, Billings SA. Output frequency response function-based analysis for nonlinear Volterra systems. Mechanical systems and signal processing 2008;22(1):102-20.
57. M Li, X Jing, A parametric frequency domain approach to analysis and design of critical design parameters of nonlinear energy harvesting systems: Parametric output spectrum and power generation function, Mechanical Systems and Signal Processing 181, 109506, 2022

Chapter 8
Case Studies: Controller Design of Vehicle Suspension Systems—A SIMO Nonlinear System Approach

Abstract To effectively implement the nCOS function method, it is essential to accurately derive the nCOS function in relation to the relevant design parameters. Although the method is originally formulated for single-input single-output (SISO) nonlinear systems, its applicability can be extended to multi-input multi-output (MIMO) systems, as well as to nonlinear systems characterized by multiple optimization variables or non-polynomial nonlinearities, such as exponential functions. The critical step lies in transforming the system of interest into an equivalent set of SISO systems. This study focuses on vehicle suspension systems exhibiting inherent nonlinear behavior, which can be reformulated as single-input, multiple-output (SIMO) systems. Under this framework, the nCOS function method can be employed to establish a direct and analytically tractable relationship between controller gains and the objective function.

Keyword Controller design · Frequency domain · Nonlinear systems · SIMO · nCOS function

8.1 Introduction

8.1.1 About This Chapter

To apply the nCOS function method, the key is to obtain the accurate nCOS function with respect to those design parameters. Although the nCOS function is established based on a single input single output (SISO) nonlinear system, similar results can be extended to any multi-input multi-output systems (MIMO) or nonlinear systems with multiple optimization variables or with other types of nonlinear functions except those polynomial forms such as exponential nonlinear functions. The essential task is to transform the system of interest into some equivalent SISO systems.

This case study discusses vehicle suspension systems with inherent nonlinearity, which can be casted into a single input, multiple output (SIMO) system. The study focuses on the parametric convergence bounds for the Volterra series expansion

X. Jing, *The Bio-Inspired X-Structure/Mechanism Approach for Exploring Nonlinear Benefits in Engineering*, https://doi.org/10.1007/978-981-96-5769-8_8

of nonlinear systems, specifically those described by a SIMO Nonlinear Auto-Regressive with eXogenous inputs (NARX) model. This analysis in the frequency domain provides a clear indication of the parametric range within which a given nonlinear system has a convergent Volterra series expansion. This range is known as the parametric bound of convergence (PBoC).

The PBoC of characteristic parameters allows for the analysis of nonlinear systems with a nonlinear multi-objective performance (MOP) function in the frequency domain. This is achieved using a nonlinear characteristic output spectrum (nCOS) method, which is based on the Volterra series expansion.

To illustrate the above findings and methodology, the study investigates a vehicle suspension system with real parameter settings borrowed from literature. This system is considered a typical SIMO nonlinear system with an MOP function that needs optimization. The results present a systematic and innovative approach for the analysis and design of nonlinear systems.

Moreover, extension to nonlinear systems with exponential nonlinear functions is also discussed with case studies in neural network analysis and a membership function optimization of a fuzzy control system.

8.1.2 Research Background

Many real-world systems, such as vehicle suspension systems [1–4] and vibration absorbers [5, 6], are multi-degree of freedom (MDoF) systems. These MDoF systems can be viewed as multiple output systems, typically characterized by inherent nonlinearity. They can always be transformed into a single input multiple output (SIMO) Nonlinear Auto-Regressive with eXogenous inputs (NARX) model using input–output experimental data. Therefore, the SIMO NARX model can represent a wide range of practical engineering systems. These include systems with multiple outputs, multi-degree-of-freedom systems, and also multi-variable systems that have a nonlinear output function or a nonlinear multi-objective-performance (MOP) function, which can be used for optimization or control.

For a Single Input Multiple Output (SIMO) nonlinear model, techniques such as harmonic balance, perturbation, and multiple scale methods [7] can be employed for its analysis and design. However, the computational cost and complexity in analysis can escalate significantly as the truncation order increases, particularly when the input is merely a harmonic input. This complexity becomes even more pronounced for multiple harmonic or arbitrary excitation signals.

The analysis and design of system nonlinear dynamics can also be conducted using frequency domain methods associated with the Volterra series. The concept of the Generalized Frequency Response Function (GFRF) [8] initiated the frequency domain analysis of nonlinear systems, which was further developed with practical GFRF and nonlinear output spectrum-based algorithms and methods [9–12].

The frequency domain method based on the Volterra series often provides straightforward insights into the analysis and design of system nonlinear dynamics [13–15].

Notably, the output frequency response of a nonlinear system can be formulated into an explicit polynomial function of model parameters, which can effectively aid nonlinear analysis, design, and optimization [9, 10, 16, 17].

When utilizing Volterra series-based methods for nonlinear analysis and design, it's crucial to ensure that the nonlinear dynamics being studied consistently have a convergent Volterra series expansion. Various attempts have been made in academic literature to tackle this issue. Some methods, designed for fading memory or nonlinear operators, are theoretically derived but might be too broad for practical application [18, 19]. Others are solely developed for specific systems, such as Duffing oscillators [20–22]. A recent study [23] presented the computation of a convergence bound for a certain class of input analytic nonlinear systems. However, the system considered in that study is limited to being affine in input, which might only represent a specific case of this study. Furthermore, the result of that study is only applicable for calculating the input convergence bound. In contrast, this study aims to address not only the input bound but also a parametric bound for any model parameters.

Nonlinearities present a complex challenge in the analysis and design of practical systems. If a nonlinear system has multiple steady states, the dynamic response can become highly complex, and even a simple harmonic excitation can result in intricate bifurcation or chaos [7]. This complexity can significantly complicate the analysis and design of nonlinear systems and potentially undermine system stability. However, if a system operates within Volterra-type nonlinearity, the system's dynamic response remains around a single steady state. This makes nonlinear analysis and design more manageable using existing methods associated with the Volterra series. The severity of nonlinearity in a given system is determined by certain characteristic parameters, such as model parameters, input magnitude, or frequency. This study aims to develop a parametric convergent criterion that ensures the underlying SIMO NARX models have a convergent Volterra series expansion. This will ensure that all characteristic parameters of the system under study remain within its parametric bound of convergence (PBoC).

Vibration control of vehicle suspensions, a typical SIMO nonlinear problem, often has several optimization objectives including body acceleration, suspension stroke, and relative tire load [2, 24]. The nonlinearity of the system often stems from the inherent nonlinearity of springs, dampers, and/or actuator saturation. Many existing methods, such as the robust control methods [25–27], are developed based on a linear suspension model for improved performance. However, these robust controllers may have limitations in their control effectiveness due to their neglect of nonlinear dynamics. With the proposed parametric convergent criterion, any SIMO nonlinear system with a nonlinear output or multi-objective performance (MOP) function can be conveniently analyzed using a method based on the nonlinear characteristic output spectrum (nCOS) function [9].

A method that is efficient and effective for the non-linear analysis and design of suspension systems is desirable. This study presents three key contributions: (a) the development of a parametric bound of convergence (PBoC) in the frequency domain for non-linear SIMO systems, which provides a definitive parametric bound for the Volterra series expansion of underlying non-linear dynamics, a result unlike any found in existing literature; (b) the transformation of the multiple objective optimization problem into an nCOS function problem within the PBoC in the frequency domain, offering a fresh perspective on non-linear analysis, design, and optimization; and (c) the introduction of a novel non-linear controller based on body acceleration for vehicle suspension, which demonstrates superior performance compared to some existing controllers in the literature.

The structure of this study is as follows. Section 8.2 begins with an introduction to the problem and preliminary findings. The main findings, regarding the parametric bound of convergence (PBoC) for SIMO NARX systems, are presented in Sect. 8.3. Section 8.4 then applies these findings to a case study of a nonlinear quarter vehicle suspension system, and formulates a multi-objective performance (MOP) function based on the nCOS. The study concludes thereafter.

8.2 Frequency Response Functions of Nonlinear Systems

8.2.1 The SIMO NARX Model

As previously stated, the single input multiple output (SIMO) NARX model can effectively represent a significant range of nonlinear systems found in practical engineering applications. This includes nonlinear vehicle suspension systems. A SIMO NARX model with M subsystems can be given as

$$y_{j_k}(k) = \sum_{p+q=1}^{N_l} \sum_{p=0}^{p+q} \sum_{j_1=1}^{M} \sum_{j_2=j_1}^{M} \cdots \sum_{j_p=j_{p-1}}^{M} \sum_{k_1,\cdots,k_{p+q}=0}^{K} \left(c_{p,q}^{j_p j_{p-1} \cdots j_1}(k_1, \ldots, k_{p+q}; j_k) \prod_{i=1}^{p} y_{j_i}(k - k_i) \prod_{i=p+1}^{p+q} u(k - k_i) \right) \tag{8.1}$$

where $y_{j_k}(k)$ means the j_kth output, $j_k = 1, 2, \ldots, M$. M is the number of subsystems, and N_l is the maximum nonlinear order in terms of output $y(k)$ and input $u(k)$. $c_{p,q}^{j_p j_{p-1} \cdots j_1}(k_1, \ldots, k_{p+q}; j_k)$ is the model parameter of the corresponding term $\prod_{i=1}^{p} y_{j_i}(k - k_i) \prod_{i=p+1}^{p+q} u(k - k_i)$ in the j_kth subsystem, which has a nonlinearity degree $p + q$ (p order in terms of the output and q order in terms of the input), and k_i is the difference order with the maximum order K. The superscript j_i, $i = 1, 2, \ldots, p$ in the model parameter $c_{p,q}^{j_p j_{p-1} \cdots j_1}(k_1, \ldots, k_{p+q}; j_k)$ means that the j_ith output, $y_{j_i}(k)$, is in the corresponding term $\prod_{i=1}^{p} y_{j_i}(k - k_i) \prod_{i=p+1}^{p+q} u(k - k_i)$. Denote $\mathrm{m} = p + q$, which is, clearly from the above, the nonlinear degree in NARX model (8.1). An example for a single input four output NARX model can be seen in Sect. 8.4.

The j_kth output of the SIMO NARX model in (8.1) can be approximated by the Volterra series (truncated up to the order N) as,

$$y_{j_k}(k) = \sum_{n=1}^{N} \int_{-\infty}^{\infty} \cdots \int_{-\infty}^{\infty} h_n^{(j_k)}(\tau_1, \ldots, \tau_n) \prod_{i=1}^{n} u(t - \tau_i) d\tau_i \tag{8.2}$$

where $h_n^{(j_k)}(\tau_1, \ldots, \tau_n)$ is the nth order Volterra kernel of the j_kth subsystem.

8.2.2 The nth Order GFRF and Nonlinear Output Spectrum

The nth order generalized frequency response function (GFRF) can be recursively calculated [28] as,

$$H_n = \begin{bmatrix} H_n^{(1)}(j\omega_1, \cdots, j\omega_n) \\ H_n^{(2)}(j\omega_1, \cdots, j\omega_n) \\ \vdots \\ H_n^{(M)}(j\omega_1, \cdots, j\omega_n) \end{bmatrix} = \begin{bmatrix} l_{1,1} & l_{1,2} & \cdots & l_{1,M} \\ l_{2,1} & l_{2,2} & \cdots & l_{2,M} \\ \vdots & & & \vdots \\ l_{M,1} & l_{M2} & \cdots & l_{M,M} \end{bmatrix}^{-1} \begin{bmatrix} A_{1,n} \\ A_{2,n} \\ \vdots \\ A_{M,n} \end{bmatrix} \tag{8.3}$$
$$= L_n(j\omega_1, \cdots, j\omega_n) A_n$$

where $H_n^{(j_k)}(j\omega_1, \ldots, j\omega_n)$ is the nth order GFRF of the j_kth subsystem, $A_{i,n}$ represents the ith element of A_n. ω_i, $i = 1, \ldots, n$ is the frequency variable in the nth order GFRF,

$$l_{j_k, j_i} = \begin{cases} -\sum\limits_{k_1=1}^{K} c_{1,0}^{j_i}(k_1; j_k) e^{-k_1(j\omega_1 + \cdots + j\omega_n)} & j_i \neq j_k, \\ 1 - \sum\limits_{k_1=1}^{K} c_{1,0}^{j_i}(k_1; j_k) e^{-k_1(j\omega_1 + \cdots + j\omega_n)} & j_i = j_k. \end{cases} \tag{8.4}$$

Equation (8.4) involves only the linear model parameters, e.g., $c_{1,0}(.)$. $A_{j_k,n}$ in (8.3) is given by,

$$A_{j_k,n} = \sum_{m=2}^{n} \sum_{p=1}^{m} \sum_{j_1=1}^{M} \sum_{j_2=j_1}^{M} \cdots \sum_{j_p=j_{p-1}}^{M} \sum_{k_1,\ldots,k_{p+q}=0}^{K} \Bigg[\sum_{k_1,\cdots,k_n=0}^{K} c_{0,n}(k_1, \ldots, k_n; j_k)$$
$$\times e^{-j(k_1\omega_1 + \cdots + k_n\omega_n)} + \sum_{k_1,\ldots,k_{p+q}=0}^{K} c_{p,q}^{j_p j_{p-1} \cdots j_1}(k_1, \ldots, k_{p+q}; j_k)$$
$$\times e^{-j(k_{p+1}\omega_{n-q+1} + \cdots + k_{p+q}\omega_n)} H_{n-q,p}^{(j_p j_{p-1} \cdots j_1)}(j\omega_1, \ldots, j\omega_{n-q}) \Bigg] \tag{8.5}$$

where $m = p + q$, and

$$H_{n-q,p}^{(j_p j_{p-1} \cdots j_1)}(j\omega_1, \ldots, j\omega_n) = \sum_{r_1,\ldots,r_p=1; r_1+\cdots+r_p=n-q}^{n-p+1} \prod_{i=1}^{p} \left[H_{r_i}^{(j_i)}(j\omega_{X_i+1}, \ldots, j\omega_{X_i+r_i}) e^{-jk_i\left(\omega_{X_i+1}+\cdots+\omega_{X_i+n}\right)} \right] \tag{8.6}$$

$\sum_{r_1,\ldots,r_p=1; r_1+\cdots+r_p=n-q}^{n-p+1}$ is the summation of all different combination of (γ_1, γ_2, …, γ_p) which satisfy, $1 \le \gamma_1, \gamma_2, \ldots, \gamma_p \le n-p+1$ and$\gamma_1 + \gamma_2 + \cdots + \gamma_p = n-q$.

$$H_{n,1}^{(j_k)}(j\omega_1, \ldots, j\omega_n) = H_n^{(j_k)}(j\omega_1, \ldots, j\omega_n) e^{-k_1(j\omega_1+\cdots+j\omega_n)} \tag{8.7}$$

$$e^{-j(k_{p+1}\omega_{n-q+1}+\cdots+k_{p+q}\omega_n)} = \begin{cases} 1 \; q = 0, p > 1 \\ 0 \; q = 0, p \le 1 \end{cases} \tag{8.8}$$

In (8.6), $X_i = \gamma_1 + \gamma_2 + \cdots + \gamma_{i-1}$, and $\gamma_1 + \gamma_2 + \cdots + \gamma_p = n - q$ for all $p \ge 2$ and positive integer γ_i. It is clear that higher order GFRFs can be recursively calculated with the lower order GFRFs, and the recursion terminates at the 1st order GFRF, i.e., $H_1^{(j_k)}(j\omega_1)$, which represents the transfer function when all the nonlinear model parameters in (8.1) are 0. To compute the 1st order GFRF via (8.3), (8.5) is then defined by,

$$A_{j_k,1} = \sum_{k_1=0}^{K} c_{0,1}(k_1; j_k) e^{-j\omega_1 k_1}. \tag{8.9}$$

With the GFRFs in (8.3)–(8.9), and when the SIMO NARX model subjects to a harmonic input

$$u(k) = |A_\omega| \cos(\omega k + \angle A_\omega) = \frac{A_\omega}{2} e^{j\omega k} + \frac{A_\omega^*}{2} e^{-j\omega k} \tag{8.10}$$

the nonlinear output spectrum of the j_kth subsystem (truncated at the order N) can be obtained as [10, 11, 14, 16]

$$Y_{j_k}(j\Omega) = \sum_{n=1}^{N} \left[\frac{1}{2^n} \sum_{\omega_1+\cdots+\omega_n=\Omega} \left(H_n^{(j_k)}(j\omega_1, \ldots, j\omega_n) \prod_{i=1}^{n} A_\omega(\omega_i) \right) \right] \tag{8.11}$$

where $|A_\omega|$ is the magnitude of the input $u(k)$, which is denoted by U later, Ω is the output frequency, ω is the input frequency, A_ω^* is the complex conjugate of A_ω, $\omega_i \in \{\omega, -\omega\}$ and

$$A_{\omega}(\omega_i) = \begin{cases} A_{\omega}, & \omega_i = \omega, \\ A_{\omega}^{*}, & \omega_i = -\omega. \end{cases} \tag{8.12}$$

8.3 The Convergence Bounds

$\mathbb{N}^{+}$ denotes all positive integers. The operator $\|\bullet\|$ means the absolute value for scalars and $\|\bullet\|_1$ norm for vectors, which is also denoted for matrix A as $\|A\| = \sup_{\|x\|=1}\|Ax\|$. Denote the upper bound of $L_n(j\omega_1, \ldots, j\omega_n)$, which is defined in (8.3), over the whole output frequency range W_{∞} as,

$$\overline{L}(\omega) = \sup_{\Omega \in W_{\infty}} \{\|L_n(j\omega_1, \ldots, j\omega_n)\|\} \tag{8.13}$$

where $W_{\infty} = \bigcup_{n=1}^{\infty} W_n = \bigcup_{n=1}^{\infty} \{\Omega | \Omega = \omega_1 + \cdots + \omega_n, \omega_i \in \{\omega, -\omega\}\}$ represents the output frequency range when the NARX model (8.1) is excited by (8.10), and W_n is the output frequency range of the nth order output (see more discussions in [11]). Define

$$C_{(p,q;j_k)}^{j_p j_{p-1} \cdots j_1} = \sum_{k_1, \ldots, k_{p+q}=0}^{K} \left\| c_{p,q}^{j_p j_{p-1} \cdots j_1}(k_1, \ldots, k_{p+q}; j_k) \right\| \tag{8.14}$$

Obviously, $C_{(p,q;j_k)}^{j_p j_{p-1} \cdots j_1}$ is a nonnegative function of the model parameters $c_{p,q}^{j_p j_{p-1} \cdots j_1}(\cdot; j_k)$.

In the following, the bound for the nonlinear output spectrum of the j_kth subsystem is given in Lemmas 1 and 2, then an analytical relationship between the bound of nonlinear output spectrum of all M subsystems in the SIMO system is built in Lemma 4, based on which the parametric convergence criterion, i.e., the PBoC, is developed in Sect. 8.3.2.

8.3.1 *Boundedness of the GFRF and Nonlinear Output Spectrum*

Lemma 1: The upper bound of the nonlinear output spectrum of the j_kth subsystem at $\Omega = k\omega$ is given by,

$$\overline{Y}_{\Omega=k\omega}^{(j_k)}(U) = \sum_{n=1}^{\infty} \left(C_{k+2(n-1)}^{n-1} \Big/ 2^{k+2(n-1)-1} \right.$$

$$\times \overline{H}^{(j_k)}_{k+2(n-1)}(j\omega_1, \ldots, j\omega_{k+2(n-1)})U^{k+2(n-1)}\Big) \; k \in \mathbb{N}^+ \tag{8.15}$$

$$\overline{Y}^{(j_k)}_{\Omega=k\omega}(U) = \sum_{n=1}^{\infty}\left(\frac{C^n_{2n}}{2^{2n}}\overline{H}^{(j_k)}_{2n}(j\omega_1, \ldots, j\omega_{2n})U^{2n}\right), k = 0. \tag{8.16}$$

where $\overline{H}^{(j_k)}_n(j\omega_1, \ldots, j\omega_n)$ is the upper bound of the nth order GFRF, which satisfies the following equation,

$$\sum_{j_k=1}^{M}\overline{H}^{(j_k)}_n(j\omega_1, \ldots, j\omega_n) = \overline{L}(\omega)\sum_{j_k=1}^{M}\Bigg[C_{(0,n;j_k)} + \sum_{m=2}^{n}\sum_{p=1}^{m}\sum_{j_1=1}^{M}\sum_{j_2=j_1}^{M}\cdots\sum_{j_p=j_{p-1}}^{M}\Bigg(C^{j_p j_{p-1}\cdots j_1}_{(p,q;j_k)} \sum_{r_1,\ldots,r_p=1,\sum r_i=n-q}^{n-m+1}\prod_{i=1}^{p}\overline{H}^{(j_i)}_{r_i}(j\omega_{X+1}, \ldots, j\omega_{X+r_i})\Bigg)\Bigg], n > 1 \tag{8.17}$$

and

$$\overline{H}^{(j_k)}_1(j\omega_1) = \left\| H^{(j_k)}_1(j\omega_1)\right\|, \quad n = 1. \tag{8.18}$$

Proof: See [29]. □

Lemma 2: The upper bound of the nonlinear output spectrum of the j_kth subsystem which involves all the frequencies in the output frequency range W_∞ is given as,

$$\begin{aligned}\overline{Y}^{(j_k)}_{\omega}(U) &= \sum_{k=0}^{\infty}\overline{Y}^{(j_k)}_{\Omega=k\omega}(U) = \sum_{\Omega\in W_\infty}\overline{Y}^{(j_k)}_{\Omega}(U)\\ &= \sum_{n=1}^{\infty}\left\|\sum_{\Omega\in W_n}\overline{Y}^{(j_k)}_n(j\overline{\omega})\right\|\\ &= \sum_{n=1}^{+\infty}\overline{H}^{(j_k)}_n(j\omega_1, \ldots, j\omega_n)U^n\end{aligned} \tag{8.19}$$

Proof: Following Lemma 3 in [30]. □

Lemma 3: When $\overline{L}(\omega) < \infty$, and the upper bound of the nonlinear output spectrum of one subsystem is divergent, the upper bounds of the nonlinear output spectrum of all the other subsystems also diverge.

Proof: See [29]. □

Remark 1: The above boundary results can be viewed as an extension of the results for the Single Input Single Output (SISO) case presented in [30, 31]. The upper limit of the nonlinear output spectrum of subsystem in (8.19) is presented as the cumulative total of an infinite power series, which can also be interpreted as the sum of the power series (bearing nonnegative coefficients) detailed in (8.15) and (8.16). If the upper limit of the nonlinear output spectrum in (8.19) is convergent, then the upper limit of the nonlinear output spectrum at any frequency for any given $k \in \mathbb{N}$ will also converge. This clearly suggests a convergent Volterra series expansion.

Lemma 4: The upper bounds of the nonlinear output spectrum of all the M subsystems satisfy,

$$\begin{aligned} &\overline{L}(\omega)\sum_{j_k=1}^{M}\sum_{m=2}^{+\infty}\sum_{p=1}^{m}\sum_{j_1=1}^{M}\sum_{j_2=j_1}^{M}\cdots\sum_{j_p=j_{p-1}}^{M}\left(U^q C_{(p,q;j_k)}^{j_p j_{p-1}\cdots j_1}\prod_{i=1}^{p}\overline{Y}_{\omega}^{(j_i)}(U)\right) \\ &\quad -\sum_{j_k=1}^{M}\overline{Y}_{\omega}^{(j_k)}(U)+\overline{L}(\omega)\sum_{j_k=1}^{M}\sum_{m=2}^{+\infty}\left(C_{(0,m;j_k)}U^m\right) \\ &\quad +\sum_{j_k=1}^{M}\left(\overline{H}_1^{(j_k)}(j\omega_1)U\right)=0. \end{aligned} \tag{8.20}$$

Proof: See [29]. □

8.3.2 *Parametric Bound of Convergence (PBoC)*

The parametric bound of convergence (PBoC) is defined as the upper bound of a characteristic parameter in which a given nonlinear system has a convergent Volterra series expansion.

Proposition 1: The PBoC can be obtained by solving,

$$\begin{vmatrix} a_{1,M_p} & a_{1,M_p-1} & \cdots & a_{1,0} & 0 & \cdots 0 \\ 0 & a_{1,M_p} & a_{1,M_p-1} & \cdots & a_{1,0} & \cdots 0 \\ & & \cdots & & & \\ 0 & 0 & \cdots & a_{1,M_p} & a_{1,M_p-1} & \cdots a_{1,0} \\ a_{2,M_p-1} & a_{2,M_p-2} & \cdots & a_{2,0} & 0 & \cdots 0 \\ 0 & a_{2,M_p-1} & a_{2,M_p-2} & \cdots & a_{2,0} & \cdots 0 \\ & & \cdots & & & \\ 0 & 0 & \cdots & a_{2,M_p-1} & a_{2,M_p-2} & \cdots a_{2,0} \end{vmatrix} = 0 \tag{8.21}$$

where M_p is the maximum nonlinear degree in terms of the output, and

$$a_{1,p} = \overline{L}(\omega)\sum_{j_k=1}^{M}\sum_{q=0}^{+\infty}\sum_{j_1=1}^{M}\sum_{j_2=j_1}^{M}\cdots\sum_{j_p=j_{p-1}}^{M}\left[U^q C_{(p,q;j_k)}^{j_p j_{p-1}\cdots j_1}\left(\prod_{i=1}^{p}\frac{\overline{H}_1^{(j_i)}(\omega)}{\overline{H}_1^{(j)}(\omega)}\right)\right], 2 \le p \le M_p \tag{8.22}$$

$$a_{1,1} = \overline{L}(\omega)\sum_{j_k=1}^{M}\sum_{q=1}^{+\infty}\sum_{j_1=1}^{M}\left(U^q C_{(1,q;j_k)}^{j_1}\frac{\overline{H}_1^{(j_1)}(\omega)}{\overline{H}_1^{(j)}(\omega)}\right) - \sum_{j_k=1}^{M}\frac{\overline{H}_1^{(j_k)}(\omega)}{\overline{H}_1^{(j)}(\omega)}, \; p = 1, p+q \ge 2 \tag{8.23}$$

$$a_{1,0} = \sum_{j_k=1}^{M}\left(\overline{H}_1^{(j_k)}(j\omega_1)U\right) + \overline{L}(\omega)\sum_{j_k=1}^{M}\sum_{m=2}^{+\infty}\left(C_{(0,m;j_k)}U^m\right), \tag{8.24}$$

$$a_{2,p-1} = \overline{L}(\omega)\sum_{j_k=1}^{M}\sum_{q=0}^{+\infty}\sum_{j_1=1}^{M}\sum_{j_2=j_1}^{M}\cdots\sum_{j_p=j_{p-1}}^{M}\left[U^q C_{(p,q;j_k)}^{j_p j_{p-1}\cdots j_1} p\left(\prod_{i=1}^{p}\frac{\overline{H}_1^{(j_i)}(\omega)}{\overline{H}_1^{(j)}(\omega)}\right)\right], \; 2 \le p \le M_p \tag{8.25}$$

$$a_{2,0} = \overline{L}(\omega)\sum_{j_k=1}^{M}\sum_{q=1}^{+\infty}\sum_{j_1=1}^{M}\left(U^q C_{(1,q;j_k)}^{j_1}\frac{\overline{H}_1^{(j_1)}(\omega)}{\overline{H}_1^{(j)}(\omega)}\right) - \sum_{j_k=1}^{M}\frac{\overline{H}_1^{(j_k)}(\omega)}{\overline{H}_1^{(j)}(\omega)}, \; p = 1, p+q \ge 2. \tag{8.26}$$

where $j \in \{1, \ldots, M\}$, which means that $\overline{H}_1^{(j)}(\omega)$ can be any of $\left\{\overline{H}_1^{(1)}(\omega), \ldots, \overline{H}_1^{(M)}(\omega)\right\}$.

Proof: The proof is sketched here with details in Appendix D of [29]. With the assumption that the solution of nonlinear differential equations or nonlinear difference equations can be seen as the perturbation of the solution of the corresponding linear equations, the result in (8.20) for all M subsystems can be cast into a new equation in terms of the magnitude bound of the first order GFRF for only one subsystem. Based on the analytic inversion lemma, the singularity condition can be further developed into a new form, such that the Sylvester resultant can then be constructed for easily solving the singularity condition without any iterative calculation. □

It's important to note that the PBoC as outlined in Proposition 1 can be calculated through any subsystem of the SIMO model. The PBoC values obtained from different subsystems will align with each other, a result that is in agreement with Lemma 3. In other words, if one subsystem isn't Volterra-type, then none of the other subsystems will be Volterra-type either.

Remark 2: If Eq. (8.21) has only one positive root, this root is considered as the computed PBoC. If there are several positive roots, the smallest one is identified as the computed PBoC. However, if there are no positive roots, the computed PBoC is 0, indicating that the nonlinear model in question has already exceeded the convergence bound, and the system is not of Volterra-type. While the PBoC estimated here is intended for harmonic input signals, it can serve as a valuable reference or guide for parameter optimization and design in real-world applications for any input signals.

Remark 3: It's important to note that the result in [32] pertains to a single input single output (SISO) case, which is just a specific instance of this study. The recursive computation of the nth order GFRFs for a SIMO NARX model is based on matrix calculations and is thus more complex than that for a SISO model. The boundary result in (8.20) is an explicit expression of the output bound for all the M subsystems. This result, the first of its kind for multi-output systems, cannot be straightforwardly extended from the result in [30]. The result in Proposition 1 applies for any j from 1 to M, aligning with Lemma 3. Using Lemma 3 and the approximation (8.40) in Appendix D, the boundary result of the nonlinear output spectrum in (8.20) for the SIMO model can then be transformed into the SISO case, as demonstrated in (8.41) in Appendix D of [29].

Remark 4: Proposition 1 only requires the bound of the first order generalized frequency response function (GFRF) and the bound $\overline{L}(\omega)$. There's no need for recursive calculation. Equation (8.21) can be converted into a nonlinear algebraic equation, which is straightforward to solve.

Algorithm 1 can be used for computation of the PBoC:

Step 1. Calculate $\overline{L}(\omega)$ according to (8.4) and (8.13); Calculate $C_{(p,q;j)}^{j_p j_{p-1} \cdots j_1}$ according to (8.14).
Step 2. Calculate $\overline{H}_1^{(j_k)}$ according to (8.3)–(8.5), (8.9), and (8.18).
Step 3. Solving (8.21), the PBoC can then be obtained.

8.4 Application for Nonlinear Analysis and Design of Vehicle Suspensions

Vehicle suspension systems often have inherent nonlinearity. For instance, a nonlinear spring force is identified with measured data as a third order polynomial $f_s = k_{s1}(\cdot) + k_{s2}(\cdot)^2 + k_{s3}(\cdot)^3$ in [33]. Current approaches, such as adaptive control [2, 3] or robust control [2, 34], often simplify the problem by treating it as a linear one or by considering the nonlinearity as a bounded uncertainty within a time-domain robust control framework. However, it is important to note that the simplified treatment of the nonlinear component in these methods may overlook its significant contribution to performance improvement [9, 15]. In engineering practice, simulation-based optimization is commonly used, but it typically requires extensive computational resources.

In the context of vehicle suspension vibration control, the performance of the suspension is typically evaluated based on three indicators: body acceleration, relative tire load, and suspension stroke. Consequently, vehicle suspension control represents a classic multi-objective optimization problem. Additionally, considering human sensitivity to vibrations, ISO2361 recommends a nominal frequency range of 4–8 Hz for assessing ride comfort. To tackle this nonlinear multi-objective optimization problem and leverage the advantages of nonlinearity in vibration control, the Volterra series-based method discussed in this study offers a powerful and straightforward tool for analyzing and designing vehicle suspensions directly in the frequency domain.

Firstly, a nonlinear controller based on the measured acceleration, i.e., $\ddot{x}_s^3$ with coefficient c_{s3}, is deliberately introduced to the suspension system for performance improvement. The PBoC of c_{s3} can then be computed to indicate whether the system is Volterra-type or has strong nonlinear dynamics. By ensuring that the system nonlinearity follows a Volterra-type structure, the Volterra series-based theory and methods discussed earlier can effectively analyze the nonlinear dynamics of the system. Additionally, a multi-objective performance (MOP) function is formulated to assess the performance of the vehicle suspension, taking into account factors such as ride comfort, suspension stroke, and relative tire load. This MOP function is considered a nonlinear output function and is expressed as a polynomial function of the model parameters of interest c_{s3} [9]. How the model parameters affect the MOP can then be studied in the frequency domain, and the design and optimization of the MOP function can be conducted directly within the computed PBoC.

In comparison to the approach used in this study, control strategies based on absolute sprung mass displacement and velocity [35–38] (such as skyhook control, ground hook control, acceleration-driven damper (ADD), mixed skyhook and ADD strategy, and state feedback control) may have a limitation. This limitation arises from the fact that the required motion signals, such as absolute displacement and velocity, are not readily available for measurement on a moving vehicle. Typically, these motion signals can be obtained through state estimation methods or filters (e.g., integrating the acceleration signal or using Kalman filters). However, these methods introduce estimation errors, control delays, and increase development or manufacturing costs. In contrast, by leveraging the advantageous nonlinearity in vibration control, a simple and nonlinear controller that utilizes measurable signals can achieve the same or even better performance. For further exploration of similar topics using Volterra series-based frequency domain methods for nonlinear vibration control, reference can be made to [14, 15].

As shown in Fig. 8.1, a quarter vehicle suspension model is given here as,

$$\begin{cases} m_t\ddot{x}_t = k_t(u - x_t) + c_t(\dot{u} - \dot{x}_t) - m_s\ddot{x}_s \\ m_s\ddot{x}_s = -c_{s3}\ddot{x}_s^3 - f_s - f_d \end{cases} \tag{8.27}$$

Fig. 8.1 A quarter vehicle suspension model

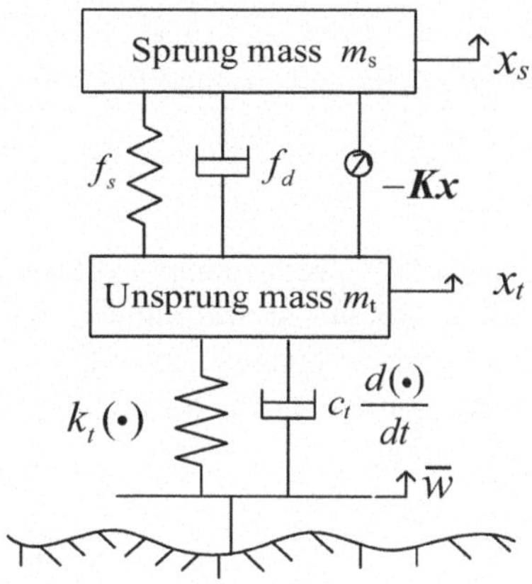

where $f_s = k_{s1}(x_s - x_t) + k_{s2}(x_s - x_t)^2 + k_{s3}(x_s - x_t)^3, f_d = c_{s1}(\dot{x}_s - \dot{x}_t) + c_{s2}(\dot{x}_s - \dot{x}_t)^2$. The model parameters are borrowed from [33], which are obtained by curve fitting with measurement data from a Hyundai Elantra front suspension and given as $m_s = 240$ kg, $m_t = 25$ kg, $k_t = 160$ kN/m, $k_{s1} = 12$ kN/m, $c_t = 10$ N · s/m, $k_{s2} = -73{,}696$ N/m^2, $k_{s3} = 3{,}170{,}400$ N/m^3, $c_{s1} = 1385.4$ N/$\left(\text{ms}^{-1}\right)$, $c_{s2} = 524$ N/$\left(\text{ms}^{-1}\right)^2$. Here only the model parameter c_{s3} is unknown, which is considered as the characteristic parameter of interest. To compute the PBoC of c_{s3}, the road input is assumed to be cosine input $u = U\cos(2\pi ft)$.

8.4.1 Compute the PBoC of c_{s3}

The suspension model in (8.27) can be transformed into a SIMO (single input four outputs) NARX model by backward discretization as,

$$\begin{cases} x_1(k) = c_{1,0}^{1}(1;1)x_1(k-1) + c_{1,0}^{1}(2;1)x_1(k-2) + c_{1,0}^{4}(0;1)x_4(k) \\ \qquad + c_{0,1}(0;1)u(k) + c_{0,1}(1;1)u(k-1) \\ x_2(k) = c_{1,0}^{2}(1;2)x_2(k-1) + c_{1,0}^{2}(2;2)x_2(k-2) + c_{1,0}^{3}(0;2)x_3(k) \\ \qquad + c_{1,0}^{5}(0;2)x_5(k) + c_{2,0}^{33}(0,0;2)x_3^2(k) + c_{2,0}^{55}(0,0;2)x_5^2(k) \\ \qquad + c_{3,0}^{333}(0,0,0;2)x_3^3(k) + c_{3,0}^{444}(0,0,0;2)x_4^3(k) \\ x_3(k) = c_{1,0}^{2}(0;3)x_2(k) + c_{1,0}^{1}(0;3)x_1(k) \\ x_4(k) = c_{1,0}^{2}(0;4)x_2(k) + c_{1,0}^{2}(1;4)x_2(k-1) + c_{1,0}^{2}(2;4)x_2(k-2) \\ x_5(k) = c_{1,0}^{3}(0;5)x_3(k) + c_{1,0}^{3}(1;5)x_3(k-1) \end{cases} \tag{8.28}$$

where $x_1 = x_t$, $x_2 = x_s$, and all the coefficients are presented in Appendix.

From (8.28), $c_{3,0}^{444}(0,0,0;2) = -c_{s3}T_s^2/m_s$, and c_{s3} is only involved in $c_{3,0}^{444}(0,0,0;2)$. Thus the PBoC of c_{s3} (denoted as C_{s3}) can be obtained as

$$C_{s3} = \left(m_s/T_s^2\right)C_{(3,0;2)}^{444} \tag{8.29}$$

where $C_{(3,0;2)}^{444}$ is the computed PBoC of $c_{3,0}^{444}(0,0,0;2)$. The computation of $C_{(3,0;2)}^{444}$ can be conducted according to the procedure in Algorithm 1. The elements in (8.21) can be obtained according to (8.22)–(8.26), which are given as, (as shown in Proposition 1, j can be any value from 1 to 4, and here j takes 4 in the following calculation),

$$a_{1,3} = \overline{L}(j\omega)\left[C_{(3,0;2)}^{333}\left(\frac{\overline{H}_1^{(3)}(j\omega)}{\overline{H}_1^{(4)}(j\omega)}\right)^3 + C_{(3,0;2)}^{444}\right],\ a_{1,0} = \sum_{j_k=1}^{4}\overline{H}_1^{(j_k)}(j\omega)U,$$

$$a_{1,2} = \overline{L}(j\omega)\left[C_{(2,0;2)}^{33}\left(\frac{\overline{H}_1^{(3)}(j\omega)}{\overline{H}_1^{(4)}(j\omega)}\right)^2 + C_{(2,0;2)}^{55}\left(\frac{\overline{H}_1^{(5)}(j\omega)}{\overline{H}_1^{(4)}(j\omega)}\right)^2\right],$$

$$a_{1,1} = -\sum_{j_k=1}^{4}\frac{\overline{H}_1^{(j_k)}(j\omega)}{\overline{H}_1^{(4)}(j\omega)},\ a_{2,2} = 3a_{1,3},\ a_{2,1} = 2a_{1,2},\ a_{2,0} = a_{1,1}$$

where $C_{(3,0;2)}^{333} = \left|c_{3,0}^{333}(0,0,0;2)\right|$, $C_{(3,0;2)}^{444} = \left|c_{3,0}^{444}(0,0,0;2)\right|$, $C_{(2,0;2)}^{33} = \left|c_{2,0}^{33}(0,0;2)\right|$ and $C_{(2,0;2)}^{55} = \left|c_{2,0}^{55}(0,0;2)\right|$. According to (8.21), the following equation can be obtained,

$$\left(27a_{1,0}^2\right)a_{1,3}^2 + \left(4a_{1,1}^3 - 18a_{1,2}a_{1,1}a_{1,0}\right)a_{1,3} + \left(4a_{1,2}^3a_{1,1} - a_{1,2}^2a_{1,3}^2\right) = 0 \tag{8.30}$$

Equation (8.30) is a quadratic equation of $a_{1,3}$, and the coefficients of (8.30) are known. Solving $a_{1,3}$ from (8.30), and then the PBoC $C_{(3,0;2)}^{444}$ can be obtained,

$$C_{(3,0;2)}^{444} = a_{1,3}/\overline{L}(j\omega) - C_{(3,0;2)}^{333}\left(\overline{H}_1^{(3)}(j\omega)/\overline{H}_1^{(4)}(j\omega)\right)^3. \tag{8.31}$$

The computed PBoC for C_{s3} is presented in Fig. 8.2 with the input amplitude given as

$$U = n_0\sqrt{2G(n_0)v}/f \tag{8.32}$$

where $G(n_0)$ is the spatial ground displacement power spectral density, which has 8 grades according to ISO/TC108/SC2N67. $G(n_0)$ is chosen as 256×10^{-6} m^3 in (8.32), which corresponds to C grade road profile. The vehicle velocity v here is 20 m/s, f is the excited frequency, and n_0 is the reference spatial frequency, which equals 0.1 m in (8.32). The input amplitude U in (8.32) is more close to actual

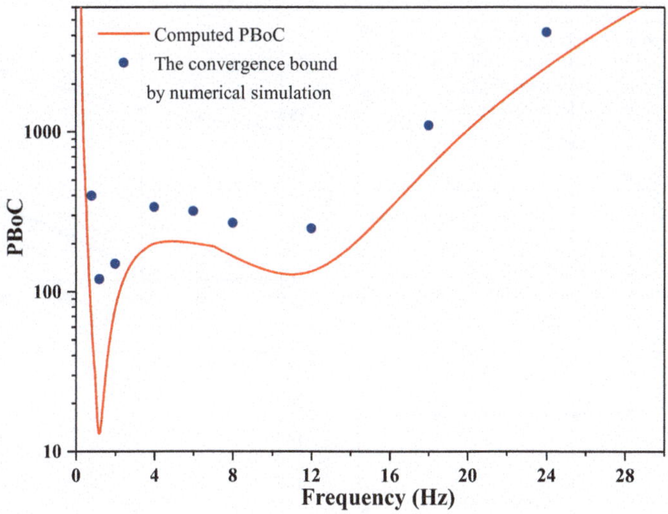

Fig. 8.2 The PBoC of Cs3

road profile over a constant amplitude in the whole frequency range because the road profile exactly has higher amplitude in lower frequency and lower amplitude in higher frequency.

From Fig. 8.2, the smallest PBoC, which locates at the first resonant frequency, is 16. The estimated convergence bound by numerical simulation is about 120. This deviation mainly comes from the large damping ratio in this example, i.e., $\xi = c_s \big/ 2\sqrt{k_{s1} m_s} = 0.4$. In other frequency range, the PBoC obtained by Proposition 1 is very close to the true convergence bound.

To validate the effectiveness of the result above, a normalized root mean square error (NRMSE) is used here to measure the closeness between the synthesized output truncated up to Nth order and the real output by ode45 as,

$$NRMSE = \sqrt{\frac{\sum \left(y_{synthesized}(k) - y_{real}(k)\right)^2}{\sum \left(y_{real}(k) - y_{mean}\right)^2}} \tag{8.33}$$

where y_{mean} is the mean value of the real output $y_{real}(k)$.

According to the results shown in Fig. 8.3, when the model parameter c_{s3} is selected to be smaller than the computed PBoC, the NRMSEs for all the outputs in Eq. (8.28) decrease significantly as the synthesized order increases. For instance, at a frequency of 4 Hz and c_{s3} value of 120 (which is smaller than the computed PBoC of 201), using a higher order approximation leads to smaller NRMSEs. This suggests that the vehicle suspension system [Eqs. (8.28) and (8.27)] exhibits a convergent Volterra series expansion. On the other hand, as the parameter c_{s3} increases, the synthesized outputs up to the same truncated order display observable growth in

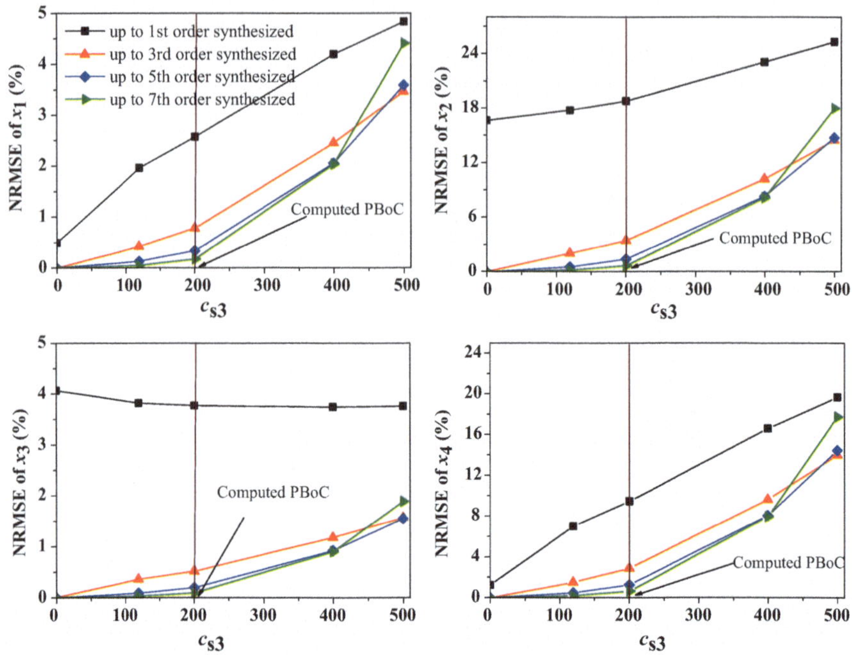

Fig. 8.3 The normalized root mean square error when f = 4 Hz

NRMSEs. When c_{s3} is much larger than the computed PBoC (e.g., $c_{s3} = 500$), the NRMSEs of the synthesized outputs continue to increase beyond a certain synthesized order (e.g., 5th order in this case), indicating divergence of the synthesized outputs. Therefore, within the range of the PBoC, the suspension system [Eqs. (8.28) and (8.27)] can be considered Volterra-type and can be effectively analyzed using Volterra-series-based methods.

8.4.2 The Multi-objective-Performance Analysis

The performance of the vehicle suspension system, for example, the ride comfort (the vertical body acceleration) $y_1 = \ddot{x}_s$, the relative dynamic tire load $y_2 = k_t(x_t - u)/(m_s + m_t)g$, and the suspension stroke $y_3 = x_s - x_t$, can be analyzed in the frequency domain using the nonlinear output spectrum if the nonlinear model (8.28) is Volterra-type. Any nonlinear output function including the MOP function can be treated with the nonlinear characteristic output spectrum (nCOS) function based method, and thus can be expressed as an explicit and analytic polynomial function of the model parameter of interest [9].

The efficacy of a vehicle's suspension system, such as ride comfort indicated by vertical body acceleration $y_1 = \ddot{x}_s$, the dynamic tire load relative to the vehicle $y_2 =$

$k_t(x_t - u)/(m_s + m_t)g$, and the suspension travel $y_3 = x_s - x_t$, can be evaluated in the frequency domain through the nonlinear output spectrum if the system adheres to a Volterra-type nonlinear model. Utilizing the nonlinear characteristic output spectrum (nCOS) method, any nonlinear output function, including the MOP function, can be articulated as a clear and analytical polynomial that is dependent on the specific model parameter of interest [9] (the model parameter is c_{s3} in this case),

$$\begin{aligned} Y_i(j\omega) = {} & Y_{i,1}(j\omega) + Y_{i,2}(j\omega) + \underbrace{Y_{i,3}(j\omega)\big|_{c_{s3}=0} + c_{s3} \cdot \varphi_{i,3}(c_{s3}; j\omega)}_{Y_{i,3}(j\omega)} \\ & + \underbrace{Y_{i,4}(j\omega)\big|_{c_{s3}=0} + c_{s3} \cdot \varphi_{i,4}(c_{s3}; j\omega)}_{Y_{i,4}(j\omega)} \\ & + \underbrace{Y_{i,5}(j\omega)\big|_{c_{s3}=0} + c_{s3} \cdot \varphi_{i,5}(c_{s3}; j\omega) + c_{s3}^2 \cdot \varphi_{i,5}(c_{s3}^2; j\omega)}_{Y_{i,5}(j\omega)} + \cdots \end{aligned} \tag{8.34}$$

where $Y_{i,m}(j\omega)$ is the mth order output spectrum for the output y_i, and $\varphi_{i,m}(c_{s3}^k; j\omega)$ is the model parameters independent part in the mth order output spectrum, which corresponds to the kth order of the model parameters c_{s3}. When c_{s3} is taken under the convergence bound, (8.34) is always valid.

In a Volterra-type nonlinear system, which operates around a stable equilibrium, there are no abrupt or chaotic behaviors. As a result, the system's dynamic response varies smoothly with its characteristic parameters. Consequently, the nonlinear characteristic output spectrum (nCOS) detailed in Eq. (8.34) at various standard frequencies effectively captures the system's output spectrum.

Under cosine wave excitation. Figure 8.4 illustrates the performance of the vertical body acceleration when the suspension system is subjected to a cosine wave excitation. In this case, cs3 is set to 120, making system (8.28) a Volterra-type across the entire frequency range. Figure 8.4 also displays the first order harmonic component of the acceleration, which is the Fourier transform of the acceleration at the first order harmonic frequency (identical to the input frequency).

It's important to note that a larger linear damping will worsen the performance in the frequencies between the two resonant frequencies, such as 4–8 Hz (which are most sensitive to the human body and are recommended as the nominal frequency range for assessing ride comfort according to ISO 2361), and also in the frequencies greater than the second resonant frequency. Conversely, a smaller linear damping will degrade the performance in/around the two resonant frequencies.

As seen in Fig. 8.4, the nonlinear controller outperforms the linear damping, i.e., it enhances performance in the frequencies between the two resonant frequencies without compromising the performance in/around the two resonant frequencies. An additional benefit of the nonlinear controller, when compared to skyhook-based methods, is that it's much simpler to measure the acceleration signal than to obtain the absolute velocity or displacement signal.

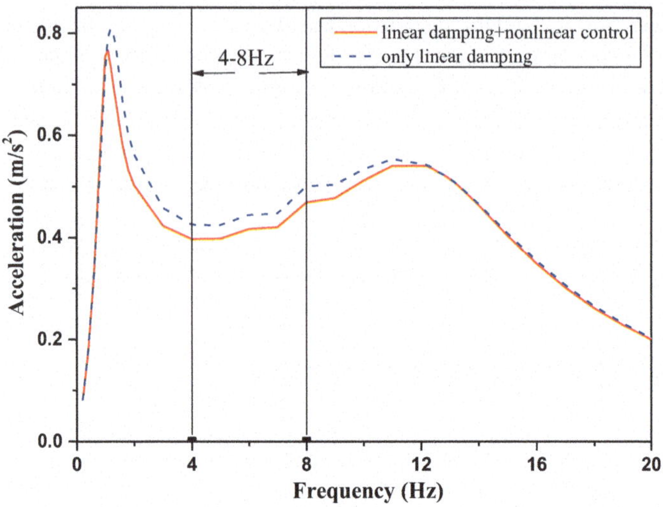

Fig. 8.4 Comparison of vertical body acceleration under cosine road excitation

The active suspension controller based on H∞ [25] are developed with linear models, and thus they are not applicable in this case. Controllers based on Fuzzy control [39, 40] and adaptive control [3, 41], etc., can deliver excellent performance involving advanced control theory, state estimation, and filtering design. In comparison to the above methods, the proposed nonlinear controller offers an alternative frequency domain approach and could be much simpler to implement.

Figure 8.4 only provides how the performance varies with frequency changes but does not present how the performance is affected by model parameters. Therefore, the nonlinear characteristic output spectrum (nCOS) function would be much better than the traditional simulation based output spectrum analysis. When it is truncated up to 5th order, (8.34) can be rearranged as,

$$Y_i(j\omega) = Y_{i0}(j\omega) + c_{s3}Y_{i1}(j\omega) + c_{s3}^2 Y_{i2}(j\omega),\ i = 1, 2, 3 \tag{8.35}$$

where $Y_{i0}(j\omega) = Y_{i,1}(j\omega) + Y_{i,2}(j\omega) + Y_{i,3}(j\omega)\big|_{c_{s3}=0} + Y_{i,4}(j\omega)\big|_{c_{s3}=0} + Y_{i,5}(j\omega)\big|_{c_{s3}=0}$, $Y_{i1}(j\omega) = \varphi_{i,3}(c_{s3}; j\omega) + \varphi_{i,4}(c_{s3}; j\omega) + \varphi_{i,5}(c_{s3}; j\omega)$ and $Y_{i2}(j\omega) = \varphi_{i,5}(c_{s3}^2; j\omega)$.

Denote the multi-objective performance (MOP) including vertical body acceleration y_1, relative dynamic tire load y_2, and suspension stroke y_3 as,

$$\begin{aligned} z(\mu_1, \mu_2, c_{s3}; \omega) = {} & \mu_1 \frac{|Y_1(j\omega)|^2}{|Y_{1l}(j\omega)|^2} + \mu_2 \frac{|Y_2(j\omega)|^2}{|Y_{2l}(j\omega)|^2} \\ & + (1 - \mu_1 - \mu_2)\frac{|Y_3(j\omega)|^2}{|Y_{3l}(j\omega)|^2} \end{aligned}$$

$$
\begin{aligned}
&= z_0(\mu_1, \mu_2; \omega) + c_{s3} z_1(\mu_1, \mu_2; \omega) \\
&\quad + c_{s3}^2 z_2(\mu_1, \mu_2; \omega) \\
&\quad + c_{s3}^3 z_3(\mu_1, \mu_2; \omega) + c_{s3}^4 z_4(\mu_1, \mu_2; \omega)
\end{aligned} \tag{8.36}
$$

where z_0, z_1, z_2, z_3, and z_4 are independent of c_{s3}, which are presented in Appendix. $Y_{il}(j\omega)$ stands for the performance when only linear damping is used, and $\mu_1, \mu_2, \mu_3 = 1 - \mu_1 - \mu_2$ stands for the weight of the vertical body acceleration, the relative tire load, and the suspension stroke in the multi-objective performance function (8.36), respectively.

To estimate the parameter-independent parts in (8.35), the following procedure is given [9].

Procedure 1:

1. Given c_{s3}, for example, $c_{s3} = 10$, then 5 simulations with different input magnitude are needed to estimate $Y_{i,m}(j\omega)$ [because it is truncated up to 5th order in (8.34)].
2. Take additionally two c_{s3}, for example, 50 and 90, repeat step 1.
3. Estimate the model parameters independent part using different $Y_{i,m}(j\omega)$ obtained by step 1 and 2.

It can be seen that only 15 simulations are needed for the estimation of the nCOS function in (8.34) and (8.35) up to the 5th order with a simple least square method. Three different $Y_{i,m}(j\omega)$s are needed in Procedure 1 because the quadratic term c_{s3}^2 exists in $Y_{i,5}(j\omega)$ in (8.34). The MOP function can then be studied with different model parameters c_{s3} and different weights (μ_1 and μ_2) according to (8.36).

To estimate the nCOS function as outlined in Eqs. (8.34) and (8.35) up to the fifth order, a mere 15 simulations are required using a straightforward least squares approach. The necessity for three distinct values of $Y_{i,m}(j\omega)$s in Procedure 1 arises due to the presence of a quadratic term c_{s3}^2 in $Y_{i,5}(j\omega)$ within Eq. (8.34). Subsequently, the MOP function can be analyzed by varying the model parameter c_{s3} and adjusting the weights (μ_1 and μ_2) as specified in Eqs. (8.36–8.37).

In (8.36–8.37), $\mu_1 = 1$, $\mu_2 = 0$ means that only vertical body acceleration is considered in the MOP function. $\mu_1 = 0$, $\mu_2 = 1$ and $\mu_1 = 0$, $\mu_2 = 0$ present the case that only relative tire load or suspension stroke is taken into account in the MOP function, respectively. $c_{s3} = 0$ stands for the case when only linear damping is used.

Figure 8.5 presents the influence of the model parameter c_{s3} on the MOP function when only vertical body acceleration y_1, relative tire load y_2, or suspension stroke y_3 is considered. The suggested nonlinear controller is observed to enhance both the vertical body acceleration and the relative tire load. These improvements are directly linked to ride comfort and vehicle handling capabilities, which are crucial for ride safety. The suspension stroke, which measures the movement between the unsprung and sprung masses, is typically kept below a certain limit. With the nonlinear controller, the suspension stroke remains comparable to that achieved with linear damping at 4 Hz, while at 8 Hz, it results in a marginally increased suspension stroke.

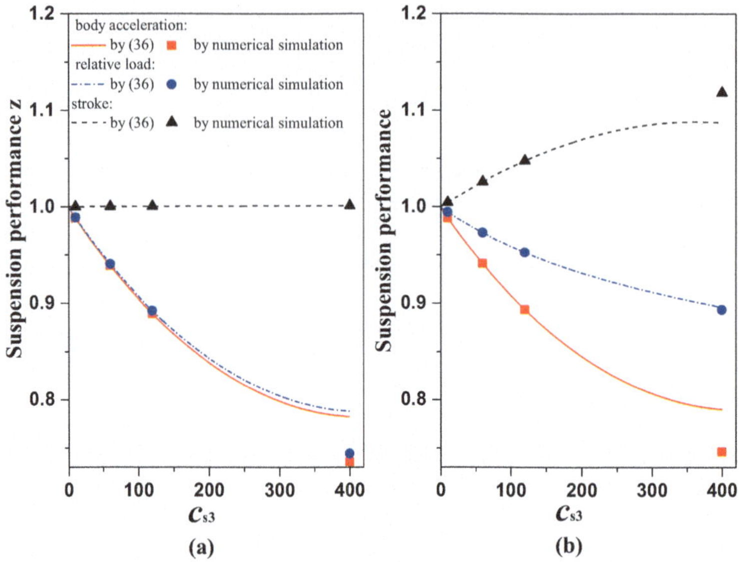

Fig. 8.5 Multi-objective performance under cosine excitation: **a** at 4 Hz, **b** at 8 Hz

The computed PBoC at $f = 4$ Hz is about 201 according to Fig. 8.3. Thus the results in Fig. 8.5 by the MOP function based on the nonlinear characteristic output spectrum (nCOS) are reliable when c_{s3} is taken smaller than the computed PBoC. If c_{s3} takes much larger than the computed PBoC, for example, 400, the MOP function based on the nCOS by (8.36–8.37) may have some prediction error when compared with those by numerical simulation, and complicated nonlinear dynamics such as bifurcation or chaos may occur in this case. This shows the significance of the results in Proposition 1 for optimization and design of nonlinear systems using the nCOS-based MOP function.

The MOP function considering both vertical body acceleration and relative tire load with different weight $\mu_1 + \mu_2 = 1$ is presented in Fig. 8.6. The MOP function takes 400,000 numerical simulations (by ode 45) when μ_1 = 0:0.01:1 and c_{s3} = 0:0.1:400 for Fig. 8.6. The nCOS method requires a mere 15 simulations for implementation. Its efficiency becomes particularly evident when analyzing multiple model parameters simultaneously, as outlined in Eq. (8.34). While the harmonic balance or perturbation methods could theoretically be applied, their computational complexity often renders them impractical. For systems that conform to the Volterra-type, the nCOS method stands out due to its computational speed and practicality. The insights from Proposition 1 are invaluable for conducting nonlinear analysis, design, and optimization grounded in Volterra series theories and methodologies.

The vehicle suspension system analysis presented here, which utilizes the MOP function at 4 Hz, can also be extended to other frequencies of interest, such as the two resonant frequencies or those beyond the second resonant frequency, although these are not detailed in this discussion. The design (optimization) of model parameters

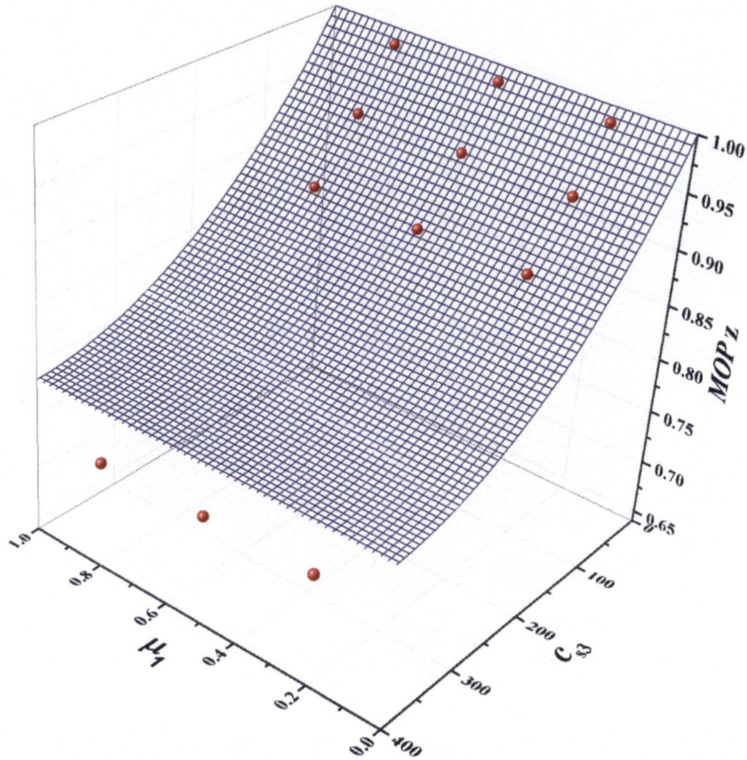

Fig. 8.6 Multi-objective performance under cosine excitation at 4 Hz with μ1 + μ2 = 1, the red ball is the MOP function by numerical simulation

c_{s3} depends on different weights μ_1, μ_2, μ_3. For the case in Fig. 8.6, a larger c_{s3} leads to a better performance improvement (with a lower MOP). Considering that the designed system would be better to be Volterra-type in the whole frequency range to avoid complicate nonlinear phenomena, the computed PBoC for c_{s3} is needed here as a constraint for the optimization, c_{s3} is thus chosen as 120 (the lowest PBoC shown in Fig. 8.2).

Under random road excitation. The Multi-Objective-Performance (MOP) function, as formulated in (8.35–8.38) and Procedure 1, is not limited to harmonic input. It can handle general or random inputs. Road disturbances are typically treated as a random process. According to ISO/TC108/SC2N67, road profiles can be categorized into eight grades. In such scenarios, the overall performance of the suspension can be evaluated using the MOP function as per (8.35–8.38) and Procedure 1. Figure 8.7 showcases the MOP function for a C grade road profile at a vehicle speed of 20 m/s. The multi-objective performance as per (8.38) aligns well with the results from numerical simulations. It's evident that the nonlinear controller is effective even under random road profiles.

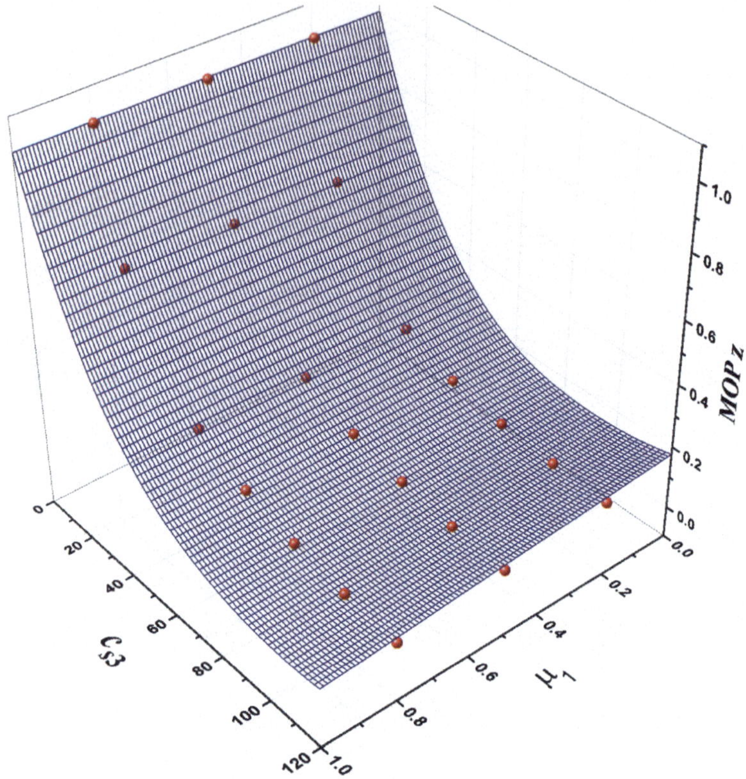

Fig. 8.7 Multi-objective performance under stochastic excitation at 4 Hz with μ1 + μ2 = 1, the red ball is the MOP function by numerical simulation

Other strategies focused on enhancing comfort include the Acceleration Driven Damper (ADD) approach [35] and the refined combination of the skyhook and ADD techniques [37]. The latter has a controller as,

$$\begin{cases} c_{in}(t) = c_{\max} & \left[(\ddot{x}_s^2 - \alpha^2\dot{x}_s^2) \le 0 \wedge \dot{x}_s(\dot{x}_s - \dot{x}_t) > 0\right] \\ & \vee\left[(\ddot{x}_s^2 - \alpha^2\dot{x}_s^2) > 0 \wedge \ddot{x}_s(\dot{x}_s - \dot{x}_t) > 0\right] \\ c_{in}(t) = c_{\min} & \left[(\ddot{x}_s^2 - \alpha^2\dot{x}_s^2) \le 0 \wedge \dot{x}_s(\dot{x}_s - \dot{x}_t) \le 0\right] \\ & \vee\left[(\ddot{x}_s^2 - \alpha^2\dot{x}_s^2) > 0 \wedge \ddot{x}_s(\dot{x}_s - \dot{x}_t) \le 0\right] \end{cases} \tag{8.37}$$

where $c_{in}(t)$ is the requested damping coefficient. The actual damping coefficient is obtained by acting a low pass filter on $c_{in}(t)$, i.e., $\dot{c}(t) = -\beta c(t) + \beta c_{in}(t)$. The comparison of the performance between the proposed nonlinear controller and that with the mixed skyhook and the ADD method is presented in Fig. 8.8. The comparison takes the following coefficients:$\alpha = 11, \beta = 30$, $c_{\min} = 300$, $c_{\max} = 4000$ [37], and $c_{s3} = 120$. Figure 8.8a showcases the performance outcomes when the suspension system

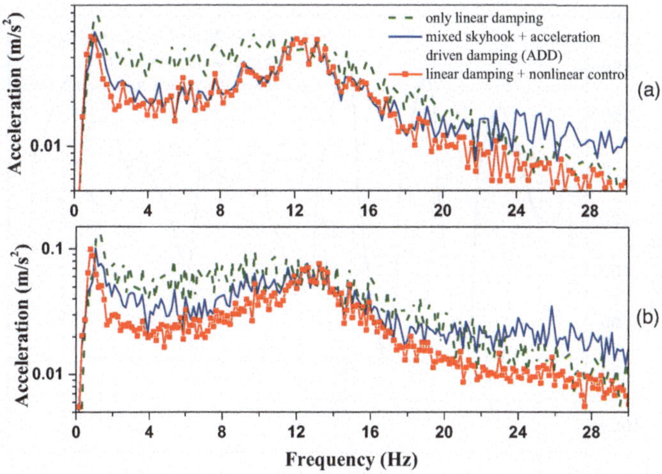

Fig. 8.8 Comparison of the performances with different strategies. **a** Subject to C level road profile with 20 m/s, **b** subject to D level road profile with 10 m/s

is tested against a C level road profile at a speed of 20 m/s, while Fig. 8.8b displays the results for a D level road profile at 10 m/s. When operating on a C level road surface, the results using the combined skyhook and ADD method are comparable to those achieved with the proposed nonlinear controller. However, under the more challenging conditions of a D level road surface, the proposed nonlinear controller significantly outperforms the mixed skyhook and ADD method.

Figure 8.9 offers a comparative analysis of the active forces produced by the proposed nonlinear controller and the hybrid skyhook and ADD method when subjected to a C level road profile. The active force generated by the proposed nonlinear controller is noticeably smoother than the hybrid method. This implies that the hybrid damping control includes high-frequency components, leading to a greater damping force at high frequencies compared to the proposed method. It's widely accepted that a larger damping force at high frequencies can adversely affect the vibration control performance in practical applications.

Based on the above results, the nCOS-based MOP analysis for various excitation inputs clearly demonstrates that: (1) the proposed nonlinear analysis method offers a clear understanding of the nonlinear relationship between characteristic parameters and system output performance, significantly aiding nonlinear analysis and design, and is also computationally efficient; (2) the performance of the suspension can be effectively enhanced with the proposed nonlinear acceleration control method, which is simpler to implement compared to many existing methods; (3) the proposed PBoC ensures that system nonlinearity consistently remains at a Volterra-type level, avoiding complex jumping or chaos phenomena.

Remark 5: If a system is verified to be of the Volterra-type, the impact of nonlinear parameters of interest on a pre-defined multi-objective output performance function

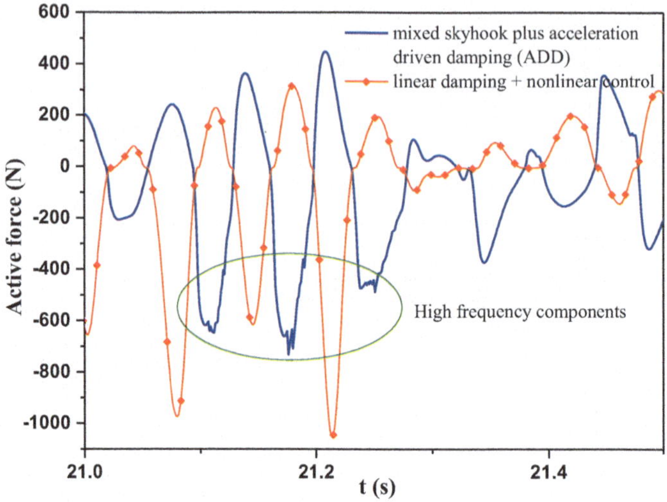

Fig. 8.9 Comparison of the active forces

can be examined using the nCOS-based MOP analysis method, as shown. Procedure 1 can be employed to identify the MOP function. It's important to note that the findings of this study are formulated for the NARX model. Numerous practical systems can be transformed into a NARX model with experimental input–output data using various algorithms found in the literature [42]. In the presence of measurement noise or other types of noise corruption, developing a reliable and reasonable parametric bound for the Volterra series expansion and robustly estimating the nonlinear MOP function would be highly pertinent. One potential approach could involve adopting a set membership concept [38]. These aspects will be systematically explored in future studies.

To facilitate a clear comprehension of the proposed nCOS-based MOP analysis method for practical nonlinear analysis and optimization design, the subsequent general procedure can be adhered to.

Procedure 2:

1. Identify the real plant into a NARX model with experimental data. The example in (8.27)–(8.28) shows how to obtain the NARX from a continuous model.
2. Determine the characteristic parameters of interest in the design and optimization, e.g., input magnitude or model/controller parameters [c_{s3} in system (8.27)].
3. Compute the PBoC of these characteristic parameters of interest according to Algorithm 1 to find parameter ranges so as to guarantee that the underlying nonlinearity is Volterra-type, example can be seen in (8.29)–(8.31).
4. Determine the nonlinear characteristic output spectrum (nCOS) based multi-objective performance (MOP) function. This can follow (8.34–8.37) and also Procedure 1.

5. Evaluate the MOP function with respect to the characteristic parameters of interest. Examples can be referred to the results in Figs. 8.5, 8.6 and 8.7.

8.5 Summary of the SIMO Approach

This study introduces the Parametric Bound of Convergence (PBoC) for a broad category of nonlinear engineering systems represented by a Single Input Multiple Output (SIMO) Nonlinear Auto-Regressive with eXogenous inputs (NARX) model. This PBoC can be computed with ease and offers valuable guidance for parameter optimization and/or system design of nonlinear systems concerning any characteristic parameters, including model parameters, input magnitude, and frequency variable.

When applied to a nonlinear vehicle suspension system, it's observed that within the PBoC of the characteristic parameter of interest, the nonlinear characteristic output spectrum (nCOS) based Multi-Objective Performance (MOP) function provides a clear understanding of the relationship between the MOP function and characteristic parameters. Furthermore, the MOP function in optimization can be effectively addressed with the proposed method.

These findings present a useful and significant method for engineering optimization related to nonlinearity. This is particularly relevant as many engineering problems are essentially nonlinear optimal design problems (e.g., nonlinear suspension systems), which are typically challenging to tackle. However, the nCOS and PBoC based MOP function introduced in this study offer a straightforward insight into these complex nonlinear relationships, thereby significantly facilitating optimization and design tasks.

8.6 Further Extensions

Similar results can be extended to analysis and design of nonlinear systems with exponential functions [43] since basis functions of ANN or fuzzy membership can be specially treated as an independent output and thus the overall system of concern can be transformed into a multi-output or multi-objective systems as shown in this study. For easy reading, the Volterra series expansion of a SIMO nonlinear system is reformulated as follows

$$
\begin{aligned}
x_k(t) &= \sum_{N=1}^{N'} x_{(k,N)}(t) \\
x_{(k,N)}(t) &= x_{(k,0)} + \int_{-\infty}^{\infty} \dots \int_{-\infty}^{\infty} h_{(k,N)}(\tau_1, \dots, \tau_N) \prod_{i=1}^{N} u(t-\tau_i) d\tau_i
\end{aligned}
\tag{8.38}
$$

where x_k and u are the kth subsystem output and single input of the system respectively; N' is the maximum order of the system nonlinearity; $h_{(k,N)}(\tau_1, ..., \tau_N)$ is the Nth order Volterra kernel; $x_{(k,0)}$ can be zero or a DC constant. The Nth order GFRF can then be defined by the multi-dimensional Fourier transform as

$$H_{(k,N)}(j\omega_1, ..., j\omega_N) = \int_{-\infty}^{\infty} ... \int_{-\infty}^{\infty} h_{(k,N)}(\tau_1, ..., \tau_N) \cdot e^{-j(\omega_1\tau_1+...+\omega_N\tau_N)} d\tau_1 ... d\tau_N \quad (8.39)$$

Equation (8.1) can also be written as [9]

$$x_k(t) = \sum_{N=1}^{N'} x_{(k,N)}(t)$$
$$x_{(k,N)}(t) = H_{(k,0)} + \frac{1}{2\pi}\int_{-\infty}^{\infty}\left[\frac{1}{(2\pi)^{N-1}}\underbrace{\int_{-\infty}^{\infty} ... \int_{-\infty}^{\infty}}_{N-1} \Xi_{(k,N)} d\omega_1 ... d\omega_{N-1}\right] e^{j\omega t} d\omega \quad (8.40a)$$

where

$$\Xi_{(k,N)} = H_{(k,N)}(j\omega_1, ..., j\omega_{N-1}, j(\omega - \omega_1 - ... - \omega_{N-1})) \cdot U_N(j\omega, j\omega_1, ..., j\omega_{N-1})$$
$$U_N(j\omega, j\omega_1, ..., j\omega_{N-1}) = U(j\omega_1)...U(j\omega_{N-1}) \cdot U(j(\omega - \omega_1 - ... - \omega_{N-1}))$$

$H_{(k,0)}$ is associated with the 0th order output $x_{(k,0)}$ which is independent of any input. Note that, $\omega_N = \omega - \omega_1 - \cdots - \omega_{N-1}$ is used in (8.40a). The output $x_{(\mathrm{k,N})}$ in the frequency domain can be written as [9]

$$X_{(k,N)}(j\omega) = \frac{1}{(2\pi)^{n-1}}\underbrace{\int_{-\infty}^{\infty} \cdots \int_{-\infty}^{\infty}}_{n-1} \Xi_{(\mathrm{k,N})} \mathrm{d}\omega_1 \cdots \mathrm{d}\omega_{n-1} \quad (8.40b)$$

or

$$X_{(k,N)}(j\omega) = \frac{1}{\sqrt{N}(2\pi)^{N-1}} \int_{\omega_1+\cdots+\omega_N=\omega} H_{(k,N)}(j\omega_1, \cdots, j\omega_n) \prod_{i=1}^{n} U(j\omega_i) d\sigma_\omega \quad (8.40c)$$

Equation (8.40b) is derived from the definition of the Fourier transform, and Eq. (8.40c) is developed from Eq. (8.40b). This provides a clear insight into the output frequency components given the input [11]. $d\sigma_\omega$ denotes the area of a unit element on the n-dimensional hyper plane $\omega = \omega_1 + \cdots + \omega_n$. The output frequency $\omega = \omega_1 + \cdots + \omega_n$ is dependent on $\omega_1 + \ldots + \omega_N$, instead of individual frequency component.

Here is to present an extension of the SIMO approach to an exponential-type nonlinear system as follows [43]:

$$g_1(\boldsymbol{x}, \boldsymbol{u}) + g_2(\boldsymbol{x}, \boldsymbol{u})e^{f_1(\boldsymbol{x}, \boldsymbol{u})} = \sum_{l=0}^{L} \frac{d^l u}{dt^l} \tag{8.41}$$

where $g_1(\boldsymbol{x}, \boldsymbol{u}), g_2(\boldsymbol{x}, \boldsymbol{u}), f_1(\boldsymbol{x}, \boldsymbol{u})$ are polynomial functions of output $\boldsymbol{x}$ and input $\boldsymbol{u}$; l is the differential order with a maximum order L; and

$$\boldsymbol{x} = \left\{x, \frac{d^1 x}{dt^1}, \frac{d^2 x}{dt^2}, \ldots, \frac{d^L x}{dt^L}\right\}, \quad \boldsymbol{u} = \left\{u, \frac{d^1 u}{dt^1}, \frac{d^2 u}{dt^2}, \ldots, \frac{d^L u}{dt^L}\right\}.$$

When $g_2(\boldsymbol{x}, \boldsymbol{u}) = 0$, System (8.41) simplifies to a polynomial nonlinear system, which is essentially the conventional polynomial-type nonlinear differential equation model. The Generalized Frequency Response Functions (GFRFs) of this model have been examined. System (8.41) can be viewed as a Single Input Single Output (SISO) system. Methods similar to those used here can also be applied to Multi-Input–Multi-Output (MIMO) systems. However, for the sake of simplicity in this discussion, only SISO or Single Input Multiple Output (SIMO) nonlinear systems are considered in this study. It is supposed that the equilibrium ($\boldsymbol{x_0} = \{x_0, 0, 0\ldots\}$ and $\boldsymbol{u} = 0$) is stable and the system has appropriate initial conditions such that a convergent Volterra series expansion as (8.1) exists for the input output relationship between u(t) and x(t); and at the equilibrium $|g_1(\boldsymbol{x_0}, \boldsymbol{0})| < \infty$, $|g_2(\boldsymbol{x_0}, \boldsymbol{0})| < \infty$, $|f_1(\boldsymbol{x_0}, \boldsymbol{0})| < \infty$ and the derivative $\frac{df_1(\boldsymbol{x}, \boldsymbol{u})}{dt}$ exists. It should be noted that there is a considerably large class of nonlinear systems in practice which have a convergent Volterra series expansion satisfying these requirements.

A. **Calculating the Generalized Frequency Response Functions (GFRFs)**

Step 1: By defining an intermediate state $x_2 = e^{f_1(\boldsymbol{x}, \boldsymbol{u})}$ and taking the derivative with respect to the time of both sides, the following equation can be obtained

$$\dot{x}_2 = e^{f_1(\boldsymbol{x}, \boldsymbol{u})} \frac{df_1(\boldsymbol{x}, \boldsymbol{u})}{dt} = x_2 \frac{df_1(\boldsymbol{x}, \boldsymbol{u})}{dt}$$

Then, the exponential-type nonlinear system (8.41) can be equivalently transformed into the following polynomial nonlinear system as,

$$\begin{cases} g_1(\boldsymbol{x}, \boldsymbol{u}) + g_2(\boldsymbol{x}, \boldsymbol{u})x_2 = \sum_{l=0}^{L} \frac{d^l u}{dt^l} \\ \dot{x}_2 = x_2 \frac{df_1(\boldsymbol{x}, \boldsymbol{u})}{dt} \end{cases} \tag{8.42}$$

with the same initial conditions as (8.41) and additioanlly $x_2(0) = e^{f_1(\boldsymbol{x}(0), \boldsymbol{u}(0))}$. Considering the assumption for (8.41), $\boldsymbol{x} = \boldsymbol{x}_0$, $\boldsymbol{u} = 0$ and $x_2 = e^{f_1(\boldsymbol{x}_0, 0)}$ is a stable equilibrium of (8.42), corresponding to the equilibrium in (8.41). Note that x_2 is the introduced system state, which is in $\mathbf{R}^+$, while the vector $\boldsymbol{x}$ is in $\mathbf{R}^{L+1}$, which is the state of the original system.

By incorporating the auxiliary equation in (8.42), the nonlinear system characterized by an exponential nonlinearity can be analytically converted into a corresponding single-input (that is, u) double-output (that is, x and x2) polynomial nonlinear system. There's no need for a Taylor series expansion in this context.

Step 2: In (8.42), the polynomial nonlinear functions can be expressed with a detailed polynomial form to include all three kinds of nonlinearities (pure input nonlinearity, pure output nonlinearity and pure cross product nonlinearity) as

$$\begin{cases} \sum_{l=0}^{L} c_{1,0}^{10}(l;1)\frac{d^l x_1}{dt^l} + \sum_{l=0}^{L} c_{1,0}^{01}(l;1)\frac{d^l x_2}{dt^l} = \sum_{l=0}^{L} c_{0,1}^{00}(l)\frac{d^l u}{dt^l} \\ \quad + \sum_{m=p+q=2}^{N} \sum_{p=0}^{m} \sum_{\substack{s_1=0 \\ s_1+s_2=p}}^{p} \sum_{l_1,\cdots,l_{p+q}=0}^{L} c_{p,q}^{s_1 s_2}(l_1,\ldots,l_{p+q};1) \\ \quad \times \prod_{i=1}^{s_1} \frac{d^{l_i} x_1}{dt^{l_i}} \prod_{i=s_1+1}^{s_1+s_2} \frac{d^{l_i} x_2}{dt^{l_i}} \prod_{i=p+1}^{p+q} \frac{d^{l_i} u}{dt^{l_i}} \\ \dot{x}_2 = \sum_{m=p+q=2}^{N} \sum_{p=0}^{m} \sum_{\substack{s_1=0 \\ s_1+s_2=p}}^{p} \sum_{l_1,\cdots,l_{p+q}=0}^{L} c_{p,q}^{s_1 s_2}(l_1,\ldots,l_{p+q};2) \\ \quad \times \prod_{i=1}^{s_1} \frac{d^{l_i} x_1}{dt^{l_i}} \prod_{i=s_1+1}^{s_1+s_2} \frac{d^{l_i} x_2}{dt^{l_i}} \prod_{i=p+1}^{p+q} \frac{d^{l_i} u}{dt^{l_i}} \end{cases} \tag{8.43}$$

where x_i is the ith output for $i = 1, 2$ with $x_1 = x$; N is the maximum nonlinearity order in terms of output $\boldsymbol{x}$ and input $\boldsymbol{u}$; $c_{p,q}^{s_1 s_2}(l_1, \ldots, l_{p+q}; j)$ is the model parameter of the corresponding term in the jth equation, which has a nonlinear degree $p + q$ (p order in terms of the outputs x_1, x_2, and q order in terms of the input u), referred to as nonlinear parameter if $p + q > 1$ otherwise as linear parameter if $p + q = 1$; l_i is the differential order with the maximum order L; $s_1, s_2 \in \mathbb{N}$, which are the nonlinearity degrees of x_1, x_2 in the nonlinear term (s_1 corresponds to the output x_1, and s_2 corresponds to the output x_2), e.g.,

$$\prod_{i=1}^{s_1} \frac{d^{l_i} x_1}{dt^{l_i}} \prod_{i=s_1+1}^{s_1+s_2} \frac{d^{l_i} x_2}{dt^{l_i}} \prod_{i=p+1}^{p+q} \frac{d^{l_i} u}{dt^{l_i}},$$

respectively. When $s < s_0$, $\prod_{i=s_0}^{s} (\cdot) = 1$.

It should be observed that the equilibrium point of the aforementioned nonlinear system is non-zero. This implies the presence of constant elements, also known as direct current (dc) components, even when the input is null. Consequently, the methodologies previously discussed for calculating the Generalized Frequency Response Functions (GFRFs) of Multi-Input–Multi-Output (MIMO) nonlinear differential systems cannot be directly used for Eqs. (8.42) or (8.43).

Step 3: Derivation of the GFRFs for (8.43) and thus the GFRFs for (8.38) are obtained. The following result is given.

Let H_0^1, H_0^2 denote the DC constant of x_1, x_2 in (8.43), respectively at the equilibrium. The nth-order GFRF can be calculated as

$$H_n = \begin{bmatrix} H_n^1(j\omega_1, \ldots, j\omega_n) \\ H_n^2(j\omega_1, \ldots, j\omega_n) \end{bmatrix} = \begin{bmatrix} l_{1,1} & l_{1,2} \\ l_{2,1} & l_{2,2} \end{bmatrix}^{-1} \begin{bmatrix} A_{1,n} \\ A_{2,n} \end{bmatrix} = L_n^{-1} \begin{bmatrix} A_{1,n} \\ A_{2,n} \end{bmatrix} \tag{8.44}$$

where

$$l_{1,1} = \sum_{l=0}^{L} c_{1,0}^{10}(l; 1)(j\omega_1 + \cdots + j\omega_n)^l - \sum_{\substack{p=2, \\ t_2=p-1}}^{N} \sum_{l_1=0}^{L} c_{p,0}^{1t_2}(l_1, 0, \ldots, 0; 1)(j\omega_1 + \cdots + j\omega_n)^{l_1} (H_0^2)^{t_2},$$

$$l_{1,2} = \sum_{l=0}^{L} c_{1,0}^{01}(l; 1)(j\omega_1 + \cdots + j\omega_n)^l - \sum_{\substack{p=2, \\ t_1=p-1}}^{N} \sum_{l_2=0}^{L} c_{p,0}^{t_1 1}(0, \ldots, 0, \ldots, l_2; 1)(j\omega_1 + \cdots + j\omega_n)^{l_2} (H_0^1)^{t_1},$$

$$l_{2,1} = - \sum_{\substack{p=2, \\ t_2=p-1}}^{N} \sum_{l_1=0}^{L} c_{p,0}^{1t_2}(l_1, 0, \ldots, 0; 2)(j\omega_1 + \cdots + j\omega_n)^{l_1} (H_0^2)^{t_2},$$

$$l_{2,2} = (j\omega_1 + \cdots + j\omega_n)$$

$$-\sum_{\substack{p=2,\\t_1=p-1}}^{N}\sum_{l_2=0}^{L} c_{p,0}^{t_1 1}(0,\ldots,0,\ldots,l_2;2)(j\omega_1+\cdots+j\omega_n)^{l_2}(H_0^1)^{t_1}, \quad (8.45)$$

and

$$A_{i,n} = \sum_{l_1,\ldots,l_n=0}^{L} c_{0,n}^{00}(l_1,\ldots,l_n;i)\prod_{i=1}^{n}(j\omega_i)^{l_i}$$
$$+\sum_{m=p+q=2}^{N}\sum_{p=1}^{m}\sum_{\substack{t=t_1+t_2=0\\t_1\le s_1,t_2\le s_2}}^{p-1}\sum_{l_1,\ldots,l_{p+q}=0}^{L} c_{p,q}^{s_1 s_2}(l_1,\ldots,l_{p+q};i)$$
$$\times (H_0^1)^{t_1}(H_0^2)^{t_2}\prod_{i=1}^{q}(j\omega_{n-q+i})^{l_{n-q+i}}H_{n-q,p-t}^{t_1' t_2'}(j\omega_1,\ldots,j\omega_{n-q}), \quad (8.46)$$

$$H_{n-q,p-t}^{t_1' t_2'}(j\omega_1,\ldots,j\omega_{n-q}) =$$
$$\sum_{i=1}^{t_1'}(j\omega_1+\cdots+j\omega_i)^{l_{t_1+1}}H_i^1(j\omega_1,\ldots,j\omega_i)$$
$$H_{n-q-i,p-t-1}^{(t_1'-i)t_2'}(j\omega_{i+1},\ldots,j\omega_{n-q})$$
$$+\sum_{i=1}^{t_2'}(j\omega_1+\cdots+j\omega_i)^{l_{s_1+t_2+1}}H_i^2(j\omega_1,\ldots,j\omega_i)$$
$$H_{n-q-i,p-t-1}^{t_1'(t_2'-i)}(j\omega_{i+1},\ldots,j\omega_{n-q}), \quad (8.47)$$

where $\sum_{i=0}^{0}(\cdot)=0$, $H_{0,0}(.)=1$, $H_{n,0}(.)=0$ for $n>0$, $H_{n,p}(.)=0$ for $n<p$, and $s_1+s_2=p$, $t_1+t_2=t$, $t_1'=s_1-t_1$, $t_2'=s_2-t_2$. The recursion in (8.47) ends when $p-t=1$(as such, $t_1'=0$ or 1 and $t_2'=1$ or 0) holds, that is,

$$\begin{aligned}H_{\overline{n},1}^{10}(j\omega_1,\cdots,j\omega_{\overline{n}}) &= (j\omega_1+\cdots+j\omega_{\overline{n}})^{l_{s_1}}H_{\overline{n}}^1(j\omega_1,\cdots,j\omega_{\overline{n}}),\\ H_{\overline{n},1}^{01}(j\omega_1,\cdots,j\omega_{\overline{n}}) &= (j\omega_1+\cdots+j\omega_{\overline{n}})^{l_{s_1+s_2}}H_{\overline{n}}^2(j\omega_1,\cdots,j\omega_{\overline{n}}),\end{aligned} \quad (8.48)$$

where $\overline{n}=1, 2, \ldots$, or n. For the first-order GFRF,

$$A_{1,1} = \sum_{l=0}^{L} c_{0,1}^{00}(l;1)(j\omega_1)^l$$

$$+ \sum_{\substack{p=1,\\ s_1+s_2=p}}^{N-1} \sum_{l_{p+1}=0}^{L} c_{p,1}^{s_1s_2}(0,\cdots,0,l_{p+1};1)\left(H_0^1\right)^{s_1}\left(H_0^2\right)^{s_2}(j\omega_1)^{l_{p+1}}$$

$$+ \sum_{p=2}^{N}\sum_{s_1=2}^{p}\sum_{i=1}^{s_1}\sum_{l_i=0}^{L} c_{p,0}^{s_1s_2}(0,l_i,0;1)\left(H_0^1\right)^{s_1-1}\left(H_0^2\right)^{s_2}(j\omega_1)^{l_i}\cdot H_1^1(j\omega_1)$$

$$+ \sum_{p=2}^{N}\sum_{s_2=2}^{p}\sum_{i=1}^{s_2}\sum_{l_{p-i+1}=0}^{L} c_{p,0}^{s_1s_2}(0,l_{p-i+1},0;1)\left(H_0^1\right)^{s_1}\left(H_0^2\right)^{s_2-1}(j\omega_1)^{l_{p-i+1}}\cdot H_1^2(j\omega_1),$$

$$A_{2,1} = \sum_{\substack{p=1,\\ s_1+s_2=p}}^{N-1} \sum_{l_{p+1}=0}^{L} c_{p,1}^{s_1s_2}(0,\ldots,0,l_{p+1};2)\left(H_0^1\right)^{s_1}\left(H_0^2\right)^{s_2}(j\omega_1)^{l_{p+1}}$$

$$+ \sum_{p=2}^{N}\sum_{s_1=2}^{p}\sum_{i=1}^{s_1}\sum_{l_i=0}^{L} c_{p,0}^{s_1s_2}(0,l_i,0;2)\left(H_0^1\right)^{s_1-1}$$

$$\left(H_0^2\right)^{s_2}(j\omega_1)^{l_i}\cdot H_1^1(j\omega_1)$$

$$+ \sum_{p=2}^{N}\sum_{s_2=2}^{p}\sum_{i=1}^{s_2}\sum_{l_{p-i+1}=0}^{L} c_{p,0}^{s_1s_2}(0,l_{p-i+1},0;2)$$

$$\left(H_0^1\right)^{s_1}\left(H_0^2\right)^{s_2-1}(j\omega_1)^{l_{p-i+1}}\cdot H_1^2(j\omega_1), \quad (8.49)$$

Please refer to [43] for the proof.

Note that the GFRFs obtained in Proposition 1 is asymmetrical. The symmetrical GFRFs can be obtained by $H_n^{sym}(j\omega_1,\ldots,j\omega_n) = \frac{1}{n!}\sum_{\text{all permutation of }\omega_1,\ldots,\omega_n} H_n^{asym}(j\omega_1,\ldots,j\omega_n)$. The L_n matrix in (8.44) can be singular when $\omega_1+\cdots+\omega_n = 0$ holds (which means that the GFRFs are calculated at dc components). For this case, the following algorithm is proposed for an approximation of the inverse matrix L_n^{-1}.

Algorithm 1:

Step a. For the nth-order GFRF, approximate ω_i by

$$\omega_i' = \omega_i + \frac{\triangle}{n}$$

for $i = 1, 2, \ldots, n$, where $\triangle$ is a small perturbation close to 0 which makes the inverse matrix $L_n^{-1}(j\omega_1'+\cdots+j\omega_n') = L_n^{-1}(j\omega_1+\cdots+j\omega_n+j\triangle) = L_n^{-1}(j\triangle)$ exists (note that $j\omega_1+\cdots+j\omega_n = 0$ as the singular point).

Step b. Approximate $H_n^i(j\omega_1,\ldots,j\omega_n)$ with $H_n^i(j\omega_1',\ldots,j\omega_n')$ □

The point $(j\omega'_n, \ldots, j\omega'_n)$ is chosen in the normal direction of the hyper plane $j\omega_1 + \cdots + j\omega_n = 0$ with a distance $d = \triangle \,/\, \sqrt{n}$ to the frequencies $(j\omega_1, \ldots, j\omega_n)$. After $\triangle$ is chosen, $L_n^{-1}(j\ \triangle)$ will be the same for different order n since $l_{1,1}, l_{1,2}, l_{2,1}$ and $l_{2,2}$ are dependent only on $\triangle$, not on n. That is, $L_1^{-1}(j\ \triangle) = L_2^{-1}(j\ \triangle) = \cdots$. Note that the distance d decreases as the order n increases, thus the approximation of $H_n^i(j\omega_1, \ldots, j\omega_n)$ with $H_n^i(j\omega'_1, \ldots, j\omega'_n)$ would be more accurate as the order n increases, i.e., $|H_n^i(j\omega_1, \ldots, j\omega_n) - H_n^i(j\omega'_1, \ldots, j\omega'_n)| \to 0$ as $n \to \infty$.

Example 1: Pure input nonlinearity

Consider the exponential nonlinear system described by

$$x = e^{au}$$

where u is the system input and x is the system output. This exponential nonlinearity can be expanded with a Taylor series as

$$x = 1 + au + \frac{a^2u^2}{2!} + \frac{a^3u^3}{3!} + \cdots \tag{8.50}$$

When the nonlinear equation is truncated up to u^3, there are 4 coefficients involved in terms of a. The GFRFs can be computed as (with the probing method),

$$H_0 = 1,\ H_1 = 1,\ H_2 = \frac{a^2}{2!},\ H_3 = \frac{a^3}{3!}, \cdots \tag{8.51}$$

With the results in the previous section, the exponential nonlinearity $x = e^{au}$ can be analytically transformed into the following single-input single-output polynomial nonlinear system (step 1–2)

$$\dot{x} = ax\dot{u} \tag{8.52}$$

with $x(0) = e^{au(0)}$. In (8.52), there exists only one coefficient a. The GFRFs can then be computed as step 3.

For the 0th order GFRF, assuming the input is zero, then it gives $H_0^{asym} = 1$.

For the first order GFRF, following (8.45–8.49) gives

$$(j\omega_1)H_1^{asym}(j\omega_1) = aH_0^{asym} \cdot (j\omega_1) \tag{8.53}$$

That is,

$$H_1^{asym}(j\omega_1) = \frac{aH_0^{asym} \cdot (j\omega_1)}{j\omega_1} = a, \tag{8.54}$$

For the second order GFRF,

$$H_2^{asym}(j\omega_1, j\omega_2)(j\omega_1 + j\omega_2) = aH_1^{asym}(j\omega_1) \cdot (j\omega_2) \tag{8.55}$$

$$H_2^{asym}(j\omega_1, j\omega_2) = \frac{aH_1^{asym} \cdot (j\omega_1)}{j\omega_1 + j\omega_2} = \frac{a^2 \cdot (j\omega_1)}{j\omega_1 + j\omega_2} \tag{8.56}$$

For the third order GFRF,

$$H_3^{asym}(j\omega_1, j\omega_2, j\omega_3)(j\omega_1 + j\omega_2 + j\omega_3) = aH_2^{asym}(j\omega_1, j\omega_2) \cdot (j\omega_3) \tag{8.57}$$

$$H_3^{asym}(j\omega_1, j\omega_2, j\omega_3) = \frac{a \cdot H_2^{asym}(j\omega_1, j\omega_2) \cdot (j\omega_1)}{j\omega_1 + j\omega_2 + j\omega_3} \tag{8.58}$$

The GFRFs in (8.56, 8.58) are asymmetrical, but can be symmetrized as,

$$\begin{aligned} H_0^{sym} &= 1, H_1^{sym}(j\omega_1) = a, \\ H_2^{sym}(j\omega_1, j\omega_2) &= \frac{a^2}{2!}, \\ H_3^{sym}(j\omega_1, j\omega_2, j\omega_3) &= \frac{a^3}{3!} \cdots \end{aligned} \tag{8.59}$$

The Generalized Frequency Response Functions (GFRFs) in (8.59) match those in (8.51), thereby confirming the validity of the result in Proposition 1. It's evident that the polynomial nonlinear system in (8.52) has fewer parameters (just one parameter) compared to the Taylor series expansion in (8.50) (which has 4 parameters). Other benefits of the proposed algorithm, such as its high computational efficiency and simplicity in analysis, will become more apparent for pure output nonlinear systems or cross input–output nonlinear systems. These aspects will be demonstrated and discussed in the subsequent section.

Example 2: Suppression of harmonic distortion

An example is borrowed from [44] as shown in Fig. 8.10. The input–output dynamics of the common emitter amplifier are governed by the exponential-type nonlinearity as

$$\frac{u - x}{R} = C\dot{x} + I_{BQ}(e^{\frac{x}{V_T}} - 1) \tag{8.60}$$

$$y = -\beta I_{BQ}(e^{\frac{x}{V_T}} - 1) \tag{8.61}$$

where $u = U \sin \omega t$, and I_{BQ} is the reverse saturation current, which always varies with temperature. It is thus necessary and of interest to study how the amplifier performance is affected by this reverse saturation current I_{BQ}.

In the literature, the exponential-type nonlinearity is often approximated by a third-order Taylor series as

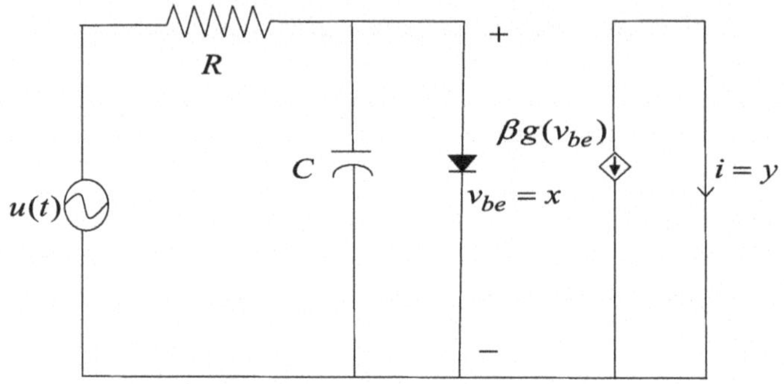

Fig. 8.10 Equivalent model of bipolar amplifier

$$\begin{aligned} g(x) = I_{BQ}(e^{x/V_T} - 1) &= \sum_{i=1}^{3} \frac{I_{BQ}}{i!V_T^i} x^i \\ &= \frac{I_{BQ}}{V_T} x + \frac{I_{BQ}}{2V_T^2} x^2 + \frac{I_{BQ}}{6V_T^3} x^3 \end{aligned} \tag{8.62}$$

It can be seen that I_{BQ} is involved in all three parameters.

Step 1: Denote

$$x_2 = I_{BQ}\left(e^{\frac{x_1}{V_T}} - 1\right),$$

and take derivative with respect to the time in both sides,

$$\dot{x}_2 = \frac{I_{BQ}}{V_T} \dot{x}_1 + \frac{1}{V_T} \dot{x}_1 x_2$$

holds.

Step 2: Equation (8.61) can then be given as $y = -\beta x_2$. Equation (8.60) can be rearranged into a single-input double-output system as

$$\begin{cases} \frac{1}{R} x_1 + C\dot{x}_1 + x_2 = \frac{1}{R} u \\ -\frac{I_{BQ}}{V_T} \dot{x}_1 + \dot{x}_2 = \frac{1}{V_T} \dot{x}_1 x_2 \end{cases} \tag{8.63}$$

Step 3: The GFRFs can then be computed according to Proposition 1.

For the 0th order, as there are no DC components in (8.60–8.61) around the zero equilibrium,

$$H_0^1 = H_0^2 = 0 \tag{8.64}$$

For the 1st order, by following (8.44–8.45, 8.49),

$$l_{1,1} = \frac{1}{R} + C(j\omega_1),\ l_{1,2} = 1,$$

$$l_{2,1} = -\frac{I_{BQ}}{V_T}(j\omega_1),\ l_{2,2} = (j\omega_1)$$

$$A_{1,1} = \frac{1}{R},\ A_{2,1} = 0$$

$$\begin{bmatrix} H_1^1(j\omega_1) \\ H_1^2(j\omega_1) \end{bmatrix} = \begin{bmatrix} \frac{1}{R} + C(j\omega_1) & 1 \\ -\frac{I_{BQ}}{V_T}(j\omega_1) & (j\omega_1) \end{bmatrix}^{-1} \begin{bmatrix} \frac{1}{R} \\ 0 \end{bmatrix} \tag{8.65}$$

For the higher orders, the GFRFs can be calculated similar to the calculation of the 1st order by following (8.46–8.47)

$$l_{1,1} = \frac{1}{R} + C(j\omega_1 + \cdots + j\omega_n),\ l_{1,2} = 1$$

$$l_{2,1} = -\frac{I_{BQ}}{V_T}(j\omega_1 + \cdots + j\omega_n),\ l_{2,2} = (j\omega_1 + \cdots + j\omega_n)$$

$$\begin{bmatrix} H_n^1(j\omega_1, \ldots, j\omega_n) \\ H_n^n(j\omega_1, \ldots, j\omega_n) \end{bmatrix} = \begin{bmatrix} \frac{1}{R} + C(j\omega_1 + \cdots + j\omega_n) & 1 \\ -\frac{I_{BQ}}{V_T}(j\omega_1 + \cdots + j\omega_n) & (j\omega_1 + \cdots + j\omega_n) \end{bmatrix}^{-1} \begin{bmatrix} 0 \\ A_{2,n} \end{bmatrix},\ n \geq 2 \tag{8.66}$$

where

$$A_{2,2} = \frac{1}{V_T}(j\omega_1)H_1^1(j\omega_1)H_1^2(j\omega_2),$$

$$A_{2,3} = \frac{1}{V_T}\left[(j\omega_1)H_1^1(j\omega_1)H_2^2(j\omega_2, j\omega_3) + (j\omega_1 + j\omega_2)H_2^1(j\omega_1, j\omega_2)H_1^2(j\omega_3)\right] \cdots$$

The calculation of the GFRFs for (8.63) is more efficient and straightforward than that by the Taylor series expansion in (8.62) because there exists only one nonlinear-related parameter, $1/V_T$, in (8.63), while at least two nonlinear-related parameters, I_{BQ}/V_T and $I_{BQ}/2V_T^2$, are involved in the calculation of GFRFs for the system approximated by the Taylor series expansion (8.62). Obviously, the computation efficiency using the proposed method is much better than the method using the Taylor series expansion.

In (8.63), it is clear that only one parameter, $-I_{BQ}/V_T$, is of interest to analyse. Denote $c = -I_{BQ}/V_T$ for ease of discussion. By using (8.65–8.66) and after some manipulations, the nCOS can be expressed as a polynomial function of c as

$$Y(j\omega) = \varphi_0(j\omega) + \varphi_1(j\omega)c + \varphi_2(j\omega)c^2 + \varphi_3(j\omega)c^3 + \varphi_4(j\omega)c^4 + \cdots \quad (8.67)$$

where the coefficients $\varphi_0(j\omega), \varphi_1(j\omega)\cdots$ are independent of the parameter c, which can be determined by following [9, 10] and partially given as follows

$$\begin{aligned}
&\varphi_0(j\omega) = 0, \\
&\varphi_1(j\omega) = -\frac{(j\omega)^2}{1 + RC(j\omega)} \\
&\quad - \int\limits_{\omega_1+\omega_2=\omega} \frac{j\omega_1}{V_T R^2} \frac{(j\omega_1 + j\omega_2)^2}{\left[1 + RC(j\omega_1 + j\omega_2)\right]^2} d\omega_1 d\omega_2 - \cdots, \\
&\cdots
\end{aligned}$$

The third-order harmonic distortion can be assessed by [45]

$$HD_3 = \frac{\varphi_0(3j\Omega) + \varphi_1(3j\Omega)c + \varphi_2(3j\Omega)c^2 + \varphi_3(3j\Omega)c^3 + \varphi_4(3j\Omega)c^4 + \cdots}{\varphi_0(j\Omega) + \varphi_1(j\Omega)c + \varphi_2(j\Omega)c^2 + \varphi_3(j\Omega)c^3 + \varphi_4(j\Omega)c^4 + \cdots} \quad (8.68)$$

where $\varphi_0(3j\Omega), \varphi_1(3j\Omega), \cdots$ are coefficients at the third-order harmonic frequency, and $\varphi_0(j\Omega), \varphi_1(j\Omega), \cdots$ are coefficients at frequency Ω which is the input frequency.

Linearization is a widely discussed topic in the literature for the analysis and design of amplifiers [44, 45]. Typically, an exponential nonlinearity is approximated using a third-order Taylor series. The Taylor series is truncated at the third order to simplify the analysis and design process, as it involves only three parameters. However, this simplification could lead to inaccuracies or incorrect results. For instance, in Fig. 8.11, the harmonic distortion produced by a third-order Taylor series significantly differs from that produced by numerical simulation. To enhance the accuracy of the result, the exponential nonlinearity should be approximated with a higher-order Taylor series expansion. However, this would increase the number of parameters involved in the analysis and design, thereby complicating the analysis and reducing computational efficiency. In the proposed method, the exponential-type nonlinear system can be analytically converted into a polynomial nonlinear system. This transformation produces fewer parameters for analysis and design (only one parameter $-I_{BQ}/V_T$ in this example), which will significantly facilitate the analysis and design of the nonlinear system.

From Fig. 8.11, it can be seen that based on the analytical transformation in the proposed method, the harmonic distortion via the nCOS function has good agreement with that by the numerical simulation. As a comparison, the harmonic distortion by the perturbation method proposed in [44] is also presented in Fig. 8.11. The result in Fig. 8.11 is studied with the following model parameters: $V_T = 26\text{mV}$, $\beta = 180$, $R = 100\Omega$, $C = 10\text{pF}$, and input frequency $\omega = 2\pi \times 5 \times 10^6 rad/\text{s}$.

B. **Summary**

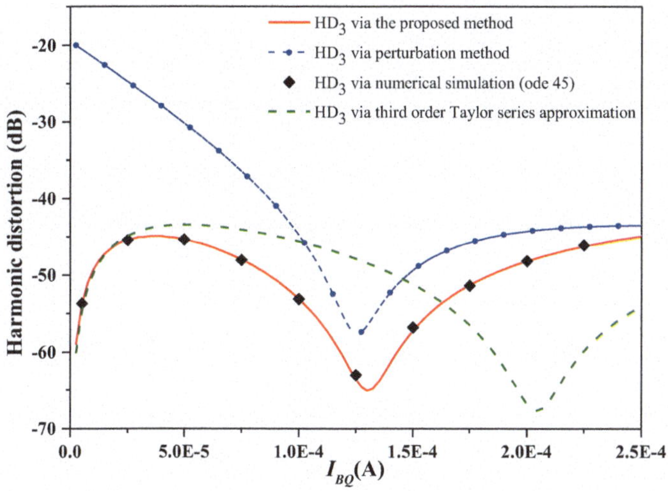

Fig. 8.11 Comparison of third-order harmonic distortion with different methods

This section explores a new method for deriving the Generalized Frequency Response Functions (GFRFs) of nonlinear systems with exponential-type nonlinearity. The proposed method has been demonstrated to be more computationally efficient and accurate compared to existing methods such as the perturbation method, Taylor series-based method, etc. The proposed algorithm enables the application of the nCOS method for the analysis and design of nonlinear systems with exponential nonlinearity in the frequency domain, offering valuable insights into the analysis and design of such nonlinearity using a novel frequency domain approach. Tasks such as feedback control, model component optimization, analysis, design, and understanding of nonlinear dynamics caused by exponential-type nonlinearity can be accomplished using the nCOS method in the frequency domain. These aspects will be further explored in future studies with specific applications in mind. While this study primarily focuses on Single Input Multiple Output (SIMO) nonlinear systems, the method can be readily extended to Multi-Input–Multi-Output (MIMO) systems, which will also be the subject of future studies.

Acknowledgements This chapter can also be referred to the published work in [29, 43] and thus for more discussions and detailed proof, please refer to the mentioned references.

Appendix

Coefficients of (8.28)

See Tables 8.1 and 8.2.

Table 8.1 Coefficients of (8.28)

$c^{1}_{1,0}(1;1) = (2m_t + c_t T_s)/D$	$c^{1}_{1,0}(2;1) = -m_t/D$
$c_{0,1}(0;1) = \left(k_t T_s^2 + c_t T_s\right)/D$	$c^{2}_{1,0}(1;2) = 2$
$c_{0,1}(1;1) = -c_t T_s/D$	$c^{2}_{1,0}(2;2) = -1$
$c^{4}_{1,0}(0;1) = -m_s T_s^2/D$	$c^{2}_{1,0}(2;4) = 1/T_s^2$
$c^{3}_{1,0}(0;2) = -k_{s1} T_s^2/m_s$	$c^{5}_{1,0}(0;2) = -c_s\, T_s^2/m_s$
$c^{33}_{2,0}(0,0;2) = -k_{s2} T_s^2/m_s$	$c^{2}_{1,0}(0;3) = 1$
$c^{333}_{2,0}(0,0,0;2) = -k_{s3} T_s^2/m_s$	$c^{1}_{1,0}(0;3) = -1$
$c^{444}_{3,0}(0,0,0;2) = -c_{s3} T_s^2/m_s$	$c^{2}_{1,0}(0;4) = -1/T_s^2$
$c^{55}_{2,0}(0,0;2) = -c_{s2} T_s^2/m_s$	$c^{2}_{1,0}(1;4) = -2/T_s^2$
$c^{3}_{1,0}(0;5) = 1/T_s$	$c^{3}_{1,0}(1;5) = -1/T_s$
$D = k_t^2 T_s^2 + c_s T_s + m_t$	$T_s = 1/2000$

Table 8.2 Coefficients of (8.35)

$$z_0(\mu_1,\mu_2;\omega) = \mu_1 \frac{|Y_{10}(j\omega)|^2}{|Y_{1l}(j\omega)|^2} + \mu_2 \frac{|Y_{20}(j\omega)|^2}{|Y_{2l}(j\omega)|^2} + (1-\mu_1-\mu_2)\frac{|Y_{30}(j\omega)|^2}{|Y_{3l}(j\omega)|^2}$$

$$z_1(\mu_1,\mu_2;\omega) = \mu_1 \frac{2\mathrm{Re}\left(Y_{10}(j\omega)Y_{11}^*(j\omega)\right)}{|Y_{1l}(j\omega)|^2} + \mu_2 \frac{2\mathrm{Re}\left(Y_{20}(j\omega)Y_{21}^*(j\omega)\right)}{|Y_{2l}(j\omega)|^2} + (1-\mu_1-\mu_2)\frac{2\mathrm{Re}\left(Y_{30}(j\omega)Y_{31}^*(j\omega)\right)}{|Y_{3l}(j\omega)|^2}$$

$$z_2(\mu_1,\mu_2;\omega) = \mu_1 \frac{|Y_{11}(j\omega)|^2 + 2\mathrm{Re}\left(Y_{10}(j\omega)Y_{12}^*(j\omega)\right)}{|Y_{1l}(j\omega)|^2} + \mu_2 \frac{|Y_{21}(j\omega)|^2 + 2\mathrm{Re}\left(Y_{20}(j\omega)Y_{22}^*(j\omega)\right)}{|Y_{2l}(j\omega)|^2} + (1-\mu_1-\mu_2)\frac{|Y_{31}(j\omega)|^2 + 2\mathrm{Re}\left(Y_{30}(j\omega)Y_{32}^*(j\omega)\right)}{|Y_{3l}(j\omega)|^2}$$

$$z_3(\mu_1,\mu_2;\omega) = \mu_1 \frac{2\mathrm{Re}\left(Y_{11}(j\omega)Y_{12}^*(j\omega)\right)}{|Y_{1l}(j\omega)|^2} + \mu_2 \frac{2\mathrm{Re}\left(Y_{21}(j\omega)Y_{22}^*(j\omega)\right)}{|Y_{2l}(j\omega)|^2} + (1-\mu_1-\mu_2)\frac{2\mathrm{Re}\left(Y_{31}(j\omega)Y_{32}^*(j\omega)\right)}{|Y_{3l}(j\omega)|^2}$$

$$z_4(\mu_1,\mu_2;\omega) = \mu_1 \frac{|Y_{12}(j\omega)|^2}{|Y_{1l}(j\omega)|^2} + \mu_2 \frac{|Y_{22}(j\omega)|^2}{|Y_{2l}(j\omega)|^2} + (1-\mu_1-\mu_2)\frac{|Y_{32}(j\omega)|^2}{|Y_{3l}(j\omega)|^2}$$

Coefficients of (8.35)

References

1. M. Zapateiro, F. Pozo, H. Karimi, and N. Luo, "Semiactive Control Methodologies for Suspension Control With Magnetorheological Dampers," Mechatronics, IEEE/ASME Transactions on, vol. 17, pp. 370-380, 2012.
2. W. Sun, H. Gao, and O. Kaynak, "Vibration Isolation for Active Suspensions With Performance Constraints and Actuator Saturation," Mechatronics, IEEE/ASME Transactions on, vol. PP, pp. 1-9, 2014.
3. W. Sun, H. Gao, and O. Kaynak, "Adaptive Backstepping Control for Active Suspension Systems With Hard Constraints," Mechatronics, IEEE/ASME Transactions on, vol. 18, pp. 1072-1079, 2013.
4. M. Fallah, R. Bhat, and W. Xie, "Optimized Control of Semiactive Suspension Systems Using H infinity Robust Control Theory and Current Signal Estimation," Mechatronics, IEEE/ASME Transactions on, vol. 17, pp. 767-778, 2012.
5. W. Kemmetmuller, K. Holzmann, A. Kugi, and M. Stork, "Electrorheological Semiactive Shock Isolation Platform for Naval Applications," Mechatronics, IEEE/ASME Transactions on, vol. 18, pp. 1437-1447, 2013.
6. M. Hosek and N. Olgac, "A single-step automatic tuning algorithm for the delayed resonator vibration absorber," Mechatronics, IEEE/ASME Transactions on, vol. 7, pp. 245-255, 2002.
7. A. Nayfeh and D. Mook, Nonlinear oscillations: Wiley-VCH, 2008.
8. D. George, "Continuous nonlinear systems," Technical report (Massachusetts Institute of Technology. Research Laboratory of Electronics) ; 355., 1959.
9. X. Jing, "Nonlinear Characteristic Output Spectrum for Nonlinear Analysis and Design," Mechatronics, IEEE/ASME Transactions on, vol. 19, pp. 171-183, 2014.
10. X. Jing, Z. Lang, and S. Billings, "Parametric Characteristic Analysis for Generalized Frequency Response Functions of Nonlinear Systems," Circuits, Systems, and Signal Processing, vol. 28, pp. 699-733, 2009.
11. X. Jing, Z. Lang, and S. Billings, "Output frequency properties of nonlinear systems," International Journal of Non-Linear Mechanics, vol. 45, pp. 681-690, 2010.
12. S. Billings and J. Peyton, "Mapping non-linear integro-differential equations into the frequency domain," International Journal of Control, vol. 52, pp. 863-879, 1990.
13. O. Boaghe and S. Billings, "Subharmonic oscillation modeling and MISO Volterra series," Circuits and Systems I: Fundamental Theory and Applications, IEEE Transactions on, vol. 50, pp. 877-884, 2003.
14. X. Jing, Z. Lang, and S. Billings, "Nonlinear influence in the frequency domain: Alternating series," Systems & Control Letters, vol. 60, pp. 295-309, 2011.
15. Z. Xiao, X. Jing, and L. Cheng, "The transmissibility of vibration isolators with cubic nonlinear damping under both force and base excitations," Journal of Sound and Vibration, vol. 332, pp. 1335-1354, 2013.
16. Z. Lang and S. Billings, "Output frequency characteristics of nonlinear systems," International Journal of Control, vol. 64, pp. 1049-1067, 1996.
17. Z. Lang, S. Billings, R. Yue, and etc., "Output frequency response function of nonlinear Volterra systems," Automatica, vol. 43, pp. 805–816, 2007.
18. S. Boyd and L. Chua, "Fading memory and the problem of approximating nonlinear operators with Volterra series," Circuits and Systems, IEEE Transactions on, vol. 32, pp. 1150-1161, 1985.
19. I. Sandberg, "Expansions for nonlinear systems," Bell Syst. Tech. J, vol. 61, pp. 159-199, 1982.
20. L. Li and S. Billings, "Analysis of nonlinear oscillators using Volterra series in the frequency domain," Journal of Sound and Vibration, vol. 330, pp. 337-355, 2011.
21. Z. Peng and Z. Lang, "On the convergence of the Volterra-series representation of the Duffing's oscillators subjected to harmonic excitations," Journal of Sound and Vibration, vol. 305, pp. 322-332, 2007.

22. G. Tomlinson, G. Manson, and G. Lee, "A Simple Criterion for Establishing an Upper Limit to the Harmonic Excitation Level of the Duffing Oscillator Using the Volterra Series," Journal of Sound and Vibration, vol. 190, pp. 751-762, 1996.
23. T. Helie and B. Laroche, "Computation of Convergence Bounds for Volterra Series of Linear-Analytic Single-Input Systems," Automatic Control, IEEE Transactions on, vol. 56, pp. 2062-2072, 2011.
24. W. Sun, Z. Zhao, and H. Gao, "Saturated Adaptive Robust Control for Active Suspension Systems," Industrial Electronics, IEEE Transactions on, vol. 60, pp. 3889-3896, 2013.
25. R. Amirifar and N. Sadati, "Low-order H infinity controller design for an active suspension system via LMIs," Industrial Electronics, IEEE Transactions on, vol. 53, pp. 554-560, 2006.
26. H. Gao, W. Sun, and P. Shi, "Robust Sampled-Data H infinity Control for Vehicle Active Suspension Systems," Control Systems Technology, IEEE Transactions on, vol. 18, pp. 238-245, 2010.
27. W. Sun, Y. Zhao, J. Li, L. Zhang, and H. Gao, "Active Suspension Control With Frequency Band Constraints and Actuator Input Delay," Industrial Electronics, IEEE Transactions on, vol. 59, pp. 530-537, 2012.
28. A. Swain and S. Billings, "Generalized frequency response function matrix for MIMO non-linear systems," International Journal of Control, vol. 74, pp. 829-844, 2001.
29. Z Xiao, X Jing, An SIMO Nonlinear System Approach to Analysis and Design of Vehicle Suspensions, IEEE/ASME Trans on Mechatronics 20 (6), 3098 – 3111, 2015
30. Z. Xiao, X. Jing, and L. Cheng, "Parameterized Convergence Bounds for Volterra Series Expansion of NARX Models," Signal Processing, IEEE Transactions on, vol. 61, pp. 5026-5038, 2013.
31. X. Jing, Z. Lang, and S. Billings, "New bound characteristics of NARX model in the frequency domain," International Journal of Control, vol. 80, pp. 140-149, 2006.
32. Z. Xiao, X. Jing, and L. Cheng, "Estimation of parametric convergence bounds for Volterra series expansion of nonlinear systems," Mechanical Systems and Signal Processing, vol. 45, pp. 28-48, 2014.
33. C. Kim and P. Ro, "A sliding mode controller for vehicle active suspension systems with non-linearities," Proceedings of the Institution of Mechanical Engineers, Part D: Journal of Automobile Engineering, vol. 212, pp. 79-92, 1998.
34. H Li, X Jing, HK Lam, P Shi, Fuzzy sampled-data control for uncertain vehicle suspension systems, IEEE Transactions on Cybernetics 44 (7), 1111 – 1126, 2014
35. S. Savaresi, E. Silani, and S. Bittanti, "On performance evaluation methods and control strategies for semi-active suspension systems," in Decision and Control, 2003. Proceedings. 42nd IEEE Conference on, 2003, pp. 2264–2269 Vol.3.
36. S. Savaresi, E. Silani, and S. Bittanti, "Acceleration-Driven-Damper (ADD): An optimal control algorithm for comfort-oriented semiactive suspensions," Journal of dynamic systems, measurement, and control, vol. 127, pp. 218-229, 2005.
37. S. Savaresi and C. Spelta, "Mixed sky-hook and ADD: Approaching the filtering limits of a semi-active suspension," Journal of dynamic systems, measurement, and control, vol. 129, pp. 382-392, 2007.
38. M. Milanese and C. Novara, "Structured Set Membership identification of nonlinear systems with application to vehicles with controlled suspension," Control Engineering Practice, vol. 15, pp. 1-16, 2007.
39. H. Li, J. Yu, C. Hilton, and H. Liu, "Adaptive Sliding-Mode Control for Nonlinear Active Suspension Vehicle Systems Using TS Fuzzy Approach," Industrial Electronics, IEEE Transactions on, vol. 60, pp. 3328-3338, 2013.
40. Hongyi Li, Honghai Liu, Huijun Gao and Peng Shi, Reliable Fuzzy Control for Active Suspension Systems with Actuator Delay and Fault, IEEE Transactions on Fuzzy Systems, vol. 20, no. 2, pp. 342-357, 2012.
41. R. Lian, "Enhanced Adaptive Self-Organizing Fuzzy Sliding-Mode Controller for Active Suspension Systems," Industrial Electronics, IEEE Transactions on, vol. 60, pp. 958-968, 2013.

42. S. Chen and S. Billings, “Representations of non-linear systems: the NARMAX model,” International Journal of Control, vol. 49, pp. 1013-1032, 1989.
43. Z Xiao, X Jing, Frequency domain analysis of exponential-type nonlinearity, 2016 American Control Conference (ACC), 7466–7471, 2016
44. Buonomo, A., & Schiavo, A. L. Perturbation analysis of nonlinear distortion in analog integrated circuits. Circuits and Systems I: Regular Papers, IEEE Transactions on, 52, 1620-1631, 2005
45. Sira, D., & Larsen, T. Process, Voltage and Temperature Compensation Technique for Cascode Modulated PAs. Circuits and Systems I: Regular Papers, IEEE Transactions on, 60, 2511-2520, 2013

Chapter 9
Case Studies: Optimization of Fuzzy Membership with the nCOS Function Method

Abstract This chapter addresses the challenge of optimizing the parameters of membership functions in fuzzy-model-based controllers, particularly under conditions of imperfect premise matching. A novel frequency-domain algorithm is proposed, which establishes an explicit relationship between membership function parameters and the desired control performance. This development facilitates a systematic optimization strategy grounded in the nonlinear characteristic output spectrum (nCOS) function. In contrast to conventional search-based optimization techniques, the proposed method offers a more efficient and insightful approach, delivering comprehensive results within a significantly reduced computational timeframe. Moreover, it provides a deeper understanding of the underlying nonlinear dynamics, extending beyond the pursuit of optimal performance alone. The application of this method leads to substantial improvements in the effectiveness of fuzzy-model-based controllers.

Keyword Fuzzy systems · Membership optimization · Frequency domain · nCOS function

9.1 Introduction

9.1.1 About This Chapter

With the SIMO nonlinear system approach and the GFRF calculation methods for exponential nonlinear systems, many nonlinear systems including neural networks and fuzzy systems can be treated easily with these methods with a unique optimization approach and novel insight into understanding of these nonlinear dynamics.

In this chapter, we look into the problem of optimizing the parameters of membership functions in fuzzy-model-based controllers, particularly when the premise matching is not perfect. We introduce a new algorithm in the frequency domain that establishes a clear connection between the parameters of the membership functions and the desired control performance. This leads to the development of an

X. Jing, *The Bio-Inspired X-Structure/Mechanism Approach for Exploring Nonlinear Benefits in Engineering*, https://doi.org/10.1007/978-981-96-5769-8_9

optimization strategy for these parameters, grounded in the nonlinear characteristic output spectrum function (nCOS). Unlike the conventional search-based optimization methods, this novel approach yields more comprehensive results in a shorter time frame, providing a deeper insight into the nonlinear effects beyond merely achieving optimal outcomes. Employing this innovative technique significantly improves the performance of the fuzzy-model-based controller. Ultimately, we apply the optimization method for fuzzy membership functions to nonlinear systems, resulting in enhanced system performance.

9.1.2 Research Background

Sugeno (T-S) fuzzy systems are capable of replicating any smooth nonlinear systems with high precision, presenting an alternative strategy for addressing nonlinear systems [1]. There is a substantial body of research on fuzzy control, as evidenced by numerous studies [2–8]. The concept of enhancing the design flexibility and robustness of a fuzzy controller by employing different membership functions from those in the T-S fuzzy model, known as imperfect premise matching, was initially suggested in Lam and Narimani [9]. This approach introduces arbitrary slack matrices to reduce the conservative nature of the design. The field of fuzzy control with imperfect premise variables has seen notable advancements in recent years [10–12]. When designing a mismatched fuzzy controller, the mismatched membership functions only need to meet a specific design criterion [9]. However, the parameters related to the mismatched controller membership functions are often chosen in a subjective or random fashion, which plays a significant role in attaining superior control performance. To date, there has been no systematic and thorough investigation into how the parameters of mismatched membership functions impact the performance of the closed-loop system within the existing design framework. It's also important to note that many studies on fuzzy-model-based control or filtering with imperfect premise matching prefer Gaussian-shaped functions for their mismatched membership functions due to their smoothness and simple representation [13–16]. Therefore, research into the optimization of parameters for mismatched membership functions, particularly Gaussian-type functions, holds significant theoretical and practical value.

Traditionally, the optimal parameters for mismatched membership functions are identified using optimization techniques that measure the closed-loop system performance directly, such as the genetic algorithm [17–19] and the swarm optimization algorithm [20, 21]. These intelligent optimization methods typically involve a global search within the available parameter space, leading to longer search times as the parameter space expands. This becomes particularly cumbersome for systems with complex dynamics and multiple dimensions, as it greatly increases computational complexity and hinders practical analysis and design. To address this issue, a new frequency domain analysis method utilizing the nonlinear Characteristic Output Spectrum (nCOS) function is introduced. This method, rooted in the Volterra series expansion theory [22], broadens the transfer function concept to encompass nonlinear systems. It offers a detailed and complete insight into the impact of nonlinear parameters on system responses [23–25]. Notable achievements have been made using the nCOS function-based method, including the analysis of vehicle suspensions with nonlinear damping and stiffness [26, 27]. In contrast to search-based optimization methods, the nCOS function-based approach precisely illustrates the influence of membership function parameters on the closed-loop system performance. Thanks to its analytical and clear expression of the nonlinear characteristic output spectrum and the parameters of mismatched membership functions, the nCOS function-based optimization method is significantly more time-efficient, even when dealing with a larger parameter space, compared to traditional intelligent optimization techniques.

The rest of the chapter is organized as follows. System description and problem formulation are given in Sect. 9.2. The nCOS function based optimization algorithm and genetic algorithm (GA) based optimization method are presented in Sect. 9.3. Applications of the proposed optimization method to nonlinear systems are given in Sect. 9.4. Conclusions are drawn in Sect. 9.5.

9.2 Problem Formulation

This section will focus on constructing the T-S fuzzy model for a nonlinear system and developing a fuzzy controller that incorporates membership functions with intentional discrepancies.

Consider a T-S fuzzy model in the following form:

Plant Rule i: IF $\theta_1(t)$ is μ_{i1} and $\theta_2(t)$ is μ_{i2} and $\cdots$ and $\theta_p(t)$ is μ_{ip}, then

$$\begin{cases} \dot{x}(t) = A_i x(t) + B_i u(t) + B_{\omega i}\omega(t) \\ z(t) = C_i x(t) + D_i u(t) \end{cases} \tag{9.1}$$

where $\mu_{i1}, \mu_{i2}, \ldots, \mu_{ip}$ are fuzzy sets, A_i, B_i, $B_{\omega i}$, C_i and D_i, $i = 1, 2, \ldots, r$, are constant matrices with compatible dimensions, $\theta(t) = [\theta_1(t), \theta_2(t), \ldots, \theta_p(t)]$ is the premise variable vector, and r is the number of fuzzy rules. The overall fuzzy system is obtained as

$$\begin{cases} \dot{x}(t) = \sum_{i=1}^{r} h_i(\theta(t))[A_i x(t) + B_i u(t) + B_{\omega i}\omega(t)] \\ z(t) = \sum_{i=1}^{r} h_i(\theta(t))[C_i x(t) + D_i u(t)] \end{cases} \tag{9.2}$$

where $h_i(\theta(t)) \geq 0$, $\sum_{i=1}^{r} h_i(\theta(t)) = 1$.

In this study, fuzzy controller that shares different premise variables with the fuzzy system, is adopted:

Controller Rule j: IF $\eta_1(t)$ is N_{j1} and $\eta_2(t)$ is N_{j2} and $\cdots$ and $\eta_p(t)$ is N_{jp}, THEN

$$u(t) = K_j x(t), \tag{9.3}$$

where $j = 1, 2, \ldots, r$. The overall mismatched fuzzy controller can be written as:

$$u(t) = \sum_{j=1}^{r} m_j(\eta(t))K_j x(t) \tag{9.4}$$

$\eta(t) = \left[\eta_1(t), \eta_2(t), \ldots, \eta_p(t)\right]$ is the imperfect premise variable vector. Then the closed-loop fuzzy system can be represented as

$$\begin{cases} \dot{x}(t) = \sum_{i=1}^{r}\sum_{j=1}^{r} h_i(\theta(t))m_j(\eta(t))\left[\overline{A}_{ij}x(t) + \overline{B}_{ij}\omega(t)\right] \\ z(t) = \sum_{i=1}^{r}\sum_{j=1}^{r} h_i(\theta(t))m_j(\eta(t))\overline{C}_{ij}x(t) \end{cases} \tag{9.5}$$

where $\overline{A}_{ij} = A_i + B_i K_j$, $\overline{C}_{ij} = C_i + D_i K_j$, $\overline{B}_{ij} = B_{\omega i}$, $m_j(\eta(t)) \geq 0$, $\sum_{j=1}^{r} m_j(\eta(t)) = 1$. For simple description, we denote $h_i(\theta(t)) = h_i$ and $m_j(\eta(t)) = m_j$.

In the context of T-S fuzzy control featuring mismatched premise variables, a range of membership functions can be utilized to improve design versatility and/or reduce the fuzzy controller's structural intricacy, provided that certain criteria [9] are met:

$$m_j - \lambda_j h_j \geq 0, 1 > \lambda_j > 0, j = 1, \ldots, r. \tag{9.6}$$

9.3 Optimization of Mismatched Memberships Based on the nCOS Function

In this section, we utilize the nCOS function to fine-tune the parameters of the membership functions for the mismatched controller. Additionally, we will present some fundamental definitions.

9.3.1 Nonlinear Output Spectrum

Based on the Volterra series expansion theory, a system characterized by a single-input multiple-output (SIMO) nonlinear differential equation can be depicted in the following manner [28]:

$$\begin{cases} y_i(t) = \sum\limits_{n=1}^{N} y_i^n(t) \\ y_i^n(t) = y_i^0(t) + \int\limits_{-\infty}^{\infty} \cdots \int\limits_{-\infty}^{\infty} h_i^n(\tau_1, \ldots, \tau_n) \prod\limits_{k=1}^{n} u(t - \tau_k) d\tau_k \end{cases} \tag{9.7}$$

where $y_i(t)$ and $u(t)$ are the ith subsystem output and input of the system, respectively, N is the maximum order of the system nonlinearity, $h_i^n(\tau_1, \ldots, \tau_n)$ is the nth order Volterra kernel, y_i^0 is zero order output or the direct current constant. Then the nth order general frequency response function (GFRF) are defined [22] as

$$H_i^n(jw_1, \ldots, jw_n) = \int\limits_{-\infty}^{\infty} \cdots \int\limits_{-\infty}^{\infty} h_i^n(\tau_1, \ldots, \tau_n) e^{-j(w_1\tau_1 + \cdots + w_n\tau_n)} d\tau_1 \cdots d\tau_n \tag{9.8}$$

Then according to Jing [29], Eq. 9.1 can be rewritten as

$$\begin{aligned} y_i(t) &= \sum_{n=1}^{N} y_i^n(t) \\ &= H_i^0 + \frac{1}{2\pi} \int\limits_{-\infty}^{\infty} \left[\frac{1}{(2\pi)^{n-1}} \int\limits_{-\infty}^{\infty} \cdots \int\limits_{-\infty}^{\infty} \Xi_i^n dw_1 \cdots dw_{n-1} \right] e^{jwt} dw \end{aligned} \tag{9.9}$$

where $U_n(jw_1, \ldots, jw_{n-1}) = U(jw_1) \cdots U(jw_{n-1}) U(j(w - w_1 - \cdots - w_{n-1}))$, $\Xi_i^n = H_i^n(jw_1, \ldots, jw_{n-1}, j(w - w_1 - \cdots - w_{n-1})) U_n(jw_1, \ldots, jw_{n-1})$, $w_n = w - w_1 - \cdots - w_{n-1}$, $U(jw)$ is the input spectrum, H_i^0 is associated with the 0th order output y_i^0, which is independent of input. Output y_i^n in the frequency domain can be written as

$$Y_i^n(jw) = \frac{1}{(2\pi)^{n-1}} \int\limits_{-\infty}^{\infty} \cdots \int\limits_{-\infty}^{\infty} \Xi_i^n dw_1 \cdots dw_{n-1} \tag{9.10}$$

According to Eq. 9.9, output response of the ith subsystem is computed by

$$Y_i(jw) = \sum_{n=1}^{N} Y_i^n(jw). \tag{9.11}$$

9.3.2 Computation of the nth Order nCOS

The nCOS, up to the nth degree, can be determined analytically from simulation or empirical data through the application of the least squares technique [23–25]. Considering input $\rho U(jw)$ with magnitude ρ, the nonlinear output spectrum in Eq. 9.10 is presented as

$$\begin{aligned} Y_i(jw)_\rho &= \sum_{n=1}^{\infty} \rho^n Y_i^n(jw) \\ &= \sum_{i=1}^{N} \rho^n Y_i^n(jw) + \sigma_{[N]}(jw) \end{aligned} \tag{9.12}$$

where $\sigma_{[N]}(jw)$ is the truncation error. To calculate $Y_i(jw)_\rho$, we can excite the system with $\rho_p U(jw)$, where $p = 1, 2, \ldots, N$, correspondingly, output $Y_i(jw)_{\rho_1}, Y_i(jw)_{\rho_2}, \ldots, Y_i(jw)_{\rho N}$ can be obtained. Then the nth order output frequency response $Y_i^n(jw)$, $n \in \{1, 2, \ldots, N\}$ is calculated as

$$\begin{bmatrix} Y_i^1(jw) \\ Y_i^2(jw) \\ \vdots \\ Y_i^N(jw) \end{bmatrix} = \begin{bmatrix} \rho_1 & \rho_1^2 & \cdots & \rho_1^N \\ \rho_2 & \rho_2^2 & \cdots & \rho_2^N \\ \vdots & \vdots & \ddots & \vdots \\ \rho_N & \rho_N^2 & \cdots & \rho_N^N \end{bmatrix}^{-1} \begin{bmatrix} Y_i(jw)_{\rho_1} \\ Y_i(jw)_{\rho_2} \\ \vdots \\ Y_i(jw)_{\rho_N} \end{bmatrix} \tag{9.13}$$

The square matrix is nonsingular if $\rho_1 \neq \rho_2 \neq \cdots \neq \rho_N$. Excitation magnitude ρ is a key factor to successfully apply aforementioned output spectrum estimated method. A ρ-selection method in Jing [25] can be adopted. U-shaped $\left|Y_i^n(jw)\right| \infty \rho$ curve can be obtained with a series of ρ, which starts from a small value. The optimal ρ should locate around the bottom of the curve which corresponds to the minimum estimation error for $\left|Y_i^n(jw)\right|$.

9.3.3 Optimization of Fuzzy Membership Functions' Parameters by nCOS Function

When utilizing a mismatched fuzzy controller with Gaussian-shaped membership functions, the closed-loop T-S fuzzy system can be redefined as a general exponential-type nonlinear system [30]. This is achieved by integrating the fuzzy controller (Eq. 9.4) into the original nonlinear plant:

$$\phi_1(\tilde{x}, \tilde{\omega}) + \phi_2(\tilde{x}, \tilde{\omega}) e^{\varphi(\tilde{x}, \tilde{\omega})} = \sum_{l=0}^{L} d^l \omega / dt^l \tag{9.14}$$

where

$$\tilde{x} = \left\{x, \frac{d^1x}{dt}, \frac{d^2x}{dt^2}, \ldots, \frac{d^Lx}{dt^L}\right\}, \tilde{\omega} = \left\{\omega, \frac{d^1\omega}{dt}, \frac{d^2\omega}{dt^2}, \ldots, \frac{d^L\omega}{dt^L}\right\} \tag{9.15}$$

and $\phi_1(\tilde{x}, \tilde{\omega})$, $\phi_2(\tilde{x}, \tilde{\omega})$ and $\varphi(\tilde{x}, \tilde{\omega})$ are polynomial functions of system state $\tilde{x}$ and input $\tilde{\omega}$, l is the differential order with maximum order L. Subsequently, optimizing the parameters of the membership functions for the mismatched controller is tantamount to refining the nonlinear parameters in Eq. (9.14) that are associated with Gaussian membership functions.

Optimization process of Gaussian membership function is summarized as Algorithm 1:

Algorithm 1: Optimization of Gaussian membership function parameters for the closed-loop fuzzy system.

1. Define new state variables $y_1 = \tilde{x}$ and $y_2 = e^{\varphi(\tilde{x},\tilde{\omega})}$, then by taking derivative of y_2 with respect to time, the following equation can be obtained:

$$\frac{dy_2}{dt} = e^{\varphi(\tilde{x},\tilde{\omega})}\frac{d\varphi(\tilde{x},\tilde{\omega})}{dt} = y_2\frac{d\varphi(y_1,\tilde{\omega})}{dt} \tag{9.16}$$

Then we can equivalently transform nonlinear system (Eq. 9.14) into the following SIMO polynomial nonlinear system with same initial conditions of (Eq. 9.14) and $y_2(0) = e^{\varphi(\hat{x}(0),\hat{\omega}(0))}$:

$$\begin{cases} \phi_1(y_1, \tilde{\omega}) + \phi_2(y_1, \tilde{\omega})y_2 = \sum_{l=0}^{L} d^l\omega/dt^l \\ \frac{dy_2}{dt} = y_2\, d\varphi(y_1, \tilde{\omega})/dt \end{cases} \tag{9.17}$$

2. Based on the probing method [28], assume the input is $\omega(t) = e^{j(w_1+\cdots+w_n)t}$, system state of (Eq. 9.17) is expressed as

$$y_i = H_i^0 + \sum_{\substack{\text{all combinations} \\ \text{of } w_{i1} \text{ in } w}} H_i^1(w_{i1})e^{jw_{i1}t} + \cdots + \sum_{\substack{\text{all permutations} \\ \text{of } (w_{i1,\ldots,}w_{in})}} \sum_{\substack{\text{all combinations} \\ \text{of } w_{i1} \text{ in } w}} H_i^n\left(w_{i1,\ldots,}w_{in}\right)e^{j(w_{i1}+\cdots+w_{in})t}$$

where $w = (w_1, \ldots, w_n)$. Substitute the equation above into (Eq. 9.17) and equate the coefficients of $e^{j(w_1+\cdots+w_n)t}$ to zero, the explicit formula of the nth order GFRFs can be obtained. Substitute GFRFs into (Eq. 9.10) with some manipulation, the output spectrum can be obtained.

3. With estimated $Y_i(jw)_\rho$ from (Eq. 9.13), the unknown frequency coefficients regarding to w in nCOS function are determined by using the least squares

method. Then the relationship between the nonlinear parameters that we are interested in and the output spectrum of (Eq. 9.14) can thus be obtained, which can clearly demonstrate the relationship between membership function parameters and system frequency response. The optimal parameters of membership function for the mismatched controller can be obtained where the output spectrum is minimized.

9.3.4 GA-Based Optimization of Mismatched Memberships

An alternative approach to solving the parameter optimization issue for the T-S fuzzy system, which employs Gaussian-shaped mismatched membership functions, is to apply a search-based optimization technique. In this research, Genetic Algorithms (GA) are utilized to confirm the reliability and efficiency of the optimization method based on the nCOS function within the frequency domain. The goal is to identify the best parameters within the fuzzy membership functions that will result in the minimization of the output frequency response $|Y(jw)|$. The process of GA optimization is summarized in Algorithm 2.

Algorithm 2: Optimization of fuzzy membership function parameters using GA.

(1) Specify eligible parameter bounds for Gaussian membership function.
(2) Use binary string to encode the parameters and generate an initial population of N chromosomes randomly.
(3) Decode the initial population into the practical values of the system's parameters.
(4) For each set of parameters, calculate the output frequency response $|Y(jw)|$ of the closed-loop fuzzy system.
(5) Perform crossover with crossover rate C and do bit mutation with mutation rate M.
(6) Retain the best chromosomes in the population.
(7) If the convergent conditions are satisfied, stop and output the result. If not, repeat Step (4)–(7).

9.4 Simulation Results

This section presents simulation outcomes to assess the effectiveness and advantages of the optimization method that relies on the nCOS function.

Consider the following nonlinear system in [31]:

$$\ddot{\xi}(t) = -0.02\xi(t) - 0.67\xi(t)^3 + u(t) + \omega(t) \tag{9.18}$$

Assume $\xi(t) \in [-d, d]$, $d \geq 0$, $x_1(t) = \xi(t)$ and $x_2(t) = \dot{\xi}(t)$, denote $x(t) = [x_1(t)\ \ x_2(t)]^T$,then the nonlinear system can be represented on $[-d, d]$ by the following fuzzy system

$$\begin{cases} \dot{x}(t) = \sum_{i=1}^{2} h_i(x(t))[A_i x(t) + B_i u(t) + B_{\omega i}\omega(t)] \\ z(t) = \sum_{i=1}^{2} h_i(x(t))C_i x(t) \end{cases} \tag{9.19}$$

where $h_1(x(t)) = 1 - \frac{x_1(t)^2}{d^2}$, $h_2(x(t)) = \frac{x_1(t)^2}{d^2}$, $d = 2$,

$$\begin{aligned} A_1 &= \begin{bmatrix} 0 & 1 \\ -0.02 & 0 \end{bmatrix}, A_2 = \begin{bmatrix} 0 & 1 \\ -0.02 - 0.67d^2 & 0 \end{bmatrix} \\ B_1 &= B_2 = \begin{bmatrix} 0 & 1 \end{bmatrix}^T, B_{w1} = B_{w2} = \begin{bmatrix} 0 & 1 \end{bmatrix}^T, \\ C_1 &= \begin{bmatrix} 0 & 0.5 \end{bmatrix}, C_2 = \begin{bmatrix} 0 & 0.5 \end{bmatrix} \end{aligned} \tag{9.20}$$

Controller membership functions are chosen as

$$m_1(x_1) = 1 - ae^{-(x_1-b)^2/2c^2}, m_2(x_1) = ae^{-(x_1-b)^2/2c^2} \tag{9.21}$$

where $a = 0.95$, $b \in [-6, 6]$, $c \in [0.84, 6]$, $w_1 = 0.01$, $w_2 = 1$ with permissible parameters $\lambda_1 = 0.05$ and $\lambda_2 = 0.0011$. Controller gains can be written as $K_1 = \begin{bmatrix} k_{11} & k_{12} \end{bmatrix}$, $K_2 = \begin{bmatrix} k_{21} & k_{22} \end{bmatrix}$, so the overall controller is

$$\begin{aligned} u(t) = \sum_{i=1}^{2} m_i(x(t))K_i x(t) &= k_{11}x_1(t) + k_{12}x_2(t) \\ &+ [(k_{21} - k_{11})x_1(t) + (k_{22} - k_{12})x_2(t)]ae^{-(x_1(t)-b)^2/2c^2} \end{aligned} \tag{9.22}$$

Assume initial condition $x(0) = 0$, $\dot{x}(0) = 0$, and fuzzy controller gains are given as $K_1 = \begin{bmatrix} -129.1940 & -334.9528 \end{bmatrix}$, $K_2 = \begin{bmatrix} -129.8766 & -339.4278 \end{bmatrix}$. By substituting (9.22) into (9.18), the closed-loop system is obtained as

$$\begin{aligned} \ddot{\xi}(t) = n_1\xi(t) + n_2\dot{\xi}(t) + n_3\xi(t)e^{-\frac{(\xi(t)-b)^2}{2c^2}} \\ + n_4\dot{\xi}(t)e^{-\frac{(\xi(t)-b)^2}{2c^2}} + n_5\xi(t)^3 + \omega(t) \end{aligned} \tag{9.23}$$

where $n_1 = k_{11} - 0.02$, $n_2 = k_{12}, n_3 = a(k_{21} - k_{11}), n_4 = a(k_{22} - k_{12})$, $n_5 = -0.67$.

It is obvious that closed-loop system (9.23) is a nonlinear system with exponential type nonlinearity: $e^{-(\xi(t)-b)^2/2c^2}$. Then following the steps in Algorithm 1, the relationship between system nCOS function and parameters b and c can be analytically calculated as follows.

First define new variables $y_1(t) = \xi(t)$ and $y_2(t) = e^{-(\xi(t)-b)^2/2c^2}$, the closed-loop system can be equivalently transformed into following polynomial nonlinear system

$$\begin{cases} \ddot{y}_1(t) = n_1 y_1(t) + n_2 \dot{y}_1(t) + n_3 y_1(t) y_2(t) \\ \qquad +n_4 \dot{y}_1(t) y_2(t) + n_5 y_1(t)^3 + \omega(t) \\ \dot{y}_2(t) = -\frac{1}{c^2} y_1(t) \dot{y}_1(t) y_2(t) + \frac{b}{c^2} \dot{y}_1(t) y_2(t) \end{cases} \tag{9.24}$$

Following the Step 2 of Algorithm 1, each order GFRF $H_i^n(jw)$ can be calculated as

$$\begin{aligned} H_1^0 &= 0, \\ H_2^0 &= e^{-b^2/2c^2}, \\ H_2^1(jw_1) &= H_2^0 H_1^1(jw_1), \\ H_1^1(jw_1) &= 1/\left[H_2^0(-m_4 jw_1 - m_3) - \left(m_2 jw_1 + w_1^2 + m_1\right)\right], \\ H_1^2(jw_1, jw_2) &= \frac{(m_4 jw_1 + m_3) H_1^1(jw_1) H_2^1(jw_2)/H_2^0}{\left[-m_4 j\varpi - m_3\right] - \left[m_2 j\varpi + \varpi^2 + m_1\right]}, \\ H_2^2(jw_1, jw_2) &= -\frac{H_1^1(jw_1)\left[H_2^0 jw_1 H_1^1(jw_2) + bjw_1 H_2^1(jw_2)\right]}{c^2 j\varpi} \\ &\quad + bH_2^0 H_1^2(jw_1, jw_2)/c^2 \\ &\quad \vdots \end{aligned} \tag{9.25}$$

where $\varpi = w_1 + w_2$. Choose $y(t) = \ddot{\xi}(t)$ as the output. Input is $\omega(t) = \rho sin(wt)$ and denote $f = -\frac{1}{c^2}$, $g = \frac{b}{c^2}$. Substitute $H_i^n(jw)$ into (9.10) with some manipulations, the output spectrum up to third order can be calculated as

$$Y_1(jw) = \rho Y_1^1(jw_1) + \rho^2 Y_1^2(jw_1, jw_2) + \rho^3 Y_1^3(jw_1, jw_2, jw_3) + \cdots \tag{9.26}$$

where

$$\begin{aligned} Y_1^1(jw_1) &= 1/\left[\varphi_1^1(jw_1) H_2^0 + \varphi_1^0(jw_1)\right], \\ Y_1^2(jw_1, jw_2) &= H_2^0/\left[\sum_{i=0}^{3} \varphi_2^i(jw_1, jw_2)\left(H_2^0\right)^i\right], \\ Y_1^3(jw_1, jw_2, jw_3) &= \frac{1}{\sum_{i=0}^{6} \varphi_3^i(w_1, w_2, w_3)(H_2^0)^i} \\ &\quad + \frac{\sum_{i=1}^{3}\left[\varphi_{3f}^i(w_1, w_2, w_3) f + \varphi_{3g}^i(w_1, w_2, w_3) g + \varphi_{3c}^i(w_1, w_2, w_3)\right](H_2^0)^i}{\sum_{i=0}^{6} \varphi_3^i(w_1, w_2, w_3)(H_2^0)^i} \end{aligned} \tag{9.27}$$

The coefficients $\varphi(\cdot)$ are independent of f and g and can be determined by using least square method.

According to the ρ-selection method aforementioned, the best excitation magnitude should be 0.6. Then choose input as $\omega(t) = 0.6\sin(\pi \cdot t)$, coefficients in output spectrum $Y_1(jw)$ are calculated as shown in Algorithm 1 and listed in Table 9.1.

To confirm the effectiveness of the suggested algorithm, we compare the output frequency response obtained from simulations with that derived from the nCOS function method, as illustrated in Fig. 9.1. The comparison reveals a strong correlation between the output spectrum computed using the nCOS function-based approach and the results from numerical simulations.

Output spectrum in terms of b and c is shown in Fig. 9.2. It can be seen from Fig. 9.2 that parameters b and c in the controller membership functions have a significant impact on system output, which has not been studied in existing research works yet. The output spectrum can provide insight guidance in choosing parameters of controller membership functions in imperfect premise variables fuzzy control.

Table 9.1 Coefficients of nCOS function functions (up to 3rd order)

Coefficient	Value
$\varphi_1^0(jw_1, jw_2)$	3.6805 + 0.4148i
$\varphi_1^1(jw_1, jw_2)$	0.7394 + 0.1926i
$\varphi_2^0(jw_1, jw_2)$	−7,225,390,291.49673 − 4,277,717,741.50991i
$\varphi_2^1(jw_1, jw_2)$	10,776,851,523.8719 + 6,359,900,317.04678i
$\varphi_2^2(jw_1, jw_2)$	−4,308,869,509.22147 − 2,537,573,509.26971i
$\varphi_2^3(jw_1, jw_2)$	525,335,424.045296 + 318,524,456.813410i
$\varphi_3^0(jw_1, jw_2, jw_3)$	−25,545,826.1335602 + 21,579,148.1058807i
$\varphi_3^1(jw_1, jw_2, jw_3)$	289,487,121.213074 − 224,697,469.113037i
$\varphi_3^2(jw_1, jw_2, jw_3)$	−1,236,823,901.44092 + 942,227,866.423584i
$\varphi_3^3(jw_1, jw_2, jw_3)$	2,671,336,546.84814 − 2,028,425,239.12524i
$\varphi_3^4(jw_1, jw_2, jw_3)$	−3,126,003,529.66870 + 2,371,476,787.70874i
$\varphi_3^5(jw_1, jw_2, jw_3)$	1,892,153,389.05457 − 1,433,259,036.73730i
$\varphi_3^6(jw_1, jw_2, jw_3)$	−464,573,681.885559 + 351,103,325.641510i
$\varphi_{3f}^1(jw_1, jw_2, jw_3)$	−56.2793221471511 − 2.15568621179773i
$\varphi_{3f}^2(jw_1, jw_2, jw_3)$	−166.064222770103 − 10.4177121317625i
$\varphi_{3f}^3(jw_1, jw_2, jw_3)$	−41.3692556868191 − 1.78030757233500i
$\varphi_{3g}^1(jw_1, jw_2, jw_3)$	40.1471021710313 + 1.63099207039340i
$\varphi_{3g}^2(jw_1, jw_2, jw_3)$	119.336723686894 + 7.28239276327076i
$\varphi_{3g}^3(jw_1, jw_2, jw_3)$	29.4474660647102 + 1.37009786867566i
$\varphi_{3c}^1(jw_1, jw_2, jw_3)$	29.6585524723414 + 1.74775303857859i
$\varphi_{3c}^2(jw_1, jw_2, jw_3)$	−19.4692646977492 − 1.12477620991194i
$\varphi_{3c}^3(jw_1, jw_2, jw_3)$	−34.6876201892883 − 1.90197813324630i

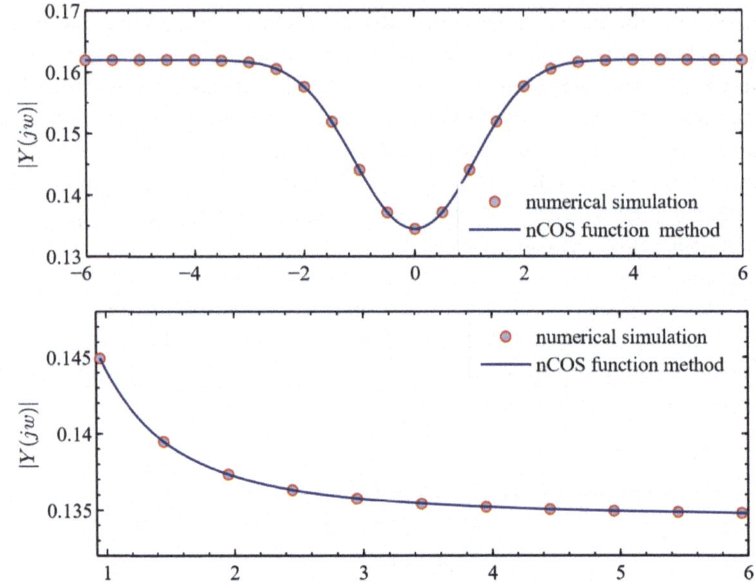

Fig. 9.1 Amplitude of Y(jw) with different b(c = 1) and c(b = 1)

For example, when $b = 2, c = 0.955, |Y(jw)| = 0.1561$. $b = -0.285, c = 5.595$, $|Y(jw)| = 0.1346$. The amplitude of output is reduced about 13.77% .

Then the genetic algorithm is applied to verify the results obtained by the nCOS function based optimization method, the GA parameters (N, C, M) are chosen as (140, 0.5, 0.04) and the parameters space of b and c are $[-6, 6]$ and $[0.84, 6]$ respectively.

Fig. 9.2 Amplitude of |Y(jw)| with different b and c

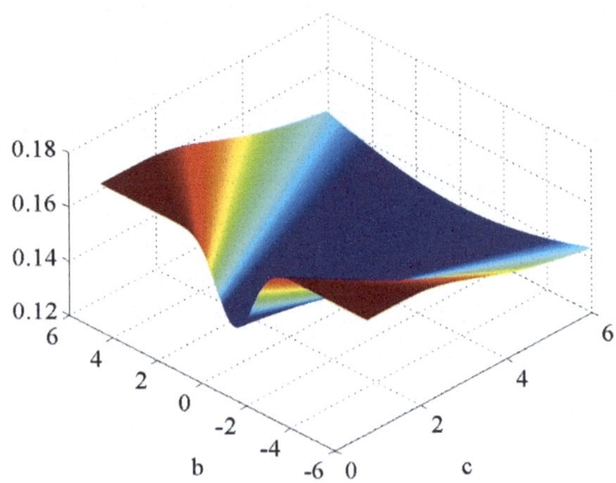

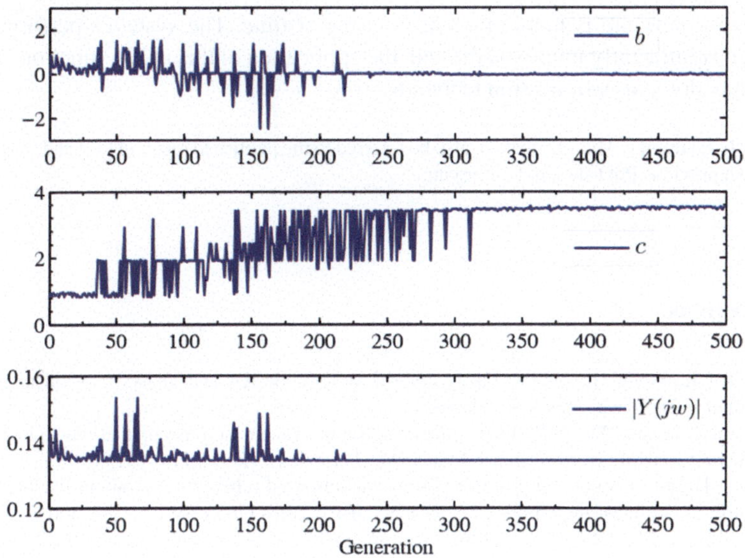

Fig. 9.3 GA-based optimization process

The GA-based optimization process is shown in Fig. 9.3. It is obtained that the final optimal output is $|Y(jw)| = 0.13448$, and the corresponding optimal parameters are approximately $b = 0$ and $c = 3.5$, which is almost identical to the result $|Y(jw)| = 0.1345$ calculated by the frequency response based method. This again verifies the effectiveness of frequency response based optimization method. It is also worth pointing out that the GA approach can only obtained optimal value of $|Y(jw)|$ and the corresponding parameters b and c. Additionally, as the parameter space expands, the time required for computation grows significantly. Conversely, the method based on the nCOS function offers a more analytical connection between parameters b and c and the targeted performance. As demonstrated in Fig. 9.2, the output response of the closed-loop system within the specified parameter space can be determined analytically, without an increase in computational load as the range of parameters broadens.

9.5 Conclusion

This research adopts Gaussian membership functions as the mismatched membership functions. It introduces an innovative optimization method centered on the nCOS function, designed to fine-tune the parameters of these Gaussian membership functions. This new frequency domain optimization strategy, when compared to the Genetic Algorithm (GA) optimization method, offers a comprehensive analysis of the nonlinear effects of fuzzy membership function parameters on the system's output response and is more efficient in terms of time. The system's performance has been significantly improved through the application of this nCOS function-based frequency domain optimization technique.

Acknowledgements This chapter can also be referred to the published result in [32] and thus gives credit to a previous PhD student LI Jingying.

References

1. Tanaka K, Wang HO (2004) Fuzzy control systems design and analysis: A linear matrix inequality approach. John Wiley & Sons
2. Takagi T, Sugeno M (1985) Fuzzy identification of systems and its applications to modeling and control. IEEE Trans Systems, Man and Cybernetics SMC-15:116–132
3. Qiu J, Ding S, Gao H, Yin S (2016) Fuzzy-model-based reliable static output feedback H_∞ control of nonlinear hyperbolic PDE systems. IEEE Trans Fuzzy Syst 24:388–400
4. Li J, Jing X, Li Z, Huang X (2019b) Fuzzy adaptive control for nonlinear suspension systems based on a bio-inspired reference model with deliberately designed nonlinear damping. IEEE Trans Ind Electron 66:8713–8723.
5. Li H, Zhang Z, Yan H, Xie X (2019a) Adaptive event-triggered fuzzy control for uncertain active suspension systems. IEEE Trans Cybern 49:4388–4397.
6. Yang X, Lam HK, Wu L (2019) Membership-dependent stability conditions for type-1 and interval type-2 T-S fuzzy systems. Fuzzy Sets Syst 356:44–62
7. Zhang M, Shi P, Ma L, et al (2019b) Quantized feedback control of fuzzy markov jump systems. IEEE Trans Cybern 49:3375–3384.
8. Gu Z, Shi P, Yue D, Ding Z (2019) Decentralized adaptive event-triggered H_∞ filtering for a class of networked nonlinear interconnected systems. IEEE Trans Cybern 49:1570–1579.
9. Lam HK, Narimani M (2009) Stability analysis and performance design for fuzzy-model-based control system under imperfect premise matching. IEEE Trans Fuzzy Syst 17:949–961.
10. Lam HK, Seneviratne LD (2011) Stability analysis of polynomial fuzzy-model-based control systems under perfect/imperfect premise matching. IET Control Theory Appl 5:1689–1697.
11. Kim SH (2013) Robust Takagi-Sugeno fuzzy control design for nonlinear networked control systems with asynchronous premise variables. IET Control Theory Appl 7:910–919.
12. Zhang C, Lam H, Qiu J, et al (2019a) A new design of membership-function-dependent controller for T-S fuzzy systems under imperfect premise matching. IEEE Trans Fuzzy Syst 27:1428–1440
13. Kim HJ, Park JB, Joo YH (2015) Decentralised H_∞ fuzzy filter for non-linear large-scale systems under imperfect premise matching. IET Control Theory Appl 9:2704–2714.
14. Kim HJ, Park JB, Joo YH (2017) H_∞ fuzzy filter for non-linear sampled-data systems under imperfect premise matching. IET Control Theory Appl 11:747–755.
15. Zhao T, Dian S (2017) Fuzzy dynamic output feedback H_∞ control for continuous-time T-S fuzzy systems under imperfect premise matching. ISA Transactions 70:248–259.
16. Li J, Huang X, Li Z (2017) Extended dissipative filtering of fuzzy systems with time-varying delay under imperfect premise matching. Int J Syst Sci 48:1731–1743
17. Nakamura E, Kehtarnavaz N (1995) Optimization of fuzzy membership function parameters. In: 1995 IEEE international conference on systems, man and cybernetics. Intelligent systems for the 21st century. pp 1–6 vol.1
18. Homaifar A, McCormick E (1995) Simultaneous design of membership functions and rule sets for fuzzy controllers using genetic algorithms. IEEE Trans Fuzzy Syst 3:129–139.

19. Larbes C, Cheikh SMA, Obeidi T, Zerguerras A (2009) Genetic algorithms optimized fuzzy logic control for the maximum power point tracking in photovoltaic system. Renewable Energy 34:2093–2100.
20. Fang G, Kwok NM, Ha Q (2008) Automatic fuzzy membership function tuning using the particle swarm optimization. In: 2008 IEEE pacific-asia workshop on computational intelligence and industrial application. pp 324–328
21. Omizegba EE, Adebayo GE (2009) Optimizing fuzzy membership functions using particle swarm algorithm. In: 2009 IEEE international conference on systems, man and cybernetics. pp 3866–3870
22. George DA (1959) Continuous nonlinear systems. Massachusetts Institute of Technology. Research Laboratory of Electronics
23. Jing X, Lang Z, Billings SA, Tomlinson GR (2006) The parametric characteristic of frequency response functions for nonlinear systems. Int J Control 79:1552–1564
24. Jing X, Lang Z, Billings SA (2008) Mapping from parametric characteristics to generalized frequency response functions of non-linear systems. Int J Control 81:1071–1088
25. Jing X (2014) Nonlinear characteristic output spectrum for nonlinear analysis and design. IEEE/ASME Trans Mechatronics 19:171–183.
26. Jing X, Lang Z (2015) Frequency domain analysis and design of nonlinear systems based on Volterra series expansion: A parametric characteristic approach. Springer
27. Xiao Z, Jing X (2016a) Frequency-domain analysis and design of linear feedback of nonlinear systems and applications in vehicle suspensions. IEEE/ASME Trans Mechatronics 21:506–517.
28. Swain AK, Billings SA (2001) Generalized frequency response function matrix for MIMO non-linear systems. Int J Control 74:829–844
29. Jing X (2012) Truncation order and its effect in a class of nonlinear systems. Automatica 48:2978–2985.
30. Xiao Z, Jing X (2016b) Frequency domain analysis of exponential-type nonlinearity. In: 2016 American control conference (ACC). pp 7466–7471
31. Tanaka K, Ikeda T, Wang HO (1996) Robust stabilization of a class of uncertain nonlinear systems via fuzzy control: Quadratic stabilizability, H_∞ control theory, and linear matrix inequalities. IEEE Trans Fuzzy Syst 4:1–13.
32. J Li, X Jing, Z Li, X Huang, Optimization of Fuzzy Membership Function based on the NCOS function method, IECON 2020 The 46th Annual Conference of the IEEE Industrial Electronics Society, 2020

Chapter 10
A Parametric Convergence Theory

Abstract A fundamental consideration in the analysis and design of nonlinear systems using Volterra series theory is ensuring that both the excitation amplitude and system parameters lie within appropriate bounds to guarantee the convergence of the Volterra series expansion. This chapter presents a systematic investigation into the parametric convergence boundaries of the Volterra series for nonlinear systems modeled by the Nonlinear AutoRegressive with eXogenous inputs (NARX) framework. The system output is expressed in a definitive analytical form, influenced not only by the input amplitude but also by the input's energy content and waveform characteristics. The proposed methodology introduces refined convergence criteria that incorporate model parameters, the maximum input amplitude, and a factor representing the total input energy or waveform shape. This advancement is particularly significant for practical applications, as a nonlinear system may exhibit chaotic behavior under a simple single-tone excitation, yet remain stable under multi-tone inputs of identical amplitude. To the best of our knowledge, this work presents the first closed-form expression for the parametric convergence bounds of Volterra-type nonlinear systems, moving beyond traditional power series-based approaches.

Keyword Convergence · Volterra series · Nonlinear systems · Closed-form expression

10.1 Introduction

10.1.1 About This Chapter

A critical aspect of analyzing and designing nonlinear systems through Volterra series theory is ensuring that the excitation magnitude and model parameters fall within suitable ranges to achieve a convergent Volterra series expansion. This chapter systematically explores the parametric convergence boundaries of the Volterra series expansion for nonlinear systems characterized by a NARX model. This exploration

X. Jing, *The Bio-Inspired X-Structure/Mechanism Approach for Exploring Nonlinear Benefits in Engineering*, https://doi.org/10.1007/978-981-96-5769-8_10

aims to determine the conditions—specifically, the excitation magnitude and parameter range—under which a NARX system can maintain a convergent Volterra series expansion for any given input signal. Typically, existing bounds are presented as dependent on the maximum input magnitude, which is adequate for single-tone harmonic inputs but overly cautious for more complex inputs, such as multi-tone or arbitrary signals. In this study, the output response of nonlinear systems is articulated in a definitive form, influenced not just by the input magnitude but also by the input's energy or waveform. These innovative methods yield more precise boundary criteria that account for model parameters, maximum input magnitude, and a factor indicative of the total input energy or waveform. This advancement holds considerable importance for practical applications, as the same nonlinear system may display chaotic behavior under a simple single-tone input but remain stable under different types of input signals, like multi-tone inputs, even if they share the same input magnitude. The findings offer valuable insights for applying Volterra series-based theories and methodologies from an engineering perspective. The Duffing equation serves as a benchmark to demonstrate the efficacy of these results.

It should be emphasized that this is the first-ever result to present a closed-form parametric convergence bound of the Volterra-type nonlinear systems, instead of a power series or others.

10.1.2 Research Background

A broad spectrum of nonlinear systems permits the establishment of an input–output relationship through a Volterra series expansion, as indicated by references [1–5]. The Volterra series, along with its analytical techniques applicable in both time and frequency domains, has been widely implemented in various practical fields including control, signal processing, and the identification, analysis, and design of systems [5–16].

The convergence of a Volterra series expansion within a nonlinear system is contingent upon the excitation input's magnitude, frequency, and energy, as well as the values of the model parameters, collectively known as characteristic parameters. A pivotal aspect of employing Volterra series-based theories or methodologies in the analysis and design of nonlinear systems is to verify that the nonlinear dynamics are representable by the Volterra series. This necessitates all characteristic parameters to be confined within specified limits. Various studies have endeavored to resolve this by concentrating solely on the bounds of maximum input magnitude. The findings presented in [1, 3, 4] offer general operator-based conclusions that theoretically affirm the feasibility of Volterra series expansions across a vast array of nonlinear systems, yet they do not readily furnish detailed parametric convergence bounds for any characteristic parameter. The bounding criteria in [17–19] are specific to certain nonlinear systems, such as Duffing oscillators. A recent study [20] dealt exclusively with input-analytic nonlinear systems. More recent and less restrictive convergence bound results are documented in [5, 21, 22], which consider input magnitude, frequency, and model parameters, rendering them superior to earlier findings. However, these newer results still represent bounds on the maximum magnitude of input excitation. While the convergence bounds in [5, 21, 22] are significantly improved for single-tone harmonic inputs, they remain conservative for multi-tone or arbitrary input signals.

The principle of superposition is inapplicable to nonlinear systems, meaning that bounds derived from single-tone inputs for a nonlinear system are not extendable to more complex inputs. A nonlinear system might respond chaotically to a simple single-tone input at a certain frequency, yet exhibit only a mildly nonlinear response to other input signals of equivalent magnitude but differing in frequency, energy, waveform, or frequency components. Mildly nonlinear systems that align with the Volterra type constitute a crucial category of nonlinear systems encountered in practical scenarios, supported by a robust theoretical framework and methodologies for signal processing, system analysis, and design. Hence, an enhanced parametric bound that signifies the presence of these nonlinearities is instrumental for comprehending or assessing the degree of nonlinearity in a practical system to facilitate further analysis and design. Consequently, less conservative convergent bounds that accommodate arbitrary input signals are vital for real-world applications. Yet, the existing literature on bound evaluation remains quite constrained.

This research introduces a novel parametric convergence boundary for the Volterra series expansion of nonlinear systems characterized by a parametric nonlinear autoregressive with exogenous input (NARX) model, applicable to various input signals. The output frequency response boundary for the NARX system is now articulated in a definitive form, rather than as an infinite power series, and is defined by the input's magnitude, frequency, and energy or waveform, in addition to the model parameters. This innovative approach allows for the development of superior parametric convergence boundaries for the Volterra series expansion, rendering many of the previous results as merely particular instances of this broader boundary result. To demonstrate the efficacy of these new findings, they are compared with prior results using the Duffing equation as a benchmark, accompanied by a thorough discussion.

10.2 Output Frequency Response of Nonlinear Systems

A considerable variety of nonlinear systems permit the application of a Volterra series expansion to define the input–output relationship in proximity to the zero equilibrium point [1, 3, 4, 23]:

$$y(t) = \sum_{n=1}^{N} \int_{-\infty}^{\infty} \cdots \int_{-\infty}^{\infty} h_n(\tau_1, \ldots, \tau_n) \prod_{i=1}^{n} u(t - \tau_i) d\tau_i \tag{10.1}$$

where $h_n(\tau_1, \ldots, \tau_n)$ is the nth order Volterra kernel. The nth order Generalized Frequency Response Function (GFRF) can be described as the multi-variable Fourier transform of the Volterra kernels, expressed as follows [2]:

$$H_n(j\omega_1, \ldots, j\omega_n) = \int_{-\infty}^{\infty} \cdots \int_{-\infty}^{\infty} h_n(\tau_1, \ldots, \tau_n) \exp(-j(\omega_1\tau_1 + \cdots + \omega_n\tau_n))d\tau_1 \cdots d\tau_n \tag{10.2}$$

In practical scenarios, it is always possible to convert a nonlinear system into a parametric form, such as the nonlinear auto-regressive model with exogenous inputs (NARX model):

$$y(t) = \sum_{m=1}^{M} y_m(t)$$
$$y_m(t) = \sum_{p=0}^{m} \sum_{k_1,k_{p+q}=0}^{K} c_{p,q}(k_1, \ldots, k_{p+q}) \prod_{i=1}^{p} y(t-k_i) \prod_{i=p+1}^{p+q} u(t-k_i) \tag{10.3}$$

where $y_m(t)$ is the mth-order output of the NARX model, $q = m - p$, $k_i = 0, =, K(\geq 1)$, $p, q \geq 0, n \geq 1$, and $\sum_{k_1,k_{p+q}=1}^{K} (\cdot) = \sum_{k_1=1}^{K} \cdots \sum_{k_{p+q}=1}^{K} (\cdot)$. $p + q$ is called the nonlinear degree of parameter $c_{p,q}(\cdot)$, and the maximum nonlinear degree is M. Nonlinear degree 1 corresponds to parameters $c_{0,1}(\cdot)$ and $c_{1,0}(\cdot)$, which are the coefficients of linear terms in the model, referred to as linear parameters, and all others with nonlinear degree >1 are referred to as nonlinear parameters. It's important to recognize that the NARX model serves as a fitting depiction for a wide array of nonlinear systems encountered in real-world applications. This includes a number of prominent nonlinear models that fall under its umbrella as particular instances [24, 25]. Additionally, there exists a variety of algorithms designed to determine a NARX model for a nonlinear system based on provided input–output data [26].

The GFRFs for System (10.3) can be obtained using the probing method [23]. A recursive algorithm to compute the nth order GFRF of System (10.3) is given as [27, 28]:

$$\begin{aligned} L_n(j\omega_1 + \cdots + j\omega_n)H_n(j\omega_1, \ldots, j\omega_n) &= \sum_{k_1,k_n=1}^{K} c_{0,n}(k_1, \ldots, k_n) \exp(-j(\omega_1 k_1 + \cdots + \omega_n k_n)) \\ &+ \sum_{q=1}^{n-1} \sum_{p=1}^{n-q} \sum_{k_1,k_{p+q}=1}^{K} c_{p,q}(k_1, \ldots, k_{p+q}) \\ &\cdot \exp(-j(\omega_{n-q+1}k_{n-q+1} + \cdots \\ &+ \omega_n k_{p+q}))H_{n-q,p}(j\omega_1, \ldots, j\omega_{n-q}) \end{aligned}$$

$$+\sum_{p=2}^{n}\sum_{k_1,k_p=1}^{K} c_{p,0}(k_1,\ldots,k_p)H_{n,p}(j\omega_1,\ldots,j\omega_n) \tag{10.4}$$

$$H_{n,p}(\cdot)=\sum_{i=1}^{n-p+1} H_i(j\omega_1,\ldots,j\omega_i)H_{n-i,p-1}(j\omega_{i+1},\ldots,j\omega_n)$$
$$\exp(-j(\omega_1+\cdots+\omega_i)k_p) \tag{10.5}$$

$$H_{n,1}(j\omega_1,\ldots,j\omega_n)=H_n(j\omega_1,\ldots,j\omega_n)\exp(-j(\omega_1+\cdots+\omega_n)k_1) \tag{10.6}$$

$$L_n(j\omega_1+\cdots+j\omega_n)=1-\sum_{k_1=1}^{K} c_{1,0}(k_1)\exp(-j(\omega_1+\cdots+\omega_n)k_1) \tag{10.7}$$

With the GFRFs above, the output frequency response of a NARX system can be obtained as follows for any input signal $u(\mathrm{t})$ with a Fourier transform $U(j\omega)$ [28, 29, 30]:

$$Y(j\omega)=\sum_{n=1}^{\infty} Y_n(j\omega) \tag{10.8a}$$

$$Y_n(j\omega)=\frac{1}{(2\pi)^{n-1}}\underbrace{\int_{-\infty}^{\infty}\cdots\int_{-\infty}^{\infty}}_{n-1}\Xi\,\mathrm{d}\omega_1\cdots\mathrm{d}\omega_{n-1} \tag{10.8b}$$

where $Y_n(j\omega)$ is the nth-order output spectrum

$$\Xi=H_n(j\omega_1,\ldots,j\omega_{n-1},j(\omega-\omega_1-\cdots-\omega_{n-1}))U_n(j\omega,j\omega_1,\ldots,j\omega_{n-1})$$
$$U_n(j\omega,j\omega_1,\ldots,j\omega_{n-1})=U(j\omega_1)\cdots U(j\omega_{n-1})U(j(\omega-\omega_1-\cdots-\omega_{n-1})) \tag{10.8c}$$

The nonlinear output frequency response detailed in Eqs. (10.8a–c) can be interpreted as a frequency-domain equivalent of the Volterra series expansion for the NARX system outlined in Eq. (10.3). This research aims to establish a more liberal parametric convergence limit, considering both model parameters and input, which ensures the convergence of the Volterra series expansion for a NARX system with any type of input signal. To achieve this, a novel method will be introduced that provides a definitive expression for the output frequency response boundary, as opposed to the infinite series presented in (10.8a), and takes into account the input's power or waveform during the evaluation of the boundary.

10.3 Parametric Convergent Bounds

10.3.1 Definitions and Notations

Prior to introducing the principal findings, we establish the following definitions and notations. Without loss of generality, this study considers the input signal denoted by $u(t)$, which has the Fourier transform written as $\rho U(j\omega)$ with $\rho > 0$ and $\max|U(j\omega)|=1$ for $\omega \in R_\omega$. Let W_ω denote the output frequency range subject to the input u(t), which can be determined with the method in [31].

For the NARX model in (10.3), define

$$C(p,q) = \begin{cases} \sum_{k_1,k_{p+q}=0}^{K} \left|c_{p,q}(k_1,\ldots,k_{p+q})\right|, & \text{for } 1 \le q \le n-1,\ 1 \le p \le n-q \\ \sum_{k_1,k_n=0}^{K} \left|c_{0,n}(k_1,\ldots,k_n)\right|, & q=n, p=0 \\ \sum_{k_1,k_p=0}^{K} \left|c_{p,0}(k_1,\ldots,k_p)\right|, & q=0, 2 \le p \le n \\ 0, & else \end{cases} \tag{10.9}$$

$C(p, q)$ is a non-negative function of parameter $c_{p,q}(\cdot)$ $(0 \le p, q \le n)$. Denote

$$\overline{H}_n = \max_{\omega_1 \cdots \omega_n \in R_\omega} (|H_n(\cdot)|) \tag{10.10}$$

Let

$$L(j\omega) = 1 - \sum_{k_1=0}^{K} c_{1,0}(k_1)\exp(-j\omega k_1) \tag{10.11}$$

For any input $U(j\omega)$ defined in R_ω, denote the minimum value of $|L(j\omega)|$ as:

$$\underline{L}_{R_\omega} = \min_{\Omega \in W_\omega} \{|L(j\Omega)|\} \tag{10.12}$$

Different input $U(j\omega)$ may lead to different $\underline{L}_{R_\omega}$.

10.3.2 The Main Results

In this study, the bounds on the magnitude of nonlinear frequency response functions, which encompass the Generalized Frequency Response Functions (GFRFs) and the output spectrum, have been previously explored in references [32–34] from a parametric perspective. Here, the magnitude bounds were articulated as polynomial expressions based on the magnitude limit of the first-order GFRF ($H_1(j\omega)$). However, the current study presents the bound of the output spectrum in Eqs. (10.8a–c) as a closed-form expression that is dependent on the input, rather than as an infinite power series focused solely on the input magnitude, as seen in prior works [5, 17–22]. This definitive expression incorporates the frequency variable, the magnitude, and the energy or waveform characteristics of the input, as well as the model parameters. It introduces a crucial method for assessing the convergence bound of the output frequency response in nonlinear systems, culminating in the establishment of new and enhanced parametric convergence bounds as detailed in Theorems 1 and 2.

Theorem 1. (Input-dependent convergence bound with respect to arbitrary inputs). *Consider the NARX system in* (10.3) *subject to any input* $u(t)$ *with a Fourier transform* $U_\rho(j\omega) = \rho U(j\omega)$, *which is defined in* $\omega \in R_\omega$ *with* $\rho > 0$ *and* $\max|U(j\omega)| = 1$ *for all* ω, *where* R_ω *is the input frequency range. The following conclusions can be made.*

(1) The magnitude of the output frequency response in (10.8a–c) is upper bounded by

$$|Y(j\omega)| \le \frac{\tilde{Y}}{\sigma} \tag{10.13}$$

where $\tilde{Y}$ is given in a closed form as

$$\tilde{Y} = \overline{H}_1\gamma + \frac{1}{L_{R_\omega}}\sum_{m=2}^{M}\sum_{\substack{p+q=m\\0\le p,q\le m}} C(p,q)\tilde{Y}^p\gamma^q \tag{10.14}$$

and

$$\gamma = \sigma\rho \text{ with } \sigma = \frac{1}{\pi}\int_{\omega_1\in R_\omega} |U(j\omega_1)| d\omega_1 \tag{10.15ab}$$

(2) The parametric convergence bound of the Volterra series expansion of the NARX system can be obtained by solving the Eq. (10.14) subject to the following constraint:

$$\frac{1}{L_{R_\omega}}\sum_{m=2}^{M}\sum_{\substack{p+q=m\\0\le p,q\le m}} pC(p,q)\tilde{Y}^{p-1}\gamma^q = 1 \tag{10.16}$$

Moreover, if the input magnitude ρ is given, solving (10.14) subject to the constraint (10.16) can give a bound for any model parameters of $p + q > 1$.

Proof. See Appendix 1. □

Remark 1. The information of the input $u(t)$ is denoted by $\gamma = \sigma\rho$, which includes two parts. One is the maximum input magnitude ρ, and the other one is the wave form information $\sigma = \frac{1}{\pi}\int_{\omega_1 \in R_\omega} |U(j\omega_1)| d\omega_1$. Therefore, the convergence bound given by solving Eqs. (10.14) and (10.16) with respect to the input is not only a simple function of the maximum input magnitude, as in all existing results, but also a function of the wave form information. Due to the wave form information $\sigma = \frac{1}{\pi}\int_{\omega_1 \in R_\omega} |U(j\omega_1)| d\omega_1$, which could be greater or less than 1, the resulting convergence bound of Theorem 1 can give more accurate evaluation of the convergence bound than using only the maximum input magnitude [5, 17–22]. This is the significant difference or contribution of this bound result compared with the existing results [5, 17–22]. Although the result in [20] is applicable to any input signal, the bound is given in terms of the maximum input magnitude in the worst case, independent of input frequency; therefore, it is more conservative. Moreover, the nonlinear system considered in [20] should be input-analytic.

Remark 2. The output frequency response of nonlinear systems given by Eqs. (10.8a–c) is bounded by $\frac{\tilde{Y}}{\sigma}$ in (10.13), which considers the waveform information σ. This approach will yield improved bounds for the output response in the frequency domain, which are less restrictive. This is because it accounts for both the maximum input magnitude and the input's waveform, as highlighted in Remark 1. The majority of existing studies [5, 17–22] only establish a bound for the worst-case scenario, which corresponds to the highest magnitude of the input signals, without considering the nature of the input—be it a single tone, multi-tone, or any other form of arbitrary signals. Such an approach results in an overly cautious assessment of the convergence boundary for multi-tone or arbitrary input signals. Notably, the bound for the magnitude (represented by $\tilde{Y}$) of the output frequency response is presented in a definitive form in Eq. (10.14), rather than as an infinite series as seen in references [5, 21, 22]. This represents another significant contribution of this research.

Remark 3. The recent findings transform the task of evaluating convergence boundaries into resolving a pair of algebraic equations, akin to the approach in references [5, 21, 22]. However, the novel methods introduced in Remarks 1–2 simplify these equations considerably, incorporating the waveform details of the input. Furthermore, the definition of $\underline{L}_{R_\omega}$ provided in Eq. (10.12) implies that varying input frequencies could lead to different outcomes for $\underline{L}_{R_\omega}$, thereby introducing a frequency-dependent aspect to the bound established by Theorem 1. This contrasts with previous methods found in [17–19, 21], which could only produce frequency-dependent bounds for single-tone harmonic inputs—a rather specific scenario.

Utilizing the aforementioned results facilitates a novel advancement in determining the convergence limit for both multi-tone and single-tone input signals. While multi-tone signals are included within the scope of the input signal outlined in Theorem 1, establishing the convergence limit for multi-tone signals is not as direct as it may seem. To this aim, consider a multi-tone signal given by:

$$u(t) = \sum_{i=1}^{\overline{K}} |F_i| \cos(\omega_i t + \angle F_i) \tag{10.17a}$$

One can see that the input frequency range is now $R_\omega = \{\omega_1, \omega_2, \ldots, \omega_{\overline{K}}\}$. Define $V_\omega = R_\omega \cup R_\omega^-$, where R_ω^- is the negative frequency set corresponding to R_ω. The multi-tone input signal can be rewritten as

$$u(t) = \sum_{i=1}^{\overline{K}} |F_i| \cos(\omega_i t + \angle F_i) = \sum_{\omega \in V_\omega} \frac{F(j\omega)}{2} e^{j\omega t} \tag{10.17b}$$

where $F(\pm\omega_i) = |F_i| e^{\pm j\angle F_i}$ for $i \in \{1, \ldots, \overline{K}\}$, which can be regarded as the frequency domain representation of the time domain signal u(t). Substituting (10.17b) into (10.1), it can be obtained that

$$\begin{aligned}
y(t) &= \sum_{n=1}^{N} y_n(t) \\
y_n(t) &= \int_{-\infty}^{\infty} \cdots \int_{-\infty}^{\infty} h_n(\tau_1, \ldots, \tau_n) \prod_{i=1}^{n} u(t - \tau_i) d\tau_i \\
&= \frac{1}{2^n} \sum_{\omega_{k_1} \in V_\omega} \cdots \sum_{\omega_{k_n} \in V_\omega} H_n(j\omega_{k_1}, \ldots, j\omega_{k_n}) \\
&\quad F(j\omega_{k_1}) \cdots F(j\omega_{k_n}) e^{j(\omega_{k_1} + \cdots + \omega_{k_n})t} \\
&= \frac{1}{2} \sum_{\omega_{k_n} = \omega - \omega_{k_1} - \cdots - \omega_{k_{n-1}}} \left\{ \frac{1}{2^{n-1}} \sum_{\omega_{k_1} \in V_\omega} \cdots \right. \\
&\quad \left. \sum_{\omega_{k_{n-1}} \in V_\omega} H_n(j\omega_{k_1}, \ldots, j\omega_{k_n}) F(j\omega_{k_1}) \cdots F(j\omega_{k_n}) \right\} e^{j\omega t} \\
&= \frac{1}{2} \sum_{\omega \in \{\omega_{k_1} + \cdots + \omega_{k_n} | \omega_{k_1}, \cdots, \omega_{k_n} \in V_\omega\}} \left\{ \frac{1}{2^{n-1}} \sum_{\omega_{k_1} \in V_\omega} \cdots \right. \\
&\quad \sum_{\omega_{k_{n-1}} \in V_\omega} H_n(j\omega_{k_1}, \ldots, j(\omega - \omega_{k_1} - \cdots - \omega_{k_{n-1}})) \\
&\quad \left. F(j\omega_{k_1}) \cdots F(j(\omega - \omega_{k_1} - \cdots - \omega_{k_{n-1}})) \right\} e^{j\omega t}
\end{aligned}$$

Equation (10.2) is used in the second equation above. Therefore, it can be obtained that, for $\omega \neq 0$,

$$Y_n(j\omega) = \frac{1}{2^{n-1}} \sum_{\omega_{k_1} \in V_\omega} \cdots \sum_{\omega_{k_{n-1}} \in V_\omega} H_n(j\omega_{k_1}, \ldots, j(\omega - \omega_{k_1} - \cdots - \omega_{k_{n-1}})) F(j\omega_{k_1}) \cdots F(j\omega_{k_{n-1}}) F(j(\omega - \omega_{k_1} - \cdots - \omega_{k_{n-1}})) \quad (10.18)$$

and the nonlinear output spectrum subject to the multi-tone input (10.17a) can be obtained via (10.8a).

Theorem 2. (Input-dependent convergence bound with respect to multi-tone inputs). *Consider the NARX system in* (10.3) *subject to a multi-tone input* $u(t) = \rho \sum_{i=1}^{\overline{K}} |F_i| \cos(\omega_i t + \angle F_i)$ *with* $\max|F_i|=1$, $\rho > 0$ *and the input frequency range* $R_\omega = \{\omega_1, \omega_2, \ldots, \omega_{\overline{K}}\}$. *Then,* (a) *the magnitude of the output frequency response in* (10.8a–c) *is upper bounded by*

$$|Y(j\omega)| \le \frac{\tilde{Y}}{\sigma} \quad (10.19)$$

with

$$\tilde{Y} = \overline{H}_1 \gamma + \frac{1}{\underline{L}_{R_\omega}} \sum_{m=2}^{M} \sum_{\substack{p+q=m \\ 0 \le p,q \le m}} C(p,q) \tilde{Y}^p \gamma^q, \quad \gamma = \sigma\rho \text{ and } \sigma = \sum_{k_1 \in \{1,2,\ldots,\overline{K}\}} \left|F_{k_1}\right| \quad (10.20abc)$$

and (b) the parametric convergence bound of the Volterra series expansion of the NARX system still can be obtained by solving the Eq. (10.20a) *with the input information index given by* (10.20bc) *and subject to the constraint given in* (10.16), *i.e.,*

$$\frac{1}{\underline{L}_{R_\omega}} \sum_{m=2}^{M} \sum_{\substack{p+q=m \\ 0 \le p,q \le m}} pC(p,q) \tilde{Y}^{p-1} \gamma^q = 1 \quad (10.20d)$$

Proof. See Appendix 2. □

With Theorems 1 and 2, the following result can be obtained.

Corollary 1. (Frequency-dependent convergence bound with respect to single-tone inputs). *Consider the NARX system in* (10.3) *subject to a single-tone input* $u(t) = \rho \cos(\omega_1 t + \angle F_1)$ *with* $\rho > 0$. . *Then, (a) the magnitude of the output frequency response in* (10.8a–c) *is upper bounded by*

$$|Y(j\omega)| \le \tilde{Y} \quad (10.21)$$

with

$$\tilde{Y} = |H_1(j\omega_1)|\rho + \frac{1}{\underline{L}_{R_\omega}} \sum_{m=2}^{M} \sum_{\substack{p+q=m \\ 0\le p,q\le m}} C(p,q)\tilde{Y}^p \rho^q \quad (10.22)$$

and (b) the parametric convergence bound of the Volterra series expansion of the NARX system can be obtained by solving the equation in (10.22) *subject to the following constraint:*

$$\frac{1}{\underline{L}_{R_\omega}} \sum_{m=2}^{M} \sum_{\substack{p+q=m \\ 0\le p,q\le m}} pC(p,q)\tilde{Y}^{p-1} \rho^q = 1 \quad (10.23)$$

Proof. See Appendix 3. □

Remark 4. Theorem 2 establishes a parameter-specific bound for multi-tone input signals, a determination not achievable with the findings in references [5, 21, 22], which only take into account the peak input magnitude. In contrast, for single-tone inputs, the bound defined by Corollary 1 surpasses the outcomes in [5, 21, 22] due to the new bound's reliance on frequency. Note that, by applying the result in [5, 21, 22], the output response bound is equivalent to solving $\tilde{Y} = \overline{H}_1\rho + \frac{1}{\underline{L}_{R_\omega}} \sum_{m=2}^{M} \sum_{\substack{p+q=m \\ 0\le p,q\le m}} C(p,q)\tilde{Y}^p \rho^q$ where $\overline{H}_1 = \sup_{\omega}(H_1(j\omega))$, while applying Corollary 1 is equivalent to solving $\tilde{Y} = |H_1(j\omega_1)|\rho + \frac{1}{\underline{L}_{R_\omega}} \sum_{m=2}^{M} \sum_{\substack{p+q=m \\ 0\le p,q\le m}} C(p,q)\tilde{Y}^p \rho^q$. One can see that the bound result for single tone inputs given by [5, 21, 22] is more conservative than our result since the result given by Corollary 1 is dependent on the magnitude of $H_1(j\omega)$ at different frequencies. For example, consider the following:

$$y(k) = c_{1,0}(1)y(k-1) + c_{1,0}(2)y(k-2) + c_{3,0}y^3(k-1) + c_{0,1}u(k-1)$$

The input is supposed to be $u(t) = \rho\cos(\omega_1 t + \angle F_1)$. . From (10.22–10.23), it can be obtained that

$$\begin{cases} c_{3,0}\tilde{Y}^3 - \underline{L}_{R_\omega}\tilde{Y} + \underline{L}_{R_\omega}|H_1(j\omega)|\rho = 0 \\ 3c_{3,0}\tilde{Y}^2 - \underline{L}_{R_\omega} = 0 \end{cases}$$

Given the model parameters, the bound for the input magnitude is obtained by solving the two equations above as:

$$\rho \le \overline{\rho} = \frac{2\sqrt{\underline{L}_{R_\omega}}}{3|H_1(j\omega)|\sqrt{3c_{3,0}}}$$

Similarly, if the input magnitude is known, a bound for the nonlinear coefficient $c_{3,0}$ can be obtained as:

$$c_{3,0} \leq \overline{c}_{3,0} = \frac{4\underline{L}_{R_\omega}}{27|H_1(j\omega)|^2\rho^2}$$

It can be seen that both bound results are frequency-dependent for the single-tone input. To apply the result in [5, 21, 22], the bounds will be $\overline{\rho} = \frac{2\sqrt{\underline{L}_{R_\omega}}}{3\overline{H}_1(j\omega)\sqrt{3c_{3,0}}} \leq \frac{2\sqrt{\underline{L}_{R_\omega}}}{3|H_1(j\omega)|\sqrt{3c_{3,0}}}$ and $\overline{c}_{3,0} = \frac{4\underline{L}_{R_\omega}}{27\overline{H}_1^2\rho^2} \leq \frac{4\underline{L}_{R_\omega}}{27|H_1(j\omega)|^2\rho^2}$, respectively. Thus, the new results are less conservative. More discussion and validation can be seen in Sect. 10.4.

Remark 5. By solving Eqs. (10.14) and (10.16) in Theorem 1 or Eqs. (10.20a, d) in Theorem 2, a convergence bound (denoted by $\overline{\gamma}$) for the input-energy index γ can be obtained. Then, through (10.15ab) or (10.20bc), the input magnitude bound for ρ can be achieved with $\overline{\rho} = \overline{\gamma}/\sigma$. For a single-tone harmonic input $u(t) = \rho\cos(\omega_1 t + \angle F_1)$, the waveform information of the input $\sigma = 1$ in (10.20c); thus, the input magnitude bound for convergence is $\overline{\rho} = \frac{\overline{\gamma}}{1} = \overline{\gamma}$, which is the same as that by applying the methods in [5, 21, 22]. For multi-tone or other inputs, due to the waveform information $\sigma = \frac{1}{\pi}\int_{\omega_1 \in R_\omega} |U(j\omega_1)| d\omega_1$ or $\sigma = \sum_{k_1 \in \{1,2,\ldots,\overline{K}\}} |F_{k_1}|$, which could be greater or less than 1, the resulting convergence bound of Theorem 1 could provide more accurate evaluation of the convergence bound than using only the maximum input magnitude [5, 17–22]. For example, a two-tone input $u(t) = \rho(\cos(\omega_1 t) + 0.2\cos(\omega_2 t))$, σ in (10.20c) is now $\sigma = 1 + 0.2 = 1.2$, thus $\overline{\rho} = \frac{\overline{\gamma}}{1.2} < \overline{\gamma}$, while the results in [21, 22] give the same bound as the single tone input $u(t) = \rho\cos(\omega_1 t)$, which is over-estimated, and the result in [20] gives the bound corresponding to the worst case over all frequencies, which is much more conservative (see Examples later). The new bound result presents more information or better evaluation of the nonlinear dynamic response than existing results, even with respect to only a single-tone input.

To encapsulate the preceding dialogue, the convergence bound outcomes derived from Theorems 1, 2, and Corollary 1 hinge on the input signals' frequency, magnitude, and shape. These results are applicable to various characteristic parameters, encompassing both input magnitude and model parameters where the sum of p and q exceeds 1. The subsequent method can be employed to calculate the parametric convergence bound.

Procedure 1 for Computation of Parametric Convergence Bounds:

A1. Compute the bounds H_1 and $\underline{L}_{R_\omega}$ with (10.4), (10.10), and (10.12) or $|H_1(j\omega)|$.

A2. Solve (10.14) and (10.16) or (10.20a, d) or (10.22–10.23) to find the bound for γ or any model parameters $c_{p,q}(\cdot)$ with $p + q > 1$.

A3. If all model parameters are known, the parametric convergence bound of the input magnitude ρ can be obtained with the computed bound of γ and (10.15ab) or (10.20bc). If input magnitude ρ is given, then γ can be computed with

(10.15ab) or (10.20bc), and a convergence bound for any specific model parameter $c_{p,q}(\cdot)$ with $p+q>1$ can be obtained from (10.16) or (10.23) as well [see example in Remark 4)]. The case where there is no real solution for (10.14) and (10.16) or (10.20a, d) or (10.22–10.23) implies that the convergence bound is infinite.

10.4 Simulations and Discussions

In this segment, we employ a discretized Duffing equation to demonstrate the novel findings. The emphasis is on multi-tone input signals due to the limitations or excessive caution of most current findings, as previously mentioned. Additional insights into single-tone input can be located in references [5, 21, 22], and the benefits of this research over [5, 21, 22] are detailed in Remarks 4–5 of this paper. We examine two scenarios for comparison:

- In the first scenario, we calculate the bound of input magnitude. It is noted that our findings provide the most precise and dependable estimation for multi-tone inputs, in contrast to other results which are either inflated [21, 22], overly cautious [20], or inapplicable [17–19].
- In the second scenario, we estimate the convergence bound for a particular model parameter. Here, our findings uniquely offer a convergence bound for model parameters relevant to any input signals, whereas the existing findings in [22] are confined to single-tone input signals.

Consider a NARX model with zero initial condition given as

$$\begin{aligned} y(k) &= c_{1,0}(1)y(k-1) + c_{1,0}(2)y(k-2) \\ &\quad + c_{3,0}(1,1,1)y^3(k-1) + c_{0,1}(1)u(k-1) \end{aligned} \tag{10.24}$$

which is obtained by discretizing backward the Duffing equation $m\ddot{y}(t)+c\dot{y}(t)+k_1y(t)+k_{30}y^3(t)=u(t)$ with parameters $m=1, c=0.01\omega_0, k_1=\omega_0^2, \omega_0=20\pi$. The model parameters in (10.24) can be obtained as: $c_{1,0}(1)=2-\frac{cT_s}{m}-\frac{kT_s^2}{m}=1.9987, c_{1,0}(2)=\frac{cT_s}{m}-1=-0.9997, c_{0,1}(1)=\frac{T_s^2}{m}=25*10^{-7}, c_{3,0}(1,1,1)=-\frac{k_{30}T_s^2}{m}$ with $T_s=\frac{1}{2000}s$.

Consider that (10.24) is subject to a multi-tone input given by:

$$u(k)=\rho(cos(\omega_1k)+\beta cos(\omega_2k)) \tag{10.25}$$

where $0<\beta<1$. Although only double-tone signals are considered here, the conclusions can be drawn for any other input signals, as revealed in Theorem 2.

10.4.1 Computation of the Convergence Bound of Input

The parametric convergence bound of input magnitude ρ can be computed according to Algorithm 1. From (10.20a, d), the following equations hold:

$$\begin{cases} C(3,0)\tilde{Y}^3 - \underline{L}_{R_\omega}\tilde{Y} + \underline{L}_{R_\omega}\overline{H}_1\gamma = 0 \\ 3C(3,0)\tilde{Y}^2 - \underline{L}_{R_\omega} = 0 \end{cases} \tag{10.26}$$

where, C(3, 0) = $c_{3,0}(1, 1, 1)$. From the second Eq. (10.26), this gives:

$$\tilde{Y} = \sqrt{\underline{L}_{R_\omega}/(3C(3,0))}$$

Substituting $\tilde{Y}$ into the first Eq. (10.26), the convergence bound of γ can be obtained. Then, according to (10.20bc), the convergence bound of ρ is straightforward, that is,

$$\overline{\rho} = \frac{\gamma}{1+\beta} = \frac{2\sqrt{\underline{L}_{R_\omega}}}{3(1+\beta)\overline{H}_1\sqrt{3C(3,0)}} \tag{10.27}$$

Given $c_{3,0}(1, 1, 1) = -153.8$, that is, $k_{30} = 0.01\omega_0{}^2$, $\beta = 0.2$, and $\omega_2 = \omega_1 + 2\pi \cdot 1 \cdot T_s$, the computed convergence bound of the input magnitude is presented in Fig. 10.1. The parameter setting here (including k_{30}, ω_2, and ω_1) is taken arbitrarily as an example. For other parameter settings, the bound results are the same and the following discussions are similar.

Figure 10.1 illustrates that the calculated convergence bound for the input magnitude ρ, when applied to model (10.24) with a multi-tone input, aligns closely with the convergence bound obtained through numerical simulation. For context, Fig. 10.1 also includes the convergence bound for the input magnitude under a single-tone input condition for model (10.24)—achieved by setting β to 0 in Eq. (10.25). This result is superior to those found in references [5, 21, 22], as elaborated in Remarks 4–5. Notably, even with a minimal β value, such as 0.2, the convergence bound for a multi-tone input remains significantly lower than that for a single-tone input, attributable to the nonlinear modulation interaction between the two input frequencies. For further details on nonlinear output frequency characteristics, refer to [31, 35].

It is important to highlight that for model (10.24) under the influence of multi-tone inputs, the findings from [5, 17–19, 21, 22] are inapplicable. Moreover, the convergence bound as per [20] is $5.3895*10^{-6}$ more conservative compared to the convergence bound established through Theorem 2 at the resonant frequency, which is 0.0032 in this instance.

It's significant to observe that the convergence bound for multi-tone inputs may be substantially lower than for single-tone inputs at lower frequencies, even when the primary input frequency is identical. This suggests that the convergence behavior of the Volterra series expansion for the same nonlinear system can vary greatly

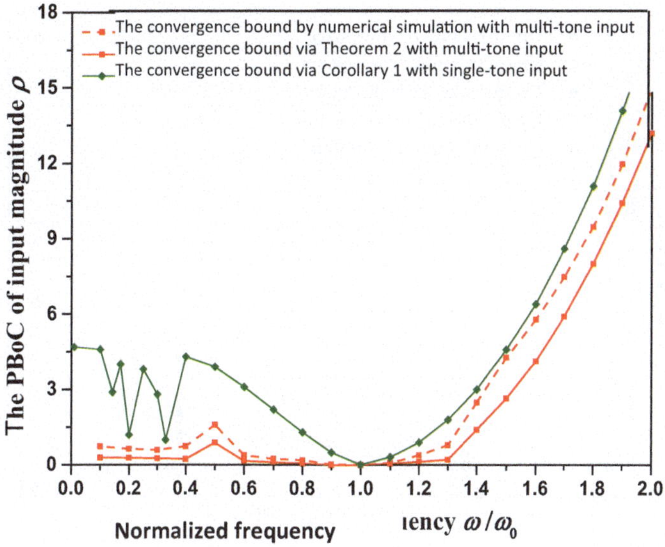

Fig. 10.1 The convergence bound of the input magnitude

depending on the input signals used. Beyond the convergence bound, the system's nonlinearity may lead to bifurcation or chaos, as noted in [23, 35]. Consequently, the convergence bound findings from this research offer valuable direction for the practical use of the Volterra series and related techniques in nonlinear analysis and design.

To validate the effectiveness of the bound result above, the normalized root mean square error is given as:

$$NRMSE = \sqrt{\frac{\sum \left(y_{synthesized}(k) - y_{real}(k)\right)^2}{\sum \left(y_{real}(k) - y_{mean}(k)\right)^2}} \tag{10.28}$$

In (10.28), $y_{real}(k)$ is obtained by numerical simulation using ode 45, $y_{mean}(k)$ is the mean value of $y_{real}(k)$, and the synthesized output response $y_{synthesized}(k)$ of a Volterra series expansion [17] up to sufficiently high orders also is computed accordingly. The *NRMSE* taken at $\omega_1 = 1.1\omega_0$ as an example is presented in Fig. 10.2. It can be seen that, when the input magnitude is smaller than the computed convergence bound, the NRMSE dramatically decreases to 0 as the synthesized order increases, but, when it is larger than the computed convergence bound, the synthesized output $y_{synthesized}(k)$ becomes slowly divergent with an observed increasing NRMSE.

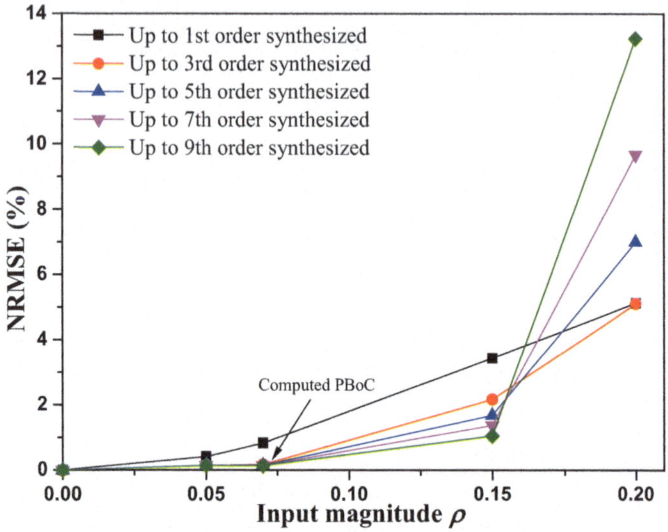

Fig. 10.2 The normalized root mean square error

10.4.2 The Convergence Bound of Model Parameters

The computation of the convergence bound of model parameter $c_{3,0}(1, 1, 1)$ is similar to the computation of the convergence bound of the input magnitude, which is given as:

$$C(3,0) = \frac{4L_{R_\omega}^2}{27(1+\beta)^2\rho^2\overline{H}_1^2}. \tag{10.29}$$

It should be emphasized that, for computation of the convergence bound of a model parameter (given the input), the methods in [17–21] are all not applicable, and the results in [22] focus on single-tone inputs.

Considering the input magnitude as $\rho = 0.5$, the computed convergence bound is presented in Fig. 10.3. In Fig. 10.3, the multi-tone input 1, multi-tone input 2, and single-tone input correspond to the input in (10.25) with $\beta = 0.1$, $\beta = 0.3$, and $\beta = 0$, respectively. The convergence bound of the model parameter with the multi-tone input also is smaller than that with the single-tone input, due to the nonlinear output frequency modulation effect. When β is changed from 0.1 to 0.3, the nonlinear degree of (10.24) also increases, which causes an even smaller convergence bound of the model parameter (see Fig. 10.3).

Remark 6. Indeed, additional examples could further validate the practicality and significance of the theoretical contributions of this research. The established parametric convergence bounds serve as a tool to determine the presence of a convergent Volterra series in a given NARX system. This can also be extended to discern whether

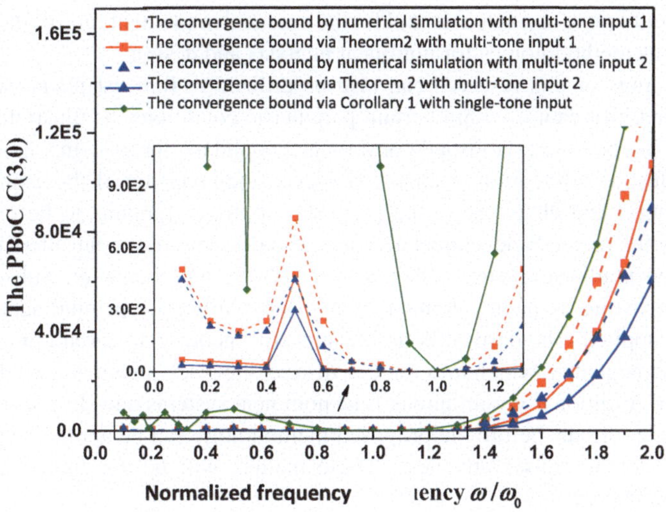

Fig. 10.3 The convergence bound of the model parameter $c_{3,0}(111)$

a nonlinear system qualifies as a Volterra-type, thus offering a criterion for designing Volterra-type nonlinear systems. It is noteworthy that Volterra-type nonlinear systems are clearly defined, making them simpler to comprehend and manage, especially with the numerous time or frequency domain methodologies available [23, 27–29, 31–35], in contrast to other complex nonlinear behaviors like bifurcation and chaos [35].

10.5 Conclusions

The Volterra series, along with its related techniques for nonlinear analysis and design, have received considerable attention and have been implemented in a range of engineering challenges. This research tackles a crucial and practical question: the parameter range within which a nonlinear system exhibits a convergent Volterra series expansion.

The suggested parametric convergence bound is straightforward to calculate for any type of input signal and enables the attainment of a convergence bound for both the input magnitude and any model parameter where the sum of p and q is greater than 1. This is an aspect not easily addressed by most existing studies. Given that the convergence bound is influenced by the input signals' frequency, magnitude, and waveform, it presents a less restrictive alternative compared to all other existing approaches (refer to the discussions in Remarks 1–5). The examples provided demonstrate that the convergence of a nonlinear system's Volterra series expansion can vary significantly based on the input signals used. Thus, the convergence bound results from this study serve as a valuable tool for the practical implementation of the Volterra series and its methodologies for nonlinear analysis and design.

The findings of this research can also be utilized to forecast phenomena such as chaos and bifurcation within certain parametric conditions, as illustrated by the examples. Beyond the established convergence boundary, the system's nonlinearity may manifest as bifurcation or chaos. However, the precision of these predictions for bifurcation and chaos using the proposed approach remains to be examined. Employing the methods developed here, one can design robust controllers to ensure that a given nonlinear system adheres to a Volterra-type behavior. Subsequently, numerous existing frequency domain strategies for Volterra-type nonlinear systems [5] can be applied. The derived bound results still incorporate certain principles of the triangle inequality, which introduces a degree of conservativeness into the bound assessment. Additionally, continuous-time nonlinear systems must be converted into discrete-time formats before the proposed methodology can be applied, which also contributes to the conservativeness. These matters will be the subject of future research endeavors.

It is also noted that several later results done in the literature actually follow the technical methods established in this study.

Acknowledgements This chapter can also be referred to the published result in [36].

Appendices

Appendix 1: Proof of Theorem 1

Proof.

(a) Since the input U(jω) is defined in a frequency range R_ω, all the frequency variables $\omega_1, \omega_2, \ldots, \omega_{n-1}$ in the output response (10.8b–c) are defined in $V_\omega = R_\omega \cup R_\omega^-$, where R_ω^- is the negative frequency set corresponding to R_ω. From (10.8a–c), it can be obtained that

$$|Y(j\omega)| \le \sum_{n=1}^{\infty} |Y_n(j\omega)| \tag{10.30}$$

$$|Y_n(j\omega)| = \frac{1}{(2\pi)^{n-1}} \Big| \underbrace{\int_{-\infty}^{\infty} \cdots \int_{-\infty}^{\infty}}_{n-1} H_n(j\omega_1, \ldots, j\omega_{n-1}, j(\omega - \omega_1 - \cdots - \omega_{n-1}))$$

$$U_n(j\omega, j\omega_1, \ldots, j\omega_{n-1}) \mathrm{d}\omega_1 \cdots \mathrm{d}\omega_{n-1} \Big|$$

$$= \frac{1}{(2\pi)^{n-1}} \Big| \underbrace{\int \cdots \int}_{\omega_1, \cdots, \omega_{n-1} \in V_\omega} H_n(j\omega_1, \ldots, j\omega_{n-1}, j(\omega - \omega_1 - \cdots - \omega_{n-1}))$$

$$U_n(j\omega, j\omega_1, \ldots, j\omega_{n-1}) \mathrm{d}\omega_1 \cdots \mathrm{d}\omega_{n-1} \Big|$$

$$\leq \frac{1}{(2\pi)^{n-1}} \underbrace{\int \cdots \int}_{\omega_1, \cdots, \omega_{n-1} \in V_\omega} |H_n(j\omega_1, \ldots, j\omega_{n-1}, j(\omega - \omega_1 - \cdots - \omega_{n-1}))$$

$$|| U_n(j\omega, j\omega_1, \ldots, j\omega_{n-1})| \mathrm{d}\omega_1 \cdots \mathrm{d}\omega_{n-1}$$

$$\leq \overline{H}_n \frac{1}{(2\pi)^{n-1}} \underbrace{\int \cdots \int}_{\omega_1, \cdots, \omega_{n-1} \in V_\omega} |U_n(j\omega, j\omega_1, \ldots, j\omega_{n-1})| \mathrm{d}\omega_1 \cdots \mathrm{d}\omega_{n-1} \tag{10.31}$$

$$\underbrace{\int \cdots \int}_{\omega_1, \cdots, \omega_{n-1} \in V_\omega} |U_n(j\omega, j\omega_1, \ldots, j\omega_{n-1})| \mathrm{d}\omega_1 \cdots \mathrm{d}\omega_{n-1}$$

$$= \rho^n \Big| \underbrace{\int \cdots \int}_{\omega_1, \cdots, \omega_{n-1} \in V_\omega} |U(j\omega - j\omega_1 - \cdots - j\omega_{n-1})|$$

$$U(j\omega_1) \cdots U(j\omega_{n-1}) | \mathrm{d}\omega_1 \cdots \mathrm{d}\omega_{n-1} \tag{10.32}$$

$$\leq \rho^n \underbrace{\int \cdots \int}_{\omega_1, \cdots, \omega_{n-1} \in V_\omega} |U(j\omega_1) \cdots U(j\omega_{n-1})| \mathrm{d}\omega_1 \cdots \mathrm{d}\omega_{n-1}$$

$$= \rho^n \left(\int_{\omega_1 \in V_\omega} |U(j\omega_1)| \mathrm{d}\omega_1 \right)^{n-1}$$

$$= \rho^n 2^{n-1} \left(\int_{\omega_1 \in R_\omega} |U(j\omega_1)| \mathrm{d}\omega_1 \right)^{n-1} \tag{10.33}$$

From (10.30–10.33), it can be derived that

$$|Y_n(j\omega)| \leq \overline{H}_n \frac{\rho^n}{(2\pi)^{n-1}} \left(2 \int_{\omega_1 \in R_\omega} |U(j\omega_1)| \mathrm{d}\omega_1 \right)^{n-1} = \frac{1}{\sigma} \overline{H}_n \gamma^n \tag{10.34}$$

From (10.34), (10.13) can be obtained with

$$\tilde{Y} = \sum_{n=1}^{\infty} \overline{H}_n \gamma^n \tag{10.35}$$

From Lemma 1 in [32], for n ≥ 2

$$|H_n(j\omega_1, \ldots, j\omega_{n-1}, j(\omega - \omega_1 - \cdots - \omega_{n-1}))| \le \bar{H}_n$$
$$= \frac{1}{\underline{L}_{R_\omega}} \sum_{m=2}^{n} \sum_{\substack{p+q=m \\ 0\le p,q\le m}} C(p,q) \left(\sum_{\substack{r_1\cdots r_p=1 \\ \sum r_i = n-q}}^{n-q-p+1} \prod_{i=1}^{p} \bar{H}_{r_i} \right)$$

Then, it can be further obtained from (10.35) that

$$\tilde{Y} = \overline{H}_1\gamma + \sum_{n=2}^{\infty} \frac{1}{\underline{L}_{R_\omega}} \sum_{m=2}^{n} \sum_{\substack{p+q=m \\ 0\le p,q\le m}} C(p,q) \left(\sum_{\substack{r_1\cdots r_p=1 \\ \sum r_i = n-q}}^{n-q-p+1} \prod_{i=1}^{p} \overline{H}_{r_i} \right) \gamma^n$$
$$= \overline{H}_1\gamma + \frac{1}{\underline{L}_{R_\omega}} \sum_{n=2}^{\infty} \sum_{m=2}^{n} \sum_{\substack{p+q=m \\ 0\le p,q\le m}} C(p,q) \left(\sum_{\substack{r_1\cdots r_p=1 \\ \sum r_i = n-q}}^{n-q-p+1} \prod_{i=1}^{p} \overline{H}_{r_i} \right) \gamma^n$$
$$= \overline{H}_1\gamma + \frac{1}{\underline{L}_{R_\omega}} \sum_{m=2}^{\infty} \sum_{\substack{p+q=m \\ 0\le p,q\le m}} C(p,q) \sum_{n=m}^{\infty} \left(\sum_{\substack{r_1\cdots r_p=1 \\ \sum r_i = n-q}}^{n-q-p+1} \prod_{i=1}^{p} \overline{H}_{r_i} \right) \gamma^n$$
$$= \overline{H}_1\gamma + \frac{1}{\underline{L}_{R_\omega}} \sum_{m=2}^{\infty} \sum_{\substack{p+q=m \\ 0\le p,q\le m}} C(p,q) \sum_{n=m}^{\infty} \left(\sum_{\substack{r_1\cdots r_p=1 \\ \sum r_i = n-q}}^{n-q-p+1} \prod_{i=1}^{p} \overline{H}_{r_i} \gamma^{n-q} \right) \gamma^q \quad (10.36)$$

Also from (10.35),

$$\tilde{Y}^p = \left(\sum_{i=1}^{\infty} \overline{H}_i \gamma^i \right)^p = \sum_{\kappa=p}^{\infty} \left(\sum_{\substack{r_1\cdots r_p=1 \\ \sum r_i = \kappa}}^{\kappa-p+1} \overline{H}_{r_1} \overline{H}_{r_2} \cdots \overline{H}_{r_p} \right) \gamma^\kappa$$

Taking $\kappa = n - q$ gives:

$$\tilde{Y}^p = \left(\sum_{i=1}^{\infty} \overline{H}_i \gamma^i \right)^p = \sum_{n=p+q}^{\infty} \left(\sum_{\substack{r_1\cdots r_p=1 \\ \sum r_i = n-q}}^{n-q-p+1} \overline{H}_{r_1} \overline{H}_{r_2} \cdots \overline{H}_{r_p} \right) \gamma^{n-q} \quad (10.37)$$

Using (10.37) in (10.36) gives:

$$\tilde{Y} = \overline{H}_1\gamma + \frac{1}{\underline{L}_{R_\omega}} \sum_{m=2}^{\infty} \sum_{\substack{p+q=m \\ 0\le p,q\le m}} C(p,q)\tilde{Y}^p\gamma^q$$

Considering the maximum nonlinear degree, i.e., m ≤ M, the equation above gives (10.14). Solving this equation leads to the output response bound for any given input signal defined in R_ω.

(b) It is known from the Analytic Inversion Lemma in [37] that an analytic function locally admits an analytic inverse near any point where the first derivative of the inverse is non-zero; however, a function cannot be analytically inverted in the neighborhood of a point where the first derivative vanishes.

Note that the output response bound in (10.35), i.e., $\tilde{Y} = \sum_{n=1}^{\infty} \overline{H}_n\gamma^n$, is analytic in its convergence region, which allows an analytic inverse, according to the analytic inversion lemma. That is, in the convergence region, there exists an inverse f^{-1} such that $\gamma = f^{-1}(\tilde{Y})$. Therefore, the minimum input magnitude bound where the first derivative of the inverse is zero, i.e., $\partial\gamma / \partial\overline{Y} = 0$, can be regarded as the convergence bound of the power series (10.35) (where the analytic inverse does not exists). Therefore, applying the first derivative operation to both sides of (10.35) with respect to $\tilde{Y}$ and setting it to be $d\gamma/d\tilde{Y} = 0$ yields (10.16). Together with (10.14), the convergence bound for γ can be obtained; thus, the bound of ρ can be computed by (10.15ab). If the input magnitude ρ is given, solving (10.14) subject to the constraint (10.16) can give a bound for any model parameters with $p + q > 1$. □

Appendix 2: Proof of Theorem 2

Proof. According to (10.8a) and (10.18), when the system (10.3) is subject to a given multi-tone input, the output frequency response is given by:

$$\begin{aligned} Y(j\omega) &= \sum_{n=1}^{\infty} Y_n(j\omega) \\ Y_n(j\omega) &= \frac{\rho^n}{2^{n-1}} \sum_{\omega_{k_1}\in V_\omega} \cdots \sum_{\omega_{k_{n-1}}\in V_\omega} H_n(j\omega_{k_1}, \ldots, \\ &\quad j(\omega - \omega_{k_1} - \cdots - \omega_{k_{n-1}}))F(j\omega_{k_1})\cdots F(j\omega_{k_n}) \\ &\quad F(j(\omega - \omega_{k_1} - \cdots - \omega_{k_{n-1}})) \end{aligned} \tag{10.38}$$

where $F(\omega_{k_i})$ can be written explicitly as [27]

$$F(\omega_{k_i}) = \left|F_{|k_i|}\right| e^{j\angle F_{|k_i|}\cdot \mathrm{sgn1}(k_i)} \quad \text{for } k_i \in \left\{\pm 1, \ldots, \pm\overline{K}\right\}, \quad \text{and}$$

$$\mathrm{sgn1}(a) = \begin{cases} 1 & a > 0 \\ 0 & a = 0 \quad for \ \ a \in \mathbb{R}. \\ -1 & a < 0 \end{cases} \tag{10.39}$$

It then can be obtained that

$$|Y(j\omega)| \le \sum_{n=1}^{\infty} |Y_n(j\omega)|$$

$$\begin{aligned}
|Y_n(j\omega)| &\le \frac{\rho^n}{2^{n-1}} \sum_{\omega_{k_1} \in V_\omega} \cdots \sum_{\omega_{k_{n-1}} \in V_\omega} \left| H_n(j\omega_{k_1}, \ldots, j(\omega - \omega_{k_1} - \cdots - \omega_{k_{n-1}})) \right| \\
&\quad \left| F(j\omega_{k_1}) \cdots F(j\omega_{k_{n-1}}) F(j(\omega - \omega_{k_1} - \cdots - \omega_{k_{n-1}})) \right| \\
&\le \frac{\rho^n}{2^{n-1}} \overline{H}_n \sum_{\omega_{k_1} \in V_\omega} \cdots \sum_{\omega_{k_{n-1}} \in V_\omega} \left| F(j\omega_{k_1}) \cdots F(j\omega_{k_{n-1}}) \right| \\
&= \frac{\rho^n}{2^{n-1}} \overline{H}_n \sum_{\omega_{k_1} \in V_\omega} \left| F(j\omega_{k_1}) \right| \cdots \sum_{\omega_{k_{n-1}} \in V_\omega} \left| F(j\omega_{k_{n-1}}) \right| \\
&= \rho^n \overline{H}_n \sum_{\omega_{k_1} \in R_\omega} \left| F(j\omega_{k_1}) \right| \cdots \sum_{\omega_{k_{n-1}} \in R_\omega} \left| F(j\omega_{k_{n-1}}) \right| \\
&= \rho^n \overline{H}_n \left(\sum_{\omega_{k_1} \in R_\omega} \left| F(j\omega_{k_1}) \right| \right)^{n-1} \\
&= \rho^n \overline{H}_n \left(\sum_{\omega_{k_1} \in R_\omega} \left| F_{|k_1|} \right| \right)^{n-1} = \rho^n \overline{H}_n \left(\sum_{\omega_{k_1} \in R_\omega} \left| F_{k_1} \right| \right)^{n-1} \\
&= \rho^n \overline{H}_n \left(\sum_{k_1 \in \{1,2,\ldots,\overline{K}\}} \left| F_{k_1} \right| \right)^{n-1} \\
&\le \frac{1}{\sigma} \overline{H}_n \gamma^n
\end{aligned}$$

Define $\tilde{Y} = \sum_{n=1}^{\infty} \overline{H}_n \gamma^n$, then $|Y(j\omega)| \le \sum_{n=1}^{\infty} |Y_n(j\omega)| = \frac{1}{\sigma} \sum_{n=1}^{\infty} \overline{H}_n \gamma^n = \frac{\tilde{Y}}{\sigma}$. Following a similar technical method as Theorem 1, it can be obtained that $\tilde{Y}$ can be given by Eq. (10.20a). Since the bound of the output response is given by $\tilde{Y} = \sum_{n=1}^{\infty} \overline{H}_n \gamma^n$, it can be obtained from Theorem 1 that the convergence bound can be achieved by solving Eq. (10.20a) with the constraint in Eq. (10.16). This completes the proof. □

Appendix 3: Proof of Corollary 1

Proof. Note that the input is a single-tone signal, the input frequency range $R_\omega = \{\omega_1\}$ and $V_\omega = \{-\omega_1, \omega_1\}$. Thus, following the proof of Theorem 2:

$$|Y(j\omega)| \le \sum_{n=1}^{\infty} |Y_n(j\omega)|$$

$$|Y_n(j\omega)| \le \frac{\rho^n}{2^{n-1}} \sum_{\omega_{k_1}=\pm\omega_1} \cdots \sum_{\omega_{k_{n-1}}=\pm\omega_1} \left|H_n(j\omega_{k_1}, \ldots, j(\omega - \omega_{k_1} - \cdots - \omega_{k_{n-1}}))\right| \left|F(j\omega_{k_1}) \cdots F(j\omega_{k_{n-1}})F(j(\omega - \omega_{k_1} - \cdots - \omega_{k_{n-1}}))\right|$$

Again, for the given single-tone input, it is clear that $|H_1(j\omega_1)|=|H_1(-j\omega_1)|$, and it can be obtained from [12, 23] that the bound of $|H_n(.)|$ can be written as

$$|H_n(\cdot)| \le \overline{H}_n(\omega_1) = a_0 + a_1|H_1(j\omega_1)| + \cdots + a_n|H_1(j\omega_1)|^n \tag{10.40}$$

where a_i for $i = 0, 1, \ldots, n$ are all positive, are functions of model parameters, and can be obtained with the method in [32–34]. Also, $\left|F(j\omega_{k_i})\right| = \rho$ for $i = 1, \ldots n$. Therefore, it can be obtained that

$$|Y(j\omega)| \le \sum_{n=1}^{\infty} |Y_n(j\omega)|$$

$$\begin{aligned}
|Y_n(j\omega)| &\le \frac{\rho^n}{2^{n-1}} \sum_{\omega_{k_1}=\pm\omega_1} \cdots \sum_{\omega_{k_{n-1}}=\pm\omega_1} \overline{H}_n(\omega_1)\left|F(j\omega_{k_1}) \cdots F(j\omega_{k_{n-1}})F(j(\omega - \omega_{k_1} - \cdots - \omega_{k_{n-1}}))\right| \\
&= \frac{\rho^n}{2^{n-1}} \overline{H}_n(\omega_1) \sum_{\omega_{k_1}=\pm\omega_1} \cdots \sum_{\omega_{k_{n-1}}=\pm\omega_1} \left|F(j\omega_{k_1}) \cdots F(j\omega_{k_{n-1}})F(j(\omega - \omega_{k_1} - \cdots - \omega_{k_{n-1}}))\right| \\
&= \frac{\rho^n}{2^{n-1}} \overline{H}_n(\omega_1) \sum_{\omega_{k_1}=\pm\omega_1} \cdots \sum_{\omega_{k_{n-1}}=\pm\omega_1} 1 \\
&= \rho^n \overline{H}_n(\omega_1)
\end{aligned}$$

Defining $\tilde{Y} = \sum_{n=1}^{\infty} \overline{H}_n(\omega_1)\rho^n$, then

$$|Y(j\omega)| \le \tilde{Y} = \sum_{n=1}^{\infty} \overline{H}_n(\omega_1)\rho^n \tag{10.41}$$

To compute the bound in (10.40), following the method in [32–34], it can be obtained that

$$|H_n(j\omega_1, \cdots, j\omega_{n-1}, j(\omega - \omega_1 - \cdots - \omega_{n-1}))| \le \bar{H}_n(\omega_1)$$

$$= \frac{1}{\underline{L}_{R_\omega}} \sum_{m=2}^{n} \sum_{\substack{p+q=m \\ 0 \le p,q \le m}} C(p,q) \left(\sum_{\substack{r_1 \cdots r_p = 1 \\ \sum r_i = n-q}}^{n-q-p+1} \prod_{i=1}^{p} \bar{H}_{r_i}(\omega_1) \right)$$

Expanding the recursive computation above will give the coefficients of (10.40), noting that $\overline{H}_1(\omega_1) = |H_1(\omega_1)|$. Hence, following the method in Theorem 1, it can be derived that $\tilde{Y} = |H_1(j\omega_1)|\rho + \frac{1}{\underline{L}_{R_\omega}} \sum_{m=2}^{M} \sum_{\substack{p+q=m \\ 0 \le p,q \le m}} C(p,q)\tilde{Y}^p \rho^q$; therefore, the parametric bound can be obtained by solving this equation subject to the constraint (10.23). This competes the proof. □

References

1. Boyd S, and L.O. Chua, "Fading memory and the problem of approximating nonlinear operators with Volterra series". *IEEE Trans. Circuits and Systems*, Vol 32, 1150–1161 (1985).
2. George DA. *Continuous nonlinear systems.* Technical report (Massachusetts Institute of Technology. Research Laboratory of Electronics), No 355 (1959)
3. Sandberg IW. "Expansions for nonlinear systems". The *Bell System Technical Journal*, Vol 61, No 2, 159-199 (1982)
4. Sandberg IW. "The mathematical foundations of associated expansions for mildly nonlinear systems". *IEEE Trans. Circuits and Systems*, Vol 30, No 7, 441-455 (1983)
5. Jing XJ and ZQ Lang, Frequency domain analysis and design of nonlinear systems based on Volterra series expansion: A parametric characteristic approach, Springer International Publishing, Switzerland, (2015)
6. Asyali, M. H., and M. Juusola, M. "Use of Meixner functions in estimation of Volterra kernels of nonlinear systems with delay". *IEEE Trans. on Biomedical Engineering*, Vol 52, No2, 229–237 (2005).
7. Prazenica R. J. and A. J. Kurdila, "Multiwavelet Constructions and Volterra Kernel Identification", *Nonlinear Dynamics*, Vol 43, Issue 3, pp 277-310 (2006).
8. Wu T, Kareem A and Ge Y, "Linear and nonlinear aeroelastic analysis frameworks for cable-supported bridges", *Nonlinear Dynamics*, Vol 74, Issue 3, pp 487-516 (2013).
9. Kotsios, S. "Transformation of finite degree discrete Volterra systems with cross products to finite input-output forms". *IEEE Trans. Automatic Control*, Vol 44, No 7, 1460-1464 (1999)
10. Mirri D, G Luculano, F Filicori, G Pasini, G Vannini, and G P Gualtieri. "A modified Volterra series approach for nonlinear dynamic systems modeling". IEEE Trans. Circuits and Systems I: Fundamental Theory and Applications, Vol 49, No 9, 1118–1128 (2002)
11. Nichols, J. M. "Frequency Distortion of Second- and Third-Order Phase-Locked Loop Systems Using a Volterra-Series Approximation". *IEEE Trans. Circuits and Systems I: Regular Papers*, Vol 56, No 2, 453–459, (2009)

12. Q, Zheng. and E.Zafiriou, "A local form of small gain theorem and analysis of feedback Volterra systems". *IEEE Trans. Automatic Control*, Vol 44, No 3, 635–640, (1999)
13. Orcioni S, "Improving the approximation ability of Volterra series identified with a cross-correlation method", *Nonlinear Dynamics*, Vol 78, Issue 4, 2861-2869 (2014)
14. Tawfiq I. and T.Vinh, "Nonlinear Behaviour of Structures Using the Volterra Series—Signal Processing and Testing Methods", *Nonlinear Dynamics,* Volume 37, Issue 2, 129-149 (2004)
15. Motato E. and C. J. Radcliffe, "Port condensation of Volterra transfer functions with cross-products", *Nonlinear Dynamics,* Vol 79, Issue 1, 593-605 (2015)
16. Guo, Y., L. Z., Guo, S. A., Billings, D., Coca, and Z. Q. Lang, "Volterra Series Approximation of a Class of Nonlinear Dynamical Systems Using the Adomian Decomposition Method". Nonlinear Dynamics, Vol 74, Issue(1–2), 359–371 (2013).
17. Li L and SA Billings. "Analysis of nonlinear oscillators using Volterra series in the frequency domain". *J. Sound and Vibration*, Vol 330, Issue 2, 337–355 (2011)
18. Peng Z and Z Lang. "On the convergence of the Volterra-series representation of the Duffing's oscillators subjected to harmonic excitations". *J. Sound and Vibration*, Vol 305, 322-332 (2007)
19. Tomlinson GR, G Manson and G Lee. "A simple criterion for establishing an upper limit to the harmonic excitation level of the Duffing oscillator using the Volterra series". *J. Sound and Vibration*, Vol 190, Issue 5 , 751–762 (1996)
20. Helie T, and B Laroche. "Computation of Convergence Bounds for Volterra Series of Linear-Analytic Single-Input Systems". *IEEE Trans. Automatic Control*, Vol 56, No9, 2062–2072 (2011)
21. Xiao Z, XJ Jing, and L Cheng. "Parameterized Convergence Bounds for Volterra Series Expansion of NARX Models". *IEEE Trans. Signal Processing*, Vol 61, No 20, 5026–5038, (2013)
22. Xiao, Z., X.J. Jing, and L Cheng. "Estimation of parametric convergence bounds for Volterra series expansion of nonlinear systems". *Mechanical Systems and Signal Processing*, Vol 45, Issue 1, 28–48 (2014)
23. Rugh WJ. *Nonlinear system theory: The Volterra/Wiener Approach*, Johns Hopkins University Press Baltimore, MD (1981)
24. Chen, S., and S. A. Billings, "Representations of non-linear systems: the NARMAX model". *Int. J. Control*, Vol 49, No3, 1013-1032(1989).
25. Chen, S., S. A., Billings, and W. Luo, "Orthogonal least squares methods and their application to non-linear system identification". *Int. J. Control*, Vol 50, No 5, 1873–1896 (1989).
26. Wei, H.L., and S. A. Billings, "Model structure selection using an integrated forward orthogonal search algorithm assisted by squared correlation and mutual information". *Int. J. Modelling, Identification and Control*, Vol 3, No 4, 341–356, (2008)
27. Jing XJ, and ZQ Lang. "On the Generalized Frequency Response Functions of Volterra Systems". *J. Dynamic Systems, Measurement, and Control*, Vol 131, No 6, 061002 (2009)
28. Peyton-Jone J and SA Billings. "Recursive algorithm for computing the frequency response of a class of non-linear difference equation models". *Int. J. Control*, Vol 50, Issue 5, 1925–1940 (1989)
29. Jing XJ. "Frequency domain analysis and identification of block-oriented nonlinear systems". *J. Sound and Vibration*, Vol 330, Issue 22, 5427-5442 (2011)
30. Lang ZQ and SA Billings. "Output frequencies of nonlinear systems". *Int. J. Control*, Vol 67, Issue 5, 713-730 (1997)
31. Jing XJ, ZQ Lang and SA Billings. "Output frequency properties of nonlinear systems". *Int. J. Non-Linear Mechanics*, Vol 45, Issue 7, 681–690 (2010)
32. Jing XJ, ZQ Lang, and SA Billings. "New bound characteristics of NARX model in the frequency domain". *Int. J. Control*, Vol 80, Issue 1, 140–149 (2007)
33. Jing XJ, ZQ Lang, and SA Billings. "Magnitude bounds of generalized frequency response functions for nonlinear Volterra systems described by NARX model". *Automatica*, 44, 838-845 (2008)
34. Jing XJ, ZQ Lang and SA Billings. "Frequency-dependent magnitude bounds of the generalized frequency response functions for NARX model". *European journal of control*, Vol 1, 68-83 (2009)

35. Nayfeh A H and D T Mook. *Nonlinear oscillations*, Wiley-VCH (2008)
36. X Jing, Z Xiao, On convergence of Volterra series expansion of a class of nonlinear systems, Asian Journal of Control 19 (3), 1089-1102, 2017
37. Flajolet P, & R Sedgewick. *Analytic combinatorics*, Cambridge Univ Pr., U.K. (2009)
38. Jing XJ, ZQ Lang, SA Billings and GR Tomlinson. "The parametric characteristic of frequency response functions for nonlinear systems". *Int. J. Control*, Vol 79, Issue 12, 1552–1564 (2006)

Chapter 11
Summary

Abstract This book focuses on a systematic frequency-domain methodology for the analysis and design of nonlinear dynamic systems. The approach is broadly applicable—not only to specific structural or mechanical configurations (such as X-structures or mechanisms)—but also to a wide range of nonlinear dynamics encountered in engineering systems. Furthermore, the methodology can be extended to nonlinear signal processing, offering enhanced tools for the analysis and design of nonlinear signals. As Part III of this book series, the present volume introduces a specialized frequency-domain technique known as the nonlinear Characteristic Output Spectrum (nCOS) function. This function provides an analytical representation of system behavior in the frequency domain and is formulated as a polynomial function of key design parameters. These parameters include coefficients associated with both linear and nonlinear model terms, as well as input signal characteristics such as frequency and amplitude. The resulting polynomial structure—referred to as the parametric characteristic—offers a clear and tractable framework for understanding and optimizing nonlinear system performance. Several benchmark applications of this method are showcased, including structural optimization, controller design, fuzzy membership selection and so on.

Keywords Frequency domain · Nonlinear systems · Optimization · Nonlinear characteristic output spectrum (nCOS)

11.1 Summary of This Book

Although most results of this book series are about the X-structure or mechanism based methods, this book is particularly focused on a systematic frequency domain method for nonlinear analysis and design involved in nonlinear dynamic systems which is not only applicable to the X-structures/mechanisms, but also to all those nonlinear dynamics of engineering systems, and can be further generalized to nonlinear signal processing for better analysis and design of nonlinear signals.

X. Jing, *The Bio-Inspired X-Structure/Mechanism Approach for Exploring Nonlinear Benefits in Engineering*, https://doi.org/10.1007/978-981-96-5769-8_11

This book is Part III of this book series and aimed to introduce a special frequency domain method, i.e., the nCOS function. This *n*onlinear *C*haracteristic *O*utput *S*pectrum function can be analytically expressed as a polynomial function of those critical design parameters including all those coefficients of model parameters (for linear terms or nonlinear terms) and also those related to input signals (frequency and amplitude etc.) in a clear polynomial structure (referred to as parametric characteristics).

Chapter 1 serves as an overview of these recently developed theory and methods, focusing on the results done by the N.D.V.C research group led by Prof XJ Jing. Chapter 2 introduces the derivation of general frequency response functions (GFRFs, which can be understood as the generalized transfer function, compared to linear systems) with examples demonstrated by apply the known probing method, and further presents the GFRFs of block oriented nonlinear systems with applications to system identifications. Chapter 3 summarizes the fundamental parametric characteristic method and related results, and the readers will see that many recent efforts in the literature are just a follow-up of this theory. A fundamental theoretic work about truncation error of all those related to unknown nonlinear identification with polynomial curve fitting is given in Chap. 4, which show light on how to make such fitting more reliable and accurate even though the truncation order is a very limited number. Applying this theory produces the method in Chap. 5—the formal formulation of the nCOS function, and also the nonlinear output regulation method in Chap. 6. These results lay down the important basis of fault detection and diagnosis of a nonlinear dynamic system (complex structures or others) by employing nonlinear features. These results related to fault detection will be summarized into another book to be published. Chapter 7 showcases the application of the nCOS function for nonlinear analysis and optimization of structural parameters in the context of X-structures for energy harvesting. The readers can see that both linear and nonlinear model parameters can all be designed with the method proposed there. Chapters 8 and 9 give some other applications with multiple optimization objectives and thus a SIMO approach is summarized there in the context of vehicle suspension control and Fuzzy membership optimization. The other fundamental issue in the field of Volterra-class nonlinear systems is the convergence of Volterra series expansion. Chapter 10 presents the first-ever closed form convergence bound with respect to model parameters, compared to all convergence results in the literature.

11.2 Results in Sister Books

This book series includes 5 books (or 5 Parts) and this is the third part, i.e., Book III, which is to introduce a frequency domain method for the analysis, design, and optimization of nonlinear systems in terms of model parameters and input excitation. The method can be successfully applied to the design of X-structures, vehicle suspension and other nonlinear systems. The first part of this book series, i.e., Book I, is about nonlinear stiffness and damping characteristics of the X-structure/mechanism.

Book II is to introduce the nonlinear inertia produced by the X-structure/mechanism, and also present the recent advance in multi-direction vibration isolation based on the X-structure method. Book III (i.e., this book) is to introduce an analysis and design method for nonlinear dynamics, i.e., the nCOS function, together with some case studies. Book IV is to discuss Energy harvesting and sensors with X-structures, which can be regarded as direct application case studies of the X-structure/mechanism method, and Book V is to provide another important application of nonlinear advantages provided by the X-structural dynamics, i.e., energy-saving robust control in active vibration control, which is a totally new approach to realization of green control systems, and some more applications of the X-structure method in robots and others will also be discussed in Book V.

The manufacturer's authorised representative in the EU is Springer Nature Customer Service Centre GmbH, Europaplatz 3, 69115 Heidelberg, Germany. If you have any concerns regarding our products, please contact ProductSafety@springernature.com

Printed and bound by CPI Group (UK) Ltd, Croydon, CR0 4YY
01/10/2026
02225850-0015